ANALYSIS AND COMPUTATION OF ELECTRIC AND MAGNETIC FIELD PROBLEMS

SECOND EDITION

Analysis and Computation of
ELECTRIC AND MAGNETIC
FIELD PROBLEMS

SECOND EDITION

by

K. J. BINNS

Reader in Electrical Machines, University of Southampton

and

P. J. LAWRENSON

Professor of Electrical and Electronic Engineering, University of Leeds

PERGAMON PRESS
OXFORD · NEW YORK · TORONTO
SYDNEY · BRAUNSCHWEIG

Pergamon Press Ltd., Headington Hill Hall, Oxford

Pergamon Press Inc., Maxwell House, Fairview Park, Elmsford,
New York 10523

Pergamon of Canada Ltd., 207 Queen's Quay West, Toronto 1

Pergamon Press (Aust.) Pty. Ltd., 19a Boundary Street,
Rushcutters Bay, N.S.W. 2011, Australia

Vieweg & Sohn GmbH, Burgplatz 1, Braunschweig

First edition 1962

Second edition 1973

Library of Congress Cataloging in Publication Data

Binns, Kenneth John.
  Analysis and computation of electric and magnetic
field problems.

  Bibliography: p.
  1. Electric fields. 2. Magnetic fields.
3. Mathematical analysis. I. Lawrenson, P. J., joint
author. II. Title. III. Title: Electric and
magnetic field problems.
QC661.B435 1973   530.1'2   73-4545
ISBN 0-08-016638-5

Printed in Hungary

CONTENTS

Part II: Direct methods

Part III: Transformation methods

Part IV: Numerical methods

Appendixes

PREFACE

IN THE first edition we attempted to provide, in a single volume, a comprehensive treatment of both analytical and numerical methods for the derivation of two-dimensional static and quasi-static electric and magnetic fields. The main objectives were to try to present the essence of each method of solution and to indicate and compare the scopes of the different methods having particular regard to the influence of digital computers. In this second edition the aim is largely the same, but the treatment has been revised to include developments which have occurred over the last ten years both in methods of solution and in new applications.

As with the first edition, the book is intended primarily for engineers, physicists, and mathematicians who are faced with problems which can only be solved by an analysis of electromagnetic fields. It is also suitable for degree students towards the end of their courses. An aim at all stages has been to emphasize the physical significance of the mathematics and, to this end, examples of practical interest have been selected wherever possible.

The main text is divided into four parts so arranged that, provided the material contained in the first of these is familiar, study can commence in any of the other three parts. Part I contains a brief introductory chapter and a chapter devoted to the fundamental theory of electric and magnetic fields. The latter has been considerably modified since the first edition so as to give, in as concise a form as possible, the background theory essential to an understanding of the methods of analysis used later in the book. A clear explanation is attempted of the derivation of quantities of physical interest such as force, inductance, and capacitance from the field solution.

Part II deals with the image and variables separation methods of solution. In addition to the topics commonly treated under these headings, the present treatment covers a wide range of field sources; and, in the chapter on images, the basic solutions are developed rigorously from considerations of surface charges and solutions are expressed in complex variable form.

Part III, the longest of the four, is devoted to transformation methods, and the authors believe that it offers the most comprehensive treatment of the subject which is available. Some of the more important topics not normally dealt with include the following: line and doublet sources, which are rarely treated in connection with electromagnetic fields; the transformation of regions exterior to finite boundaries; and the powerful numerical methods which have been developed to enlarge the scope of conformal transformation.

Part IV deals with finite difference methods which can be used to solve any problem relevant to this book. All classes of boundary shape and condition are discussed and Chapter 2 has been enlarged to take account of recent computational developments. It should provide a useful introduction in a particularly important and rapidly developing area.

For their helpful comments we are most grateful to Dr. E. M. Freeman of Brighton Polytechnic and Professor P. Hammond of the University of Southampton.

K.J.B.   P.J.L.

PART I

INTRODUCTION

PART I

INTRODUCTION

CHAPTER 1

INTRODUCTION

Types of field discussed. All static electric and magnetic fields in a uniform medium are described by Poisson's equation or its particular form, Laplace's equation. Poisson's equation applies within regions of distributed current or charge, and Laplace's equation applies in all other regions of the field. In Chapter 2 the properties of fields described by these equations are reviewed, and the whole of the remainder of the book is devoted to different methods for the solution of the field equations.

In addition to the above static fields, which they describe exactly, Laplace's and Poisson's equations also describe, to a high degree of accuracy, several types of time-varying field. The commonest of these occurs when the frequency and boundaries are such that the effect of eddy currents is negligible. However, Laplacian solutions can also be used when the eddy currents are so strong that negligible flux penetrates a boundary surface. Electromagnetic radiation phenomena are described by the wave equation, but for certain problems, such as the determination of the characteristic impedance of transmission lines, Laplacian solutions are applicable.

All physical fields are, of course, three-dimensional, but for most cases of practical interest exact analytical solutions are not available, and numerical solutions often involve a prohibitive amount of computation. However, approximate solutions of quite sufficient accuracy can be obtained by using a two-dimensional treatment, i.e. by neglecting the variation of the field in one direction. As a result, analysis becomes possible in very many cases, and in the others the labour of numerical solution is greatly reduced. Two examples of two-dimensional treatment occur in the calculation of the magnetic fields in rotating electrical machines. Firstly, the distribution of the main field within the air gap can be found with negligible error by analysing the field at a cross-section perpendicular to the axis (the variation along the length of the machine being neglected). Secondly, the field outside the machine ends can be found, though rather less accurately than in the previous example, by analysing the field in an axial plane (neglecting the peripheral variations).

Types of solution. Most of this book is concerned with solutions of Laplace's equation, though the more general form, Poisson's equation, is discussed in Chapters 5 and 11. There are two reasons for giving more attention to Laplace's equation: firstly, the majority of fields of practical importance are of this simpler type, and, secondly, since Poissonian fields are the more difficult to solve, advantage is frequently taken of the relatively small importance of the Poissonian region to replace it by an equivalent filament, so effectively making the whole field Laplacian. For example, in calculating the inductance of a transmission line, the field is solved for a current concentrated in a central filament of the line.

All solutions fall into one of two classes, analytical or numerical. In the first class a solution is in the form of an algebraic equation in which values of the parameters defining the field can be substituted. A solution in the second class takes the form of a set of numerical values of the function describing the field for one particular set of values of the parameters. All analytical methods have been in common use for at least sixty years, but it is only within the last thirty years or so that numerical methods have come into prominence. The recent development of numerical methods has been greatly stimulated by the advent of fast digital computing machines which have made possible routine solutions, to a high degree of accuracy, of many types of problems which would otherwise be extremely or even prohibitively laborious.

Where either analytical or numerical methods can be employed for the solution of a particular field, the choice of the most suitable method can sometimes be difficult to make. Analytical methods have the advantage that a general solution can be derived, from which it is possible to gain an overall picture of the effect of the various parameters. In contrast, with numerical methods it is necessary to calculate separately for each set of values of the parameters; a consequent disadvantage is that an overall picture can often be achieved only at the expense of a great amount of computation. However, for some problems for which analytical methods are possible, the determination of an analytical solution can be so involved and the computation so lengthy that numerical methods are simpler and quicker.

Analogous fields. In many aspects of engineering and physics there are physical phenomena which are directly analogous to electric and magnetic field phenomena. Amongst these are the flow of heat in conducting media and the flow of an inviscid liquid. For example, the temperature distribution between two boundaries having a constant temperature difference between them, or the distribution of the stream function of an ideal fluid passing between these boundaries, is identical in form with the voltage distribution between the same boundaries having a constant electric potential difference. Thus a solution to one problem

TABLE 1.1. ANALOGOUS QUANTITIES IN SCALAR POTENTIAL FIELDS

Quantity	Electrostatic	Electric current	Magneto-static	Heat flow	Fluid flow	Gravita-tional
Potential	Potential V	Potential V	Potential Ω	Temperature	Velocity potential	Newtonian potential
Potential gradient	Electric field strength E	Electric field strength E	Magnetic field strength H	Temperature gradient	Velocity	Gravitation force
Constant of medium	Permittivity ε	Conductivity σ	Permeability μ	Thermal conductivity	Density	Reciprocal of gravitation constant
Flux density	Electric flux density D	Current density J	Magnetic flux density B	Heat flow density	Flow rate	
Source strength	Charge density ϱ_e	Current density J	Pole density ϱ_m	Heat source density	Density of efflux	Mass density
Field conductance	Capacitance C	Conductance G	Permeance Λ	Thermal conductance		

of a particular physical type is directly applicable to other problems of different types, and methods developed in this book for electric and magnetic fields apply equally to the other fields mentioned above. Table 1.1 shows the equivalence of quantities in the different types of scalar potential field. In addition to the ones tabulated, consideration is given in the book to magnetic fields within regions of distributed current, and it is of interest to note that this type of field is analogous, for example, to that of fluid flow with vorticity.

CHAPTER 2

BASIC FIELD THEORY

THIS chapter provides a very brief review of the basic concepts of stationary electric and magnetic fields in just sufficient detail to cover the background theory required for the methods of analysis described in the book. Initially, the development is based on the point sources of field, but thereafter attention is given primarily to the line sources, the charge, the pole, and the current which are basic to the two-dimensional fields considered in this book.

2.1. Electric fields

2.1.1. *The electrostatic field vectors*

The concept of electric charge is of fundamental importance in the study of electric fields. A charge of magnitude q coulombs is considered to emit a total electric flux of q units; hence, an electric flux q emanates from any closed surface containing a charge q.

The *electric flux density* at a point is the vector **D**, and its direction is that of the flux. Considering a spherical surface of radius r, with its centre at the position of a point charge, it is evident from considerations of symmetry that the direction of the flux is radially outward and that the density of flux crossing the surface is equal to $q/4\pi r^2$, i.e. the magnitude of the flux density is given by

$$D = \frac{q}{4\pi r^2}. \tag{2.1}$$

The force exerted on unit charge placed at a point, a distance r from a charge q, is proportional to q/r^2, and so to the value of the vector **D** at that point due to the charge q. Thus if a vector **E**, known as the *electric field strength*, is defined to describe the force acting on the unit charge, then **E** is proportional to **D** for a given medium and may be expressed as

$$\mathbf{D} = \varepsilon_0 \varepsilon \mathbf{E}, \tag{2.2}$$

where ε_0 is the primary electric constant and ε is the relative permittivity of the surrounding medium. So combining eqns. (2.1) and (2.2) gives

$$E = \frac{q}{4\pi \varepsilon_0 \varepsilon r^2}. \tag{2.3}$$

In free space this becomes

$$E = \frac{q}{4\pi \varepsilon_0 r^2},$$

which, because of the nature of the variation of E with r, is called the inverse square law.

Consider now a charge distributed over a volume. As the volume tends to zero, the limit, at a point, of the outward flux per unit volume is called the *divergence* of the vector **D**, and is a scalar. Thus the divergence of **D** at any point within the volume is equal to the charge density ϱ_c, i.e.

$$\text{div } \mathbf{D} = \varrho_c. \tag{2.4}$$

The field of a line charge. When charge is uniformly distributed along an infinite straight line, the direction of the flux leaving the charge is everywhere perpendicular to the line, and the flux emitted per unit length of the line is equal to the linear charge density q. At a radius r about the charge, the flux density **D** is given in magnitude by

$$D = \frac{q}{2\pi r}, \tag{2.5}$$

and so

$$E = \frac{p}{2\pi\varepsilon_0\varepsilon r}. \tag{2.6}$$

Thus the field strength varies inversely as the distance from the line charge.

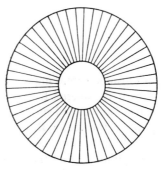

FIG. 2.1

This field is two-dimensional, and in all such fields a quantity of flux may be represented by a *number* of *flux lines*. At any point the direction of such a line is that of the flux density, and the concentration of the lines is a direct indication of the flux density there. A simple example of the distribution of flux lines is provided by the field of two charged conducting concentric cylinders (Fig. 2.1). From symmetry it is seen that the flux passes radially between the two cylinders and, since the quantity of flux passing each surface is the same, the flux densities on the surfaces of the cylinders are inversely proportional to their circumferences and therefore to their radii.

2.1.2. *Electric potential*

The scalar quantity, called the *electric potential V*, is a point function defined as the the work done in moving unit charge from infinity to the point. Now the work done dV in moving unit charge a small distance dl is given by

$$dV = -\mathbf{E}\cdot\mathbf{dl}, \tag{2.7}$$

since **E** is the force on unit charge. The negative sign means that the potential decreases with the positive direction of **E**. Equation (2.7) may be written in the form

$$E = -\frac{dV}{dl},$$

which expresses the fact that the component of electric field strength in any direction is equal to the potential gradient in that direction. This relationship is also expressed by the vector equation

$$\mathbf{E} = -\operatorname{grad} V. \tag{2.8}$$

The work done in moving a charge between two points in an electrostatic field is independent of the path taken, and the work done in moving it round any closed path (returning

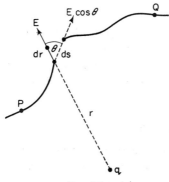

FIG. 2.2

to the starting point) is zero (Fig. 2.2). A field having this property is said to be conservative. Therefore eqn. (2.7) may be written in the form

$$V_P - V_Q = -\int_c E \, dl, \tag{2.9}$$

which is used below.

For the field of a line charge, since E is directed radially outward, the potential difference between points P and Q is

$$V_P - V_Q = -\int \frac{q \, dr}{2\pi\varepsilon_0\varepsilon r}, \tag{2.10}$$

and the difference in potential between points at radius r_1 and points at radius r_2 is

$$\frac{-q}{2\pi\varepsilon_0\varepsilon}\left(\log r\right)_{r_1}^{r_2} = \frac{-q}{2\pi\varepsilon_0\varepsilon} \log \frac{r_2}{r_1}.$$

Since the potential is a point function, it is possible to draw a line which passes through points of the same potential. Such a line is called an *equipotential line*, and no work is done in moving a charge along one. Since no work is done in moving a charge in a direction perpendicular to the field strength, equipotential lines are perpendicular to flux lines.

As an example, consider the field between two charged circular concentric boundaries. The equipotential lines are circles (Fig. 2.3); this is because the boundaries are conducting (and therefore equipotential) and because the field has circular symmetry. Flux lines

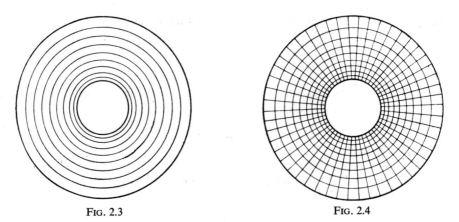

FIG. 2.3   FIG. 2.4

and equipotential lines are shown together in Fig. 2.4. They form a network of orthogonal lines, and this network is called a *field map*. Such maps provide useful, general pictures of fields, giving not only qualitative but also approximate quantitative information about the distribution of flux. They are widely used in the book to aid the reader's understanding of the various problems examined.

2.1.3. *Potential function and flux function*

To facilitate analysis, a *potential function* ψ is defined such that the *change* in this function between any two points is proportional to the *change in potential* between them. Its value at any point, with respect to some origin (of potential), is a direct measure of the value of the potential there and, in addition, a line joining points having the same value of potential function is an equipotential line.

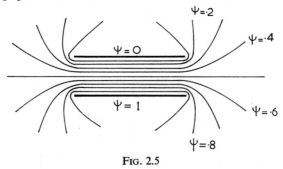

FIG. 2.5

Consider the field of two charged conducting plates (Fig. 2.5). Let $\psi = 0$ represent the value of the potential on one plate and $\psi = 1$ that on the other, so that there is unit difference of potential between the plates. Equipotential lines can be drawn in the space round the plates, representing $\psi = $ constant, for values of ψ between 0 and 1. For example, $\psi = 0.1$ represents a line joining points differing in potential from that of the lower potential plate by one tenth of the potential difference between the plates.

In a similar manner a *flux function* φ is defined such that $\varphi = $ constant defines a flux line; and two lines $\varphi = \varphi_0$ and $\varphi = \varphi_0 + n$ have n units of flux passing between them.

From the above definitions it is seen that differences in flux and potential function represent quantities of flux and potential differences (the zeros of both functions being chosen arbitrarily). Lines drawn for constant values of potential and flux functions, so chosen that ψ and φ change in equal steps, form a field map in which the regions enclosed between the intersecting lines are curvilinear squares. In a uniform field these curvilinear squares become exactly square.

Since the potential and flux functions are orthogonal one function can be derived from the other by use of the equation

$$\left(\frac{dy}{dx}\right)_{\varphi=\text{constant}} = -1 \bigg/ \left(\frac{dy}{dx}\right)_{\psi=\text{constant}}. \tag{2.11}$$

However, the equations relating ψ and φ directly are most simply developed, not from the above equation or its equivalent in other coordinate forms, but by the use of complex variable theory, see section 2.4.

As an example of the forms of flux and potential functions consideration is again given to the field of a line charge. It is shown [eqn. (2.9)] that, in general, the difference in potential between the ends of a contour C is $\int_C E\,dl$, and so

$$\psi = K_1 \int_C E\,dl + K_2,$$

where K_1 and K_2 are chosen arbitrarily. For a line charge $E = q/2\pi\varepsilon_0\varepsilon r$ and so

$$\psi = K_1 \int_C \frac{q\,dr}{2\pi\varepsilon_0\varepsilon r} + K_2$$

$$= \frac{K_1 q}{2\pi\varepsilon_0\varepsilon}\log r + K_2.$$

Thus choosing $K_1 = 1$ and $K_2 = 0$ for simplicity gives

$$\psi = \frac{q}{2\pi\varepsilon_0\varepsilon}\log r. \tag{2.12}$$

The flux function is derived by expressing the fact that flux is emitted from the charge equally in all radial directions. The flux leaving in a radial wedge is proportional to the angle of the wedge and so the flux function varies linearly with θ, and has the form

$$\varphi = K_3\theta + K_4.$$

The total flux emitted from a charge is equal to q, and so as θ changes by 2π, φ changes by q. Hence, $K_3 = q/2\pi$ and choosing $K_4 = 0$ the flux function may be written

$$\varphi = \frac{q\theta}{2\pi}. \tag{2.13}$$

It will be noted that the flux and potential functions for the field of two charged concentric circular conductors are the same as those for the field of a line charge, both fields having circular equipotential lines and straight radial flux lines.

2.1.4. *Capacitance*

The capacitance between two conducting surfaces is given by the ratio of total flux common to the surfaces to the potential difference between them. Consequently it may be evaluated conveniently from the flux and potential functions. If ψ_1 and ψ_2 are the potential functions of the two conductors, and if φ' and φ'' are the values of the flux functions for the lines bounding the mutual flux, then the capacitance C is given by

$$C = \frac{\varphi' - \varphi''}{\psi_1 - \psi_2}. \tag{2.14}$$

This relationship is valid even when more than two conductors are present. When there are only two conductors (one of which may be at infinity) the flux between them is equal to the charge q on either, and

$$C = \frac{q}{\psi_1 - \psi_2}. \tag{2.15}$$

In the case of two charged concentric cylinders the potential function has been shown to vary as

$$\frac{q}{2\pi\varepsilon_0\varepsilon} \log r,$$

and so, if the boundaries have radii of r_1 and r_2, the difference in potential function between the boundaries is

$$\frac{q}{2\pi\varepsilon_0\varepsilon} \log r_1 - \frac{q}{2\pi\varepsilon_0\varepsilon} \log r_2.$$

Therefore the capacitance between the cylinders is

$$q \left/ \left(\frac{q}{2\pi\varepsilon_0\varepsilon} \log \frac{r_1}{r_2} \right) \right. ,$$

which equals

$$\frac{2\pi\varepsilon_0\varepsilon}{\log r_1/r_2}. \tag{2.16}$$

2.1.5. *Laplace's and Poisson's equations*

From the basic theory so far developed it is simple to establish Poisson's equation and its particular form, Laplace's equation. These equations describe the flux and potential distributions for all the fields considered in this book, and it is with their solution that the remaining chapters are concerned. They are developed first in terms independent of a particular coordinate system by use of the concepts of divergence and gradient, and then in terms of cartesian coordinates by considering the net flux leaving an elemental rectangular box (Fig. 2.6).

In a region containing charge distributed with uniform density ϱ_c, the divergence of the flux is everywhere equal to ϱ_c, and this may be expressed in terms of **E** by combining eqns. (2.2) and (2.4) to give

$$\operatorname{div}(\varepsilon_0\varepsilon\mathbf{E}) = \varrho_c.$$

Using eqn. (2.8) this may be written in terms of potential as

$$\text{div}\,(-\varepsilon_0\varepsilon\,\text{grad}\,V) = \varrho_c$$

or

$$\text{div}\,(\text{grad}\,V) = -\varrho_c/\varepsilon_0\varepsilon. \qquad (2.17)$$

For a region containing no charge,

$$\varrho_c = 0$$

and

$$\text{div}\,(\text{grad}\,V) = 0. \qquad (2.18)$$

Equation (2.17) is Poisson's equation, and eqn. (2.18) is Laplace's equation.

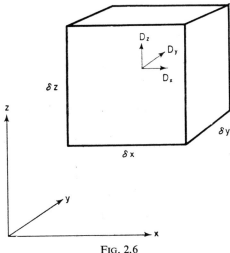

FIG. 2.6

These equations are now derived in cartesian form by considering a small cube with sides of length δx, δy, and δz parallel to the axes of x, y, and z (Fig. 2.6). Let the vector $\mathbf{D}$, with components D_x, D_y, and D_z, be the flux density at the centre of the cube and consider the two faces of the volume element perpendicular to the axis of x. The flux entering the cube through the left-hand face is

$$\left(D_x - \frac{1}{2}\frac{\partial D_x}{\partial x}\,\delta x\right)\delta y\,\delta z,$$

and that leaving the cube through the right-hand face is

$$\left(D_x + \frac{1}{2}\frac{\partial D_x}{\partial x}\,\delta x\right)\delta y\,\delta z.$$

Thus, the net flux leaving the cube in the x-direction is

$$\frac{\partial D_x}{\partial x}\,\delta x\,\delta y\,\delta z.$$

There are similar expressions for the y- and z-directions, and so the total flux leaving the element is

$$\left(\frac{\partial D_x}{\partial x}+\frac{\partial D_y}{\partial y}+\frac{\partial D_z}{\partial z}\right)\delta x\,\delta y\,\delta z.$$

But this is equal to the total charge enclosed, $\varrho_c\,\delta x\,\delta y\,\delta z$, and so

$$\frac{\partial D_x}{\partial x}+\frac{\partial D_y}{\partial y}+\frac{\partial D_z}{\partial z}=\varrho_c. \tag{2.19}$$

Now if E_x, E_y, and E_z are the components of field strength,

$$\left.\begin{aligned}D_x &= \varepsilon_0\varepsilon E_x, \\ D_y &= \varepsilon_0\varepsilon E_y, \\ D_z &= \varepsilon_0\varepsilon E_z.\end{aligned}\right\} \tag{2.20}$$

and

Combining these with eqn. (2.19) yields

$$\frac{\partial E_x}{\partial x}+\frac{\partial E_y}{\partial y}+\frac{\partial E_z}{\partial z}=\frac{\varrho_c}{\varepsilon_0\varepsilon}, \tag{2.21}$$

but, since field strength is equal to the potential gradient [cf. eqn. (2.8)],

$$\left.\begin{aligned}E_x &= -\frac{\partial V}{\partial x}, \\[1mm] E_y &= -\frac{\partial V}{\partial y}, \\[1mm] E_z &= -\frac{\partial V}{\partial z};\end{aligned}\right\} \tag{2.22}$$

and

so that substituting, in (2.21), in terms of the potential gradients, gives

$$\frac{\partial^2 V}{\partial x^2}+\frac{\partial^2 V}{\partial y^2}+\frac{\partial^2 V}{\partial z^2}=-\frac{\varrho_c}{\varepsilon_0\varepsilon}. \tag{2.23}$$

This is Poisson's equation in cartesian form, and when $\varrho_c = 0$ it becomes Laplace's equation

$$\frac{\partial^2 V}{\partial x^2}+\frac{\partial^2 V}{\partial y^2}+\frac{\partial^2 V}{\partial z^2}=0. \tag{2.24}$$

In two-dimensional problems the variation of potential in one direction is zero, so that $\partial^2 V/\partial z^2 = 0$, and thus the two-dimensional form of Poisson's equation is

$$\frac{\partial^2 V}{\partial x^2}+\frac{\partial^2 V}{\partial y^2}=-\frac{\varrho_c}{\varepsilon_0\varepsilon}. \tag{2.25}$$

This is written in terms of cylindrical polar coordinates as

$$\frac{\partial^2 V}{\partial r^2}+\frac{1}{r}\frac{\partial V}{\partial r}+\frac{1}{r^2}\frac{\partial^2 V}{\partial\theta^2}=-\frac{\varrho_c}{\varepsilon_0\varepsilon}. \tag{2.26}$$

All these equations have been expressed in terms of the potential V, but clearly they apply equally to the potential function ψ. Hence, in cartesian form,

$$\frac{\partial^2 \psi}{\partial x^2} + \frac{\partial^2 \psi}{\partial y^2} = -\frac{K_1 \varrho_c}{\varepsilon_0 \varepsilon}. \tag{2.27}$$

By virtue of the fact that the flux and potential functions are orthogonal, it can be shown (see section 2.4) that the flux function also obeys Poisson's equation. Hence

$$\frac{\partial^2 \varphi}{\partial x^2} + \frac{\partial^2 \varphi}{\partial y^2} = -\frac{K_2 \varrho_c}{\varepsilon_0 \varepsilon}. \tag{2.28}$$

All these field equations have been derived from the principle of the conservation of flux and are thus an expression of this physical principle.

2.1.6. Principle of superposition

The total flux entering or leaving any surface from any distribution of n charges is the algebraic sum of the separate fluxes due to each charge. Also, since work is a scalar quantity, the work done in moving unit charge through a field due to many charges is the algebraic sum of the separate amounts of work done against the field of each charge separately. Thus the flux and potential functions of the fields of charges taken separately may be added as scalars to give the resultant values of these functions at a point; that is, the principle of superposition may be applied to them.

2.1.7. Electric fields of currents

When an electric field is applied across a conducting medium there is a movement of electric charge, constituting an electric current, in the material. This current is analogous to the electric flux in a dielectric medium, and the electric current density corresponds to the electrostatic flux density.

The current density $\mathbf{J}$, due to a line source of current i, is given in magnitude by

$$J = \frac{i}{2\pi r}. \tag{2.29}$$

The magnitude of the electric current density $\mathbf{J}$ depends on the *conductivity* σ of the material in such a way that

$$\mathbf{J} = \sigma \mathbf{E}, \tag{2.30}$$

the direction of $\mathbf{J}$ being in the direction of $\mathbf{E}$. The potential V is therefore related to the current source i by

$$V = \text{constant} - \frac{i}{2\pi\sigma} \log r.$$

It is often convenient to consider the current flow as caused by the potential difference between the boundaries, and the conductance of the medium between them is defined as the ratio of the current flow to their potential difference. The value at a point of the outward current per unit volume is div $\mathbf{J}$ and is zero once the volume has become charged. Therefore in the steady state

$$\text{div } \mathbf{J} = 0. \tag{2.31}$$

Because of the relationship between the current field and the electrostatic field, it can be seen that the current field may also be described in terms of flux and potential functions which obey Laplace's equation,[†] but since the divergence of **J** is always zero Poisson's equation does not apply.

2.2. Magnetic fields

Magnetic fields can be considered to be due to poles or currents or both, but, whatever their assumed source, they are described in terms of the same field vectors.

The unit of current, the ampere, is defined in terms of the force between parallel current filaments. The magnetic field strength and magnetic flux density can be defined in terms either of currents or of poles because of the equivalence of the current loop and the magnetic shell. Because of the analogy with the electric field of a point charge, it is convenient to express the magnetic field vectors first in terms of the magnetic pole.

2.2.1. The field of magnetic poles

When a ferromagnetic material is magnetized, magnetic dipoles are produced in it and they give rise to field components superimposed on those causing the magnetization. Although poles exist only in pairs as dipoles, it is useful to consider first the isolated magnetic pole.

A point pole, of magnitude m, is considered to emit m units of flux, and so the *flux density* **B** at a distance r from it is given in magnitude by

$$B = \frac{m}{4\pi r^2}.$$
(2.32)

The *magnetic field strength* **H** is such that unit pole placed in the field experiences a force of H units in the same direction as the flux density. The ratio of B to H depends on the capacity of the material to produce dipoles and it is known as the permeability of the material. μ_0 is the *primary magnetic constant* and μ is the permeability of the material *relative* to that of free space, and hence

$$\mathbf{B} = \mu\mu_0\mathbf{H}.$$
(2.33)

From these relationships it is evident that the vectors **B** and **H** are analogous to the electric field vectors **D** and **E**. Hence the field of magnetic poles is directly analogous to the electrostatic field. Consequently, *magnetic potential Ω*, at a point, defined as the work pone in moving unit pole from infinity to the point, obeys Poisson's equation,

$$\operatorname{div}(\operatorname{grad}\Omega) = -\varrho_m/\mu\mu_0,$$
(2.34)

where ϱ_m is the *magnetic pole density*. This implies that

$$\operatorname{div}\mathbf{B} = \varrho_m$$

and

$$\mathbf{H} = -\operatorname{grad}\Omega.$$
(2.35)

[†] The analogy between this type of field and other fields is used in the electrolytic tank[1] to obtain the flux distribution for problems in which direct measurement is difficult or impossible.

The above assumes the existence of free poles, but since magnetic poles occur only in pairs as dipoles, the net flux crossing any surface is zero. Consequently

$$\text{div } \mathbf{B} = 0 \tag{2.36}$$

and

$$\text{div (grad } \Omega) = 0, \tag{2.37}$$

and Ω satisfies Laplace's equation.

2.2.2. *The magnetic field of line currents*

The field strength due to a line current acts tangentially to circles centred on the current, and the work done in moving a magnetic pole once round the current in any closed path is constant. The work done is the product of force and the distance moved, which is $2\pi r$ for a circular path of radius r centred on the current; hence the field strength, the force per unit pole, is given in magnitude by

$$H = \frac{i}{2\pi r}. \tag{2.38}$$

The flux density B is given by

$$B = \frac{\mu\mu_0 i}{2\pi r}, \tag{2.39}$$

and, since it also acts tangentially, the magnetic flux lines for a line current are circles having the conductor as centre.

Flux function. From the last equation it is seen that the flux passing between two points at radii r_1 and r_2 is

$$\int_{r_1}^{r_2} \frac{\mu\mu_0 i}{2\pi r} \, dr,$$

which equals

$$\frac{\mu\mu_0 i}{2\pi} [\log r]_{r_1}^{r_2}.$$

The flux function is proportional to this, and is usually expressed in a form independent of $\mu\mu_0 i$ as

$$\varphi = \frac{1}{2\pi} \log r. \tag{2.40}$$

By comparing eqns. (2.12) and (2.40) it is seen that the flux function for a line current varies with radius in the same manner as does the potential for a line pole or charge.

Magnetic potential of a line current. The gradient of the potential is in the direction of the vector $\mathbf{B}$, and so equipotential lines are radial. Therefore, because of the symmetry of the field, as a point moves round the current its potential changes in direct proportion to the change in its angular position θ with respect to the current. The change in potential for one complete revolution is i, and so the change in potential for movement through an angle θ is given by $\theta i/2\pi$. Because the potential changes continuously with rotation about the current, it is multi-valued at a point.

Potential function. A single-valued potential function can be defined for $0 \leqslant \theta \leqslant 2\pi$ and for a specified direction of rotation (counter-clockwise), so that

$$\psi = \frac{\theta}{2\pi}. \tag{2.41}$$

With the above restrictions the potential function for the field of several line currents may be derived by superposition (see, for example, section 10.2). It is to be noted that the potential function for a line current, though somewhat artificial, still obeys Laplace's equation by virtue of the conservation of flux (see section 2.1.5). By comparing eqns. (2.13) and (2.41) it is evident that the potential function for a line current varies in the same manner as does the flux function for a line pole or charge.

2.2.3. *The magnetic field of distributed currents*

The concept of curl. The work done in moving unit pole once round a path enclosing a current *i* is *i* units, so that

$$\oint H \, dl = i. \tag{2.42}$$

The value, at a point, of the line integral of H per unit area enclosed by the path, as the path length approaches zero, is called the *curl* of **H**. Hence eqn. (2.42) may be written

$$\text{curl } \mathbf{H} = \mathbf{J}, \tag{2.43}$$

where **J** is the current density at the point to which the path reduces. For a line current, curl **H** is zero everywhere except at the position of the current where it is infinite (since the current density there is infinite). It is important to appreciate also that for the field of magnetic poles, curl **H** is always zero, since it has a value only within current-carrying regions. (It is to be noted that the electric field also has the property of zero curl of the field strength. Further, like the magnetic field of poles it is, of course, conservative.)

However, consideration of the work done as a pole moves once round a closed path inside a *current-carrying region* shows an important difference when compared with that in non-curl fields. The amount of work depends on the current enclosed, but, since an infinite number of paths may be described which start and finish at the same point, it can have an infinite number of values. The concept of scalar potential, therefore, has no application to the field inside current-carrying regions. In such a field the lines perpendicular to the flux lines are called *lines of no work* (since no work is done as a pole moves along them), and the meeting point of these lines inside the region is called the *kernel*.

Vector potential. To facilitate the calculation of the field inside current-carrying regions, a vector **A**, called the *vector potential*, is defined so that its line integral round any closed path is equal to the total flux enclosed by that path. As the area of a surface bounded by the path tends to zero, the limiting value of the line integral per unit area, curl **A**, is equal to the flux density **B**. Hence,

$$\text{curl } \mathbf{A} = \mathbf{B}, \tag{2.44}$$

the direction of **A** being in the plane perpendicular to the direction of **B**. This equation defines only the gradients of **A**, and to specify **A** completely its divergence must be defined also. For simplicity of analysis it is customary to make

$$\text{div } \mathbf{A} = 0 \tag{2.45}$$

(Note that the relation between **A** and **B** is analogous to that between **H** and **J**.)

In any such two-dimensional field, since it is in a plane normal to the flux, **A** is parallel to the current. Hence, since the change in **A** between two flux lines, per unit length in the direction of the current, is equal to the flux enclosed (considering a plane parallel to the current), **A** has the properties of the flux function, and lines joining points having the same value of **A** are flux lines.

The field equation. The equations, in terms of **A**, describing the magnetic field of distributed currents are now established, first, in vector form. From eqns. (2.33) and (2.43),

$$\text{curl } \mathbf{B} = \mu\mu_0\mathbf{J}, \tag{2.46}$$

and, therefore, from the definition of **A**, eqn. (2.44),

$$\text{curl (curl } \mathbf{A}) = \mu\mu_0\mathbf{J}.$$

Now it is shown in any book on vector analysis that

$$\text{curl (curl } \mathbf{A}) = \text{grad (div } \mathbf{A}) - \nabla^2\mathbf{A},$$

and so, from eqn. (2.45),

$$\nabla^2\mathbf{A} = -\mu\mu_0\mathbf{J}. \tag{2.47}$$

This means that the components of vector potential satisfy Poisson's equation or Laplace's equation when $J = 0$.

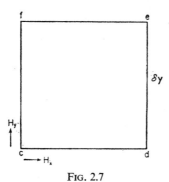

FIG. 2.7

In cartesian coordinates the two-dimensional form of eqn. (2.47) is established by considering the field inside an infinitely small rectangle with sides δx and δy (Fig. 2.7). If H_x is the component of field in the x-direction at the corner c, the components at the other corners are:

$$\text{at } d, \quad H_x + \frac{\partial H_x}{\partial x}\delta x,$$

$$\text{at } e, \quad H_x + \frac{\partial H_x}{\partial x}\delta x + \frac{\partial H_x}{\partial y}\delta y,$$

and

$$\text{at } f, \quad H_x + \frac{\partial H_x}{\partial y}\delta y.$$

There are similar expressions for the field components in the y-direction in terms of the component H_y at the corner c. The mean value of a component of field strength along one side is the average of the components of field at the ends of that side, and so, for instance, the mean value of H_x along the side *de* is

$$H_x + \frac{\partial H_x}{\partial x} \delta x + \frac{1}{2} \frac{\partial H_x}{\partial y} \delta y.$$

Now the work done in moving unit pole round the path *cdefc* is given by the algebraic sum of the products of the field component along each side and the length of the side. This is equal to the amount of current enclosed by the path (eqn. 2.43) and, therefore,

$$\left(H_x + \frac{1}{2} \frac{\partial H_x}{\partial x} \delta x \right) \delta x + \left(H_y + \frac{\partial H_y}{\partial x} \delta x + \frac{1}{2} \frac{\partial H_y}{\partial y} \delta y \right) \delta y$$

$$- \left(H_x + \frac{\partial H_x}{\partial y} \delta y + \frac{1}{2} \frac{\partial H_x}{\partial x} \delta x \right) \delta x - \left(H_y + \frac{1}{2} \frac{\partial H_y}{\partial y} \delta y \right) \delta y$$

$$= J_z \, \delta x \, \delta y,$$

where J_z is the current density in the z-direction. This equation reduces to

$$\left(\frac{\partial H_y}{\partial x} - \frac{\partial H_x}{\partial y} \right) = J_z. \tag{2.48}$$

The expression on the left-hand side of this equation, being the line integral of $H \, dl$ round the path, per unit area enclosed, is one component of the curl of **H** in cartesian form. Similarly, from eqn. (2.44) the components of flux density are given by

$$B_y = \frac{\partial A_x}{\partial z} - \frac{\partial A_z}{\partial x} \tag{2.49}$$

and

$$B_x = \frac{\partial A_z}{\partial y} - \frac{\partial A_y}{\partial z}. \tag{2.50}$$

But, since the field does not vary in the z-direction, the terms $\partial A_x/\partial z$ and $\partial A_y/\partial z$ are zero, and so

$$B_x = -\frac{\partial A_z}{\partial x} \tag{2.51}$$

and

$$B_y = \frac{\partial A_z}{\partial y}. \tag{2.52}$$

Remembering that $\mathbf{B} = \mu\mu_0\mathbf{H}$, substitution from eqn. (2.51) and (2.52) in (2.48) gives

$$\frac{\partial}{\partial x} \left(-\frac{\partial A_z}{\partial x} \right) - \frac{\partial}{\partial y} \left(\frac{\partial A_z}{\partial y} \right) = \mu\mu_0 J_z,$$

and so, finally,

$$\frac{\partial^2 A_z}{\partial x^2} + \frac{\partial^2 A_z}{\partial y^2} = -\mu\mu_0 J_z. \tag{2.53}$$

This equation, in terms of vector potential, describes the two-dimensional magnetic field of a current flowing in the z-direction. Putting $J_z = 0$ in this equation reduces it to

Laplace's equation which describes the field in a region carrying no current. Since A has the property of a flux function, eqn. (2.53) may be written

$$\frac{\partial^2 \varphi}{\partial x^2} + \frac{\partial^2 \varphi}{\partial y^2} = -\mu\mu_0 J. \tag{2.54}$$

It should be noted that, although **A** is introduced to facilitate analysis of fields of distributed current, it is frequently used for the field of line currents.

The field of a current-carrying conductor of circular section. To demonstrate the use of vector potential in a simple case, the field of an infinitely long straight conductor of relative permeability μ and of circular section, carrying current of uniform density J, is examined. This field has circular symmetry, and consequently the field vectors do not vary with movement at constant radius r about the centre of the conductor. Because of this it may be described without reference to **A**, and this is first done.

When a unit pole is moved round a circle concentric with the conductor, the work done is equal to the current enclosed. Therefore, when the circle is inside the conductor, the field strength there, H_1, is expressed by

$$2\pi r H_1 = \pi r^2 J, \tag{2.55}$$

and so

$$H_1 = \frac{Jr}{2}. \tag{2.56}$$

Again, when the circle is outside the conductor, the field strength there, H_2, is given by

$$2\pi r H_2 = \pi a^2 J,$$

a being the radius of the conductor, and so

$$H_2 = \frac{Ja^2}{2r}. \tag{2.57}$$

Thus it is seen that the field strength is proportional to the radius inside the conductor and inversely proportional to the radius outside, where it is identical with that of a line current situated at the centre of the conductor.

To determine the form of the vector potential function describing this field, consider a rectangular path in an axial plane, having two sides at radius r and $(r+\delta r)$ parallel to the direction of the current, and two sides unit distance apart. By definition the line integral of **A** taken once round this rectangle is equal to the flux linking the rectangle. Thus as the vector potential is constant at any given radius, the change in the vector potential function from radius r to $(r+\delta r)$ is equal to $B\delta r$. Inside the conductor, where $A = A_1$, the flux density is, from eqn. (2.56), $\frac{1}{2}\mu\mu_0 Jr$, and so

$$dA_1 = \frac{1}{2}\mu\mu_0 Jr \, dr, \tag{2.58}$$

which, when integrated, gives

$$A_1 = \frac{1}{4}\mu\mu_0 Jr^2 + C_1.$$

Since only changes in vector potential are specified by eqn. (2.44), the origin of A_1 is arbitrary and, for convenience, the constant C_1 can be made zero to give

$$A_1 = \frac{1}{4}\mu\mu_0 Jr^2. \tag{2.59}$$

AC 3

Outside the conductor, where $A = A_2$, the flux density is, from eqn. (2.57), $\mu_0 Ja^2/2r$, and so

$$dA_2 = \frac{1}{2}\mu_0 \frac{Ja^2}{r}\, dr,$$

which, when integrated, makes

$$A_2 = \tfrac{1}{2}\mu_0 Ja^2 \log r + C_2. \tag{2.60}$$

Now the vector potential is continuous across the boundary of the conductor and so, when $r = a$, $A_1 = A_2$. Hence

$$\tfrac{1}{4}\mu\mu_0 Ja^2 = \tfrac{1}{2}\mu_0 Ja^2 \log a + C_2, \tag{2.61}$$

giving

$$C_2 = \tfrac{1}{4}\mu_0 Ja^2(\mu - 2 \log a).$$

Thus substituting for C_2 in eqn. (2.60) yields

$$A_2 = \frac{1}{4}\mu_0 Ja^2\left[\mu + 2 \log\left(\frac{r}{a}\right)\right]. \tag{2.62}$$

In this simple example the boundary is a flux line, and an expression for the field strength is simply obtained. In general, however, it is necessary first to obtain a solution in terms of vector potential, satisfying Poisson's equation and the boundary conditions (see section 5.1).

2.2.4. *Inductance*

The concept of inductance is most useful in analysing the effects in a circuit of the magnetic fields of currents changing in magnitude with time. Though a discussion of its significance involves a consideration of induced e.m.f. and stored energy and so is beyond the scope of this book,[†] a definition is given in mathematical form and its use is demonstrated.

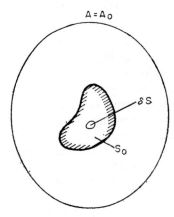

FIG. 2.8

Consider a conductor of arbitrary cross-section, as shown in Fig. 2.8, in a magnetic field described by the vector potential function A. The field may be due to current in the conductor itself, giving rise to what is termed *self*-inductance, or to currents in other conductors,

† For a full discussion the reader should consult the book by Carter (see Appendix IV, Bibliography).

giving rise to *mutual* inductance. Let δS be any elemental area of the conductor and let the vector potential function of the field at the position of the element be A. Then the inductance L, associated with all the flux linking the conductor but bounded by the flux line, $A = A_0$, is defined by

$$L = \frac{1}{IS_0} \int (A_0 - A) \, dS, \tag{2.63}$$

where I is the current giving rise to the field and where S_0 is the total area of the conductor, the integration being performed over this area.

For a rectangular conductor the integration is conveniently performed using cartesian coordinates with axes chosen to be parallel to the sides of the conductor. Then

$$L = \frac{1}{IS_0} \int_{x_1}^{x_2} \int_{y_1}^{y_2} (A_0 - A) \, dx \, dy, \tag{2.64}$$

where x_1 and x_2 are the limits of the conductor on the x-axis and y_1 and y_2 are those on the y-axis.

In evaluating inductance analytically it is simply necessary to determine the function A for the field, to substitute in eqn. (2.63) and to perform the integration. If the field solution is known in the form of a map of flux lines, the conductor is divided up into finite elements, each bounded by flux lines; and then taking a mean value of A for each element the sum of the terms $(A_0 - A) \, \delta S$, one for each element, is obtained.

Equation (2.64) has been used by Billig[2] and is used, in section 5.6.3, to determine analytically the inductance of a rectangular conductor in a slot.

2.3. Boundary conditions

When a field has regions with different electric or magnetic properties, separate functions are often used to describe the fields in the different regions. It is simple to establish the relationship between these functions at the boundaries, and this is done here for an interface between two regions in a magnetic field. Directly analogous equations exist for electric fields.

Consider an elemental length bd of the interface between two regions of relative permeability μ_1 and μ_2. Flux passes from one region to the other and a pair of parallel flux lines passing through b and d are considered to bend at the boundary (Fig. 2.9). The lines ab and cd are normal to the flux lines φ' and φ'' respectively. Now unless the current density on the boundary itself is infinite (a case discussed later), the work done in taking unit pole round the typical infinitesimally small path $abcda$ is zero. But ab and dc are equipotential lines, and so the work done from b to c plus the work done from d to a is zero. Hence if H_1 and H_2 are the field strengths on either side of the interface,

$$-H_1(bc) + H_2(da) = 0, \tag{2.65}$$

the negative sign denoting movement in a direction opposite to that of H_1. If θ_1 and θ_2 are the angles made by dc and ab with db, then

$$da = db \sin \theta_2 \quad \text{and} \quad bc = db \sin \theta_1,$$

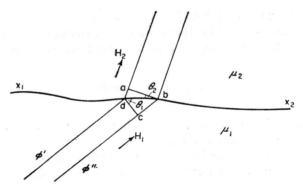

FIG. 2.9

and combining these with (2.65) gives

$$H_1 \sin \theta_1 = H_2 \sin \theta_2. \tag{2.66}$$

Hence, in the absence of an interface current, the tangential components of H on each side of an interface are equal. Again, since flux is continuous, if B_1 and B_2 are the densities on each side of the boundary,

$$B_1(dc) = B_2(ab),$$

and so

$$B_1 \cos \theta_1 = B_2 \cos \theta_2. \tag{2.67}$$

Thus the normal components of flux density are equal on each side of the boundary.
It can be seen by combining eqns. (2.66) and (2.67) that

$$\frac{H_1}{B_1} \tan \theta_1 = \frac{H_2}{B_2} \tan \theta_2,$$

and so

$$\frac{\tan \theta_1}{\tan \theta_2} = \frac{\mu_1}{\mu_2}. \tag{2.68}$$

If the permeability is infinite on one side of the boundary, then θ is zero on the other side, which means that flux enters an infinitely permeable (equipotential) surface at right angles.
In the case when current of infinite density (as in a current sheet) is distributed along the boundary surface, the normal components of flux density on the two sides of the boundary are still equal, but the tangential components of field strength are discontinuous by an amount equal to the magnitude of the surface density S; i.e.,

$$-H_1 \sin \theta_1 + H_2 \sin \theta_2 = S. \tag{2.69}$$

For any given problem the relationship connecting the field vectors on the two sides of an interface, eqn. (2.67) and eqn. (2.66) or (2.69), are expressed in terms of the appropriate derivatives of the functions ψ and φ (or A) defining the fields. In terms of the potential functions ψ_1, and ψ_2, eqns. (2.66) and (2.67) become

$$\frac{\partial \psi_1}{\partial s} = \frac{\partial \psi_2}{\partial s}, \tag{2.70}$$

where s is distance measured tangential to the interface, and

$$\mu_1 \frac{\partial \psi_1}{\partial n} = \mu_2 \frac{\partial \psi_2}{\partial n}, \tag{2.71}$$

where n is distance measured normal to the interface. In terms of φ they become

$$\frac{1}{\mu_1} \frac{\partial \varphi_1}{\partial n} = \frac{1}{\mu_2} \frac{\partial \varphi_2}{\partial n} \tag{2.72}$$

and

$$\frac{\partial \varphi_1}{\partial s} = \frac{\partial \varphi_2}{\partial s}. \tag{2.73}$$

These equations are used in terms of the particular coordinate system appropriate to a particular problem.

When current is distributed on one or both sides of the interface, the above boundary conditions are conveniently expressed in terms of vector potential. Then, since the normal components of flux density are equal on the two sides of a boundary, from eqn. (2.51), the tangential gradients of the vector potential function are of equal magnitude and A replaces φ in eqn. (2.73). Also, as mentioned in the example of a cylindrical conductor, the vector potential function is made continuous across a boundary.

2.4. Conjugate functions

The application of complex variable theory to the analysis of fields makes possible the simple solution of many problems which would otherwise be difficult or impossible. The real and imaginary parts of any continuous, regular[†] function of a complex variable are called *conjugate functions* and it is shown that both are solutions of Laplace's equation. Further, the flux and potential function for any field are conjugate functions, and they may be combined together in a single function of a complex variable and so handled with facility.

2.4.1. *Laplace's equation*

Consider a complex variable z defined by

$$z = x + jy. \tag{2.74}$$

Let a complex variable t, defined by

$$t = u + jv, \tag{2.75}$$

be any continuous regular function of z; that is, let

$$t = F(z). \tag{2.76}$$

Differentiating this equation partially with respect to x gives

$$\frac{\partial t}{\partial x} = \frac{\partial F(z)}{\partial z} \frac{\partial z}{\partial x};$$

[†] A continuous single-valued function is regular if the partial derivatives $\partial u/\partial x$, $\partial v/\partial x$, $\partial u/\partial y$, $\partial v/\partial y$ exist, are continuous and satisfy the Cauchy–Riemann equations (see section 2.4.2).

but

$$\frac{\partial z}{\partial x} = 1,$$

and so

$$\frac{\partial t}{\partial x} = F'(z). \tag{2.77}$$

Further, differentiating this equation again with respect to x gives

$$\frac{\partial^2 t}{\partial x^2} = \frac{\partial}{\partial x} F'(z)$$

$$= \frac{\partial F'(z)}{\partial z} \frac{\partial z}{\partial x},$$

and so

$$\frac{\partial^2 t}{\partial x^2} = F''(z). \tag{2.78}$$

Also, differentiating eqn. (2.76) partially with respect to y gives

$$\frac{\partial t}{\partial y} = F'(z) \frac{\partial z}{\partial y};$$

but

$$\frac{\partial z}{\partial y} = j,$$

and so

$$\frac{\partial t}{\partial y} = jF'(z). \tag{2.79}$$

Differentiating again with respect to y gives

$$\frac{\partial^2 t}{\partial y^2} = \frac{\partial}{\partial y} jF'(z)$$

$$= jF''(z) \frac{\partial z}{\partial y},$$

and so

$$\frac{\partial^2 t}{\partial y^2} = -F''(z). \tag{2.80}$$

Finally, combining eqns. (2.78) and (2.80)[†] gives

$$\frac{\partial^2 t}{\partial x^2} + \frac{\partial^2 t}{\partial y^2} = 0. \tag{2.81}$$

Thus any regular function of a complex variable obeys eqn. (2.81), which will be recognized as Laplace's equation.

---

† This assumes that $F'(z)$ and $F''(z)$ are unique; that is, that $F(z)$ is regular.

Equation (2.81) may be written as two equations, in terms of the real and imaginary parts of t, by noting that differentiation of eqn. (2.75) gives

$$\frac{\partial^2 t}{\partial x^2} = \frac{\partial^2 u}{\partial x^2} + j\frac{\partial^2 v}{\partial x^2}$$

and

$$\frac{\partial^2 t}{\partial y^2} = \frac{\partial^2 u}{\partial y^2} + j\frac{\partial^2 v}{\partial y^2}.$$

The resulting equations are

$$\frac{\partial^2 u}{\partial x^2} + \frac{\partial^2 u}{\partial y^2} = 0 \tag{2.82}$$

and

$$\frac{\partial^2 v}{\partial x^2} + \frac{\partial^2 v}{\partial y^2} = 0, \tag{2.83}$$

and they show that not only a function, but also the real and imaginary parts of any regular function of a complex variable obey Laplace's equation.

2.4.2. Cauchy–Riemann equations

Consider again eqn. (2.75). When differentiated with respect to x it gives

$$\frac{\partial t}{\partial x} = \frac{\partial u}{\partial x} + j\frac{\partial v}{\partial x},$$

and combining this with eqn. (2.77) gives

$$F'(z) = \frac{\partial u}{\partial x} + j\frac{\partial v}{\partial x}. \tag{2.84}$$

Similarly it is seen from eqns. (2.75) and (2.79) that

$$jF'(z) = \frac{\partial u}{\partial y} + j\frac{\partial v}{\partial y}. \tag{2.85}$$

Then eliminating $F'(z)$ between eqns. (2.84) and (2.85)[†] yields

$$j\frac{\partial u}{\partial x} - \frac{\partial v}{\partial x} = \frac{\partial u}{\partial y} + j\frac{\partial v}{\partial y}, \tag{2.86}$$

and, since the real and imaginary parts on both sides of the equation must be equal, then

$$\frac{\partial u}{\partial x} = \frac{\partial v}{\partial y} \tag{2.87}$$

† See previous footnote.

and

$$\frac{\partial u}{\partial y} = -\frac{\partial v}{\partial x}.$$

(2.88)

There are the Cauchy–Riemann equations and they are satisfied by the real and imaginary parts of any regular function of a complex variable.

2.4.3. *Flux and potential functions as conjugate functions*

A constant value of t defines two curves in the z-plane, $u(x, y) = $ constant and $v(x, y) = $ constant; and since t may take an infinite number of values these equations define two families of curves. Let $u = u_0$ and $v = v_0$ define one curve from each family and consider the intersection of the curves at point $z = z_0$. Taking first the curve for $u = u_0$, the slope is dy/dx and, since u remains constant as x and y change,

$$\frac{\partial u}{\partial x} \delta x + \frac{\partial u}{\partial y} \delta y = 0,$$

and, in the limit,

$$\left(\frac{dy}{dx}\right)_{z_0} = \left(\frac{-\partial u/\partial x}{\partial u/\partial y}\right)_{z_0}.$$

(2.89)

Similarly, the slope of the curve $v = v_0$ is given by

$$\left(\frac{dy}{dx}\right)_{z_0} = \left(\frac{-\partial v/\partial x}{\partial v/\partial y}\right)_{z_0}.$$

(2.90)

From eqns. (2.89) and (2.90) the product of the slopes is

$$\left(\frac{\partial u}{\partial x} \middle/ \frac{\partial u}{\partial y}\right)_{z_0} \left(\frac{\partial v}{\partial x} \middle/ \frac{\partial v}{\partial y}\right)_{z_0},$$

which, from the Cauchy–Riemann equations (2.87) and (2.88), is equal to -1. Therefore the two curves intersect at right angles and this proves that families of curves corresponding to constant values of conjugate functions are orthogonal.

Because conjugate functions are solutions of Laplace's equation and are orthogonal functions they may be used to represent flux and potential functions. Consider the flux and potential functions describing a field in the (x, y) plane, and let

$$\varphi = f_1(x, y) = f_1(z) \quad \text{and} \quad \psi = f_2(x, y) = f_2(z).$$

Then, since these can be represented as conjugate functions, they may be combined together in a single function of a complex variable, $w(z)$, where

$$w(z) = f_1(z) + jf_2(z).$$

This function, w, is called the *complex potential function* and it is of fundamental importance in the use of complex variable theory for the solution of field problems; in terms of φ and ψ it is

$$w = \varphi + j\psi.$$

(2.91)

It should be emphasized that all the above quantities are merely numbers having no dimensions. However, in a particular problem it is often convenient to choose a scale constant so that the value of either φ or ψ gives directly a quantity of flux or a value of potential difference. Note that φ and ψ can be interchanged in eqn. (2.91) when convenient.

It is pointed out earlier (see section 2.2) that φ and ψ can be derived from each other, and it is now evident that, being conjugate functions, they are related by the Cauchy–Riemann equations. Hence, from these equations, the relationships are

$$\frac{\partial \varphi}{\partial x} = \frac{\partial \psi}{\partial y} \tag{2.92}$$

and

$$\frac{\partial \psi}{\partial x} = -\frac{\partial \varphi}{\partial y}. \tag{2.93}$$

These are based on the expression of the field in cartesian coordinates, but they may be expressed equally in the forms appropriate to other coordinate systems. For example, in circular cylinder coordinates the equations are

$$\frac{\partial \varphi}{\partial r} = \frac{1}{r} \frac{\partial \psi}{\partial \theta} \tag{2.94}$$

and

$$\frac{\partial \psi}{\partial r} = -\frac{1}{r} \frac{\partial \varphi}{\partial \theta}. \tag{2.95}$$

2.4.4. *Simple examples of the use of conjugate functions*

To demonstrate the use of conjugate functions and the complex potential function, two fields treated earlier in terms of purely real functions are considered.

Charged concentric cylinders. It is shown earlier, eqn. (2.12), that, for the field between charged concentric cylinders, the potential function is given by

$$\psi = \frac{q}{2\pi\varepsilon_0\varepsilon} \log r \tag{2.96}$$

and the flux function by

$$\varphi = \frac{q}{2\pi} \theta. \tag{2.97}$$

In order that φ and ψ can be combined to give the complex potential function w, it is necessary that they be expressed with appropriate scale factors by multiplying one of them by a constant factor; in this case it is convenient to multiply the potential function by $\varepsilon_0\varepsilon$ to give

$$\psi = \frac{q}{2\pi} \log r. \tag{2.98}$$

The complex potential function then becomes [interchanging ψ and φ as compared with eqn. 2.91)]

$$w = \psi + j\varphi = \frac{q}{2\pi}(\log r + j\theta), \tag{2.99}$$

which may be expressed in terms of the complex variable t, where $t = r \exp(j\theta)$, as

$$w = \frac{q}{2\pi} \log t. \tag{2.100}$$

This represents the field of two charged concentric cylinders (or of a line charge), centred about the origin of the t-plane, and substitution for the coordinates of a point in the t-plane gives the values of flux and potential function there, to a scale determined by the form of eqn. (2.100). This equation gives quantities of flux directly in the m.k.s. system of units, but to make the solution dimensionless it is convenient to use the form

$$w = \frac{1}{2\pi} \log t. \tag{2.101}$$

The field map can be obtained by writing this equation in the form

$$t = \exp(2\pi\psi + 2\pi j\varphi), \tag{2.102}$$

and by substituting values of ψ and φ.

The field of a line current. To show the way in which the flux and potential functions may be derived from each other, the field of a line current in complex-variable form is considered. From eqn. (2.40) the flux function is

$$\varphi = \frac{1}{2\pi} \log r, \tag{2.103}$$

and the Cauchy–Riemann equation (2.94) may be rewritten

$$\psi = \int r \frac{\partial \varphi}{\partial r} \, \delta\theta.$$

Hence, differentiating eqn. (2.103) and substituting gives, in general,

$$\psi = \frac{1}{2\pi} \theta + f(r).$$

From symmetry, it is apparent that ψ is a function of θ only and so $f(r)$ is equal to an arbitrary constant, which may be ignored, and so the potential function becomes

$$\psi = \frac{1}{2\pi} \theta, \tag{2.104}$$

which is seen to be identical with eqn. (2.41). φ and ψ can, of course, be combined to form

$$w = \frac{1}{2\pi} \log t, \tag{2.105}$$

which is identical with eqn. (2.101). Therefore the complex potential functions for the fields of a line charge and a line current can be expressed by identical functions, for the fields of a line charge and a line current vary in identical ways, but with the flux and potential functions interchanged.

2.5. Equivalent pole and charge distributions

For a number of purposes—in particular the calculation of forces on boundaries (see the next section) and the derivation of image solutions (see Chapter 3)—it is best to consider the effect of a boundary as being due simply the charges, poles, or currents which lie along the boundary line. The following discussion of such surface distributions is restricted to magnetic fields and pole distributions, but, of course, electric fields and charge distributions are analogous. Surface currents can also be used as an alternative to poles, but the two representations are equivalent and surface poles are easier to handle.

Consider a region 1 of permeability $\mu_1\mu_0$ separated from a region 2 of permeability $\mu_2\mu_0$ by a boundary of arbitrary shape (Fig. 2.10). The distribution of poles, which when lying along the boundary line, gives the same effect on the field in region 1 as does the presence of the interface, is to be found. It is convenient to consider the boundary as consisting of an infinitely thin region of permeability μ_0 in which the pole distribution lies. At any point on the boundary, let H_n be the normal component of the applied field strength, i.e. the field strength due to all the field sources which may be in either or both of the regions in the absence of polarized media. The effect of the polarized media can be accounted for by a normal component of field H'_n, at the boundary, and considered to act in the same direction as H_n in region 1.

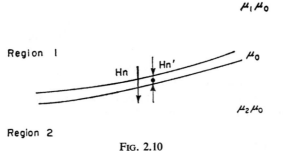

FIG. 2.10

Thus the resultant normal field at a point on the boundary is $H_n+H'_n$ in region 1 and $H_n-H'_n$ in region 2, and since the normal component of flux is continuous across the boundary it is necessary that

$$\mu_1(H_n+H'_n) = \mu_2(H_n-H'_n)$$

or

$$H'_n = \left(\frac{\mu_2-\mu_1}{\mu_2+\mu_1}\right)H_n. \tag{2.106}$$

The surface pole density ϱ_s which gives rise, in a region of permeability μ_0, to the component H'_n, and so to the effect of the boundary, is found simply. Since flux passes equally in each of the two directions normal to the boundary,

$$\varrho_s = 2\mu_0 H'_n, \tag{2.107}$$

and so, eliminating H'_n between eqns. (2.106) and (2.107), ϱ_s is related to the applied field by

$$\varrho_s = 2\mu_0\left(\frac{\mu_2-\mu_1}{\mu_2+\mu_1}\right)H_n. \tag{2.108}$$

For the calculation of force discussed in the next section, it is convenient to express the pole density in terms of the *resultant* field strength in region 1, H_{n1}. This equals $H_n+H'_n$, and from eqns. (2.107) and (2.108) it is seen that

$$\varrho_s = \frac{\mu_0}{\mu_2}(\mu_2-\mu_1)H_{n1}. \tag{2.109}$$

2.6. Forces

The force acting on a line source in a field is found simply, but the calculation of forces acting on boundaries is often difficult. There are several methods available for calculating *total* force on a boundary, and two of the most useful ones are discussed. These are reasonably simple to apply and give solutions for most fields of practical interest. For general reading on the calculation of force the authors recommend the recent papers by Carpenter[3] and the books by Moullin, Stratton, and Carter,[4] and Maxwell. The calculation of force *distribution* is briefly mentioned in Maxwell and reference (3), but the application to practical problems is not discussed, except by Carter[4] and Hammond[5]. In section 2.6.4 is given a simple method applicable to equipotential boundaries.

2.6.1. *Line sources*

In any physical problem sources have a finite cross-sectional area, but in most calculations it is sufficiently accurate to treat them as line sources. The force per unit length experienced by a line source in a field (of the same kind as that of the source) is given by the product of the field strength (or in the case of a line current flux density) at the position of the source due to all the other influences, and the strength of the source. For a line charge q per unit length in an electric field of strength E, the force is Eq; for a line pole in a magnetic field it is Hm; and for a line current in a magnetic field it is Bi.

2.6.2. *Distributed sources*

The force experienced by a field source, which is distributed over an area, can be calculated by considering the source as being made up of line elements and by summing the forces on these elements (see section 5.2.3). The field strength at the position of an element is calculated as that due to all the field influences except the source considered. If the field of the other elements of the source is taken into account, the force calculated includes a contribution from the internal forces in the source and these sum to zero when the total external force is calculated.

2.6.3. *Total force acting on a boundary*

Simple expressions for the total force acting on a boundary are derived in terms of the magnetic field; analogous relations exist for the electric field.
Equivalent pole distribution. The first method is based on a consideration of the force exerted on the surface pole distribution which accounts for the influence of the boundary on the external field. The method does not give the force *distribution* for a (practical) piece of iron[4] but, since the external field is truly represented, it does give the correct value for the total force. The required surface pole distribution is given by eqn. (2.109) in terms of H_{n1}.

Now the normal component of force per unit length F_n acting on the surface poles is given by

$$F_n = H_n \varrho_s, \qquad (2.110)$$

where H_n is the applied normal field strength. The tangential component of force per unit length F_T is given by

$$F_T = H_{T1} \varrho_s, \qquad (2.111$$

where H_{T1} is the tangential field strength. From eqn. (2.106), remembering that $H_{n1} = H_n + H'_n$ it is simply shown that

$$H_n = H_{n1}\left(\frac{\mu_2 + \mu_1}{2\mu_2}\right).$$

Hence, substituting for ϱ_s from eqn. (2.109) in eqn. (2.110) gives

$$F_n = \tfrac{1}{2}\mu_0\left(1 - \mu_1^2/\mu_2^2\right)H_{n1}^2, \tag{2.112}$$

and in eqn. (2.111) gives

$$F_T = \mu_0(1 - \mu_1/\mu_2)H_{n1}H_{T1}. \tag{2.113}$$

The total force on the iron is given by integration of these functions over the whole of the boundary surface.

Virtual work. The mechanical force acting on a magnetic material can be determined from the effect, on the energy balance, of a small displacement dx in the direction of the force. The equations given below are derived, for example, in Chapter 2 of the book by Seely.

If the potential difference between two iron surfaces is assumed constant and the flux passing between them changes by an amount $d\varphi$ for a displacement dx, the force in the x-direction is given by

$$f_x = \frac{1}{2}\psi\frac{d\varphi}{dx}. \tag{2.114}$$

If the flux is assumed constant and the potential changes by an amount $d\psi$, the force is then given by

$$f_x = \frac{1}{2}\varphi\frac{d\psi}{dx}. \tag{2.115}$$

The assumption of constant potential difference or flux is made as convenient for the analysis. Equation (2.114) is used later (see section 8.2.8) in the analysis of the force on the armature of a contactor.

2.6.4. *Force distribution over a boundary*

The evaluation of force distribution is in general extremely difficult but, when the boundaries are so highly permeable that it can be assumed that H is zero inside the iron boundary, a simple method may be used. The magnetic forces act only on the surface poles, and since $\mu_2 = \infty$, eqns. (2.112) and (2.113) become

$$F_n = \tfrac{1}{2}\mu_0 H_n^2 \tag{2.116}$$

and

$$F_T = 0.$$

Hence the force, which is normal to the boundary at any point, is obtained by integrating the square of the field strength along the boundary.

References

1. G. LIEBMANN, Electrical analogues, *Br. J. Appl. Phys.* **4**, 193 (1953).
2. E. BILLIG, The calculation of the magnetic field of rectangular conductors in a closed slot and its application to the reactance of transformer windings, *Proc. Instn. Elect. Engrs.*
3. C. J. CARPENTER, Surface-integral methods of calculating forces in magnetized iron parts, *Proc. Instn. Elect. Engrs.* **107** C, 19 (1960).
4. G. W. CARTER, Distribution of mechanical forces in magnetized material, *Proc. Instn. Elect. Engrs.* **112**, 1771 (1965).
5. P. HAMMOND, Forces in electric and magnetic fields, *Bull. Elect. Engng. Educ.* **25**, 17 (1960).

PART II

DIRECT METHODS

CHAPTER 3

IMAGES

3.1. Introduction

The method of images can be used to give solutions to some important problems involving straight-line or circular boundaries and in a particularly simple manner; for it offers certain ready-made solutions which eliminate the need for formal solutions of Laplace's and Poisson's equations. The idea of images for field problems is due to Lord Kelvin, but Maxwell, Lodge,[1] and Searle[2] extended the scope of the method.

The essence of the method consists in replacing the effects of a boundary on an applied field by simple distributions of currents or charges *behind* the boundary line (called images), the desired field being given by the *sum* of the *applied* and the *image* fields. A different system of images is required for the field on each side of a boundary, but a knowledge of one group of images quickly leads to the other, since the solutions for the two regions are connected by the boundary conditions.

In the following discussion, the magnitudes and positions of images for the single straight-line and circular boundaries are established, for convenience, in terms of the electric field of a line charge using a method first indicated by Hammond;[3] the distribution of surface charge (or polarity) representing the influence of the boundary is first found and is then replaced by a simple equivalent array (the images). (As shown in section 2.2.2, the electric field of a line charge and the magnetic field of a line current are analogous, and so, from a knowledge of the images of the electric one, the images of the magnetic one may be deduced directly.) There is no general method of deriving the images for any given problem with multiple boundaries, though a method of successive approximation (see Maxwell, article 315), gives correct results in certain cases.

It is possible to derive image solutions from a solution to Poisson's equation obtained by use of the complex Fourier transform. This method is described in a recent paper by Mullineux and Reed.[4] However, the mathematical manipulation involved in transforming the field equation is considerable compared to the elementary trigonometry needed in the method to be described. Also some image systems can be derived using theorems developed originally for the solution of problems in hydrodynamics.[5] Thus the images for a circular boundary can be deduced from Milne-Thomson's circle theorem (see Milne-Thomson, p. 154). It is possible to check all solutions obtained (by the above or other means) by considering the boundary conditions. If a set of images is proposed for each region of a field, their validity can be tested directly. If a set is proposed for one region only, it can be verified by consideration of surface charge distribution. A simpler check is possible when the boundaries are equipotential or flux line, for then the boundary conditions are easily seen to be satisfied by the system of images which is symmetrically disposed about all the boundary lines.

It is convenient here to consider images under two headings: firstly those due to plane boundaries, and, secondly, those due to circular boundaries. For both of these, consideration is given to the field of line charges (or currents), but the effect of a single circular boundary on an applied uniform field (images of doublets) is also discussed. The image representation of the fields of distributed currents is discussed in section 3.4.

Images are also very useful for solving *three-dimensional* fields, involving plane boundaries with *any shape of conductor*, for example, in problems involving windings near iron boundaries,[3, 6, 13, 14] but a discussion of these does not fall within the scope of the book.

3.2. Plane boundaries

The images for a line charge near an infinite plane boundary are first established, and then the results for this boundary are extended to give solutions for various combinations of plane boundaries. These include two parallel and up to four intersecting plane boundaries, the angles of intersection being submultiples of π. For the single boundary the solution for finite permittivity is given but, for multiple-boundary problems the discusssion is restricted to equipotential and flux-line boundaries.[†] In the cases considered the images are symmetrically disposed about all of the boundaries. Their positions are those of the optical images, in reflecting surfaces coincident with the boundaries, of an object coincident with the charge.

3.2.1. *Single plane boundary*

Consider the derivation of the image charges which give the solution for the field of a line charge q near to an infinite plane boundary (Fig. 3.1). The charge lies in region 1 of permittivity $\varepsilon_0\varepsilon_1$, region 2 having permittivity $\varepsilon_0\varepsilon_2$. Let the two regions be in the (x, y) plane, the boundary coinciding with the x-axis and the charge being at the point $y = a$.

At any point P on the boundary line the normal component of applied field E_n is given by

$$E_n = \frac{q \cos \alpha}{2\pi\varepsilon_0\varepsilon_1 \sqrt{x^2+a^2}}, \qquad (3.1)$$

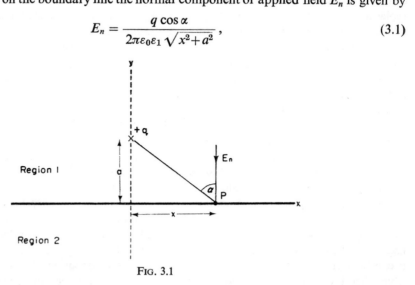

FIG. 3.1

[†] Maxwell, article 317, has given solutions for fields due to a point charge involving parallel boundaries of different conductivities, and these solutions can be extended by superposition to apply to line charges.

where α is as shown in Fig. 3.1. Substituting for α in terms of a and x gives

$$E_n = \frac{qa}{2\pi\varepsilon_0\varepsilon_1(x^2+a^2)},\tag{3.2}$$

and this component of the applied field can be considered to induce charge along the boundary. As shown in section 2.5, the normal component of field strength E_n' due to this equivalent distribution of surface charge is simply related to the applied normal field by the equation

$$E_n' = E_n\left(\frac{\varepsilon_2-\varepsilon_1}{\varepsilon_2+\varepsilon_1}\right).\tag{3.3}$$

Thus the effect of the boundary on the field in region 1 can be completely specified by the component E_n' along the boundary, and this is given, by substitution from eqn. (3.2) in eqn. (3.3), as

$$E_n' = \frac{q}{2\pi\varepsilon_0\varepsilon_1}\left(\frac{\varepsilon_2-\varepsilon_1}{\varepsilon_2+\varepsilon_1}\right)\frac{a}{x^2+a^2}.\tag{3.4}$$

This is the same field distribution, however, as would result either from a charge $-q(\varepsilon_2-\varepsilon_1)/(\varepsilon_2+\varepsilon_1)$ at a distance a behind the boundary line, or, equally, from a charge $+q(\varepsilon_2-\varepsilon_1)/(\varepsilon_2+\varepsilon_1)$ at a distance a from the boundary in region 1. Hence either of these charges can be used to account for the effect of the boundary. However, only one of them can be used in the representation of the field in each of the two regions. Thus, considering first the field in region 1, it is seen that the second of the charges would cause a change in the total flux entering or leaving any curve enclosing the charge q in region 1 [that is eqn. (2.4) for the divergence of flux would not be obeyed], and so it cannot be used to represent the field in region 1. The other charge, however, lies *outside* region 1, and so it does not affect the divergence of flux there. This charge, called an image charge, gives, in conjunction with the actual charge q, the field solution in region 1. (It should be emphasized that this image charge does not apply to the field in region 2.)

In order to derive the image charge for region 2, it is simplest to consider again E_n'. In region 2 this opposes the applied field, and so the equivalent point charges, giving the correct distribution of the normal component of field in region 2, have opposite signs to those above. Of these two, the image charge must again be outside the field region, and so the required one is that of strength $-q(\varepsilon_2-\varepsilon_1)/(\varepsilon_2+\varepsilon_1)$ lying at the point $y=+a$.[†]

Image of line currents. In a similar way to that above, the images of a line current, of strength i, in a medium of permeability μ_1, near an infinite straight-line boundary behind which is a medium of permeability μ_2, can be shown to be similar to the above images. For the field in region 1, the required image is of strength $i(\mu_2-\mu_1)/(\mu_2+\mu_1)$ and it lies at $y=-a$ in region 2; for the field in region 2 the image is of strength $-i(\mu_2-\mu_1)/(\mu_2+\mu_1)$ and lies at the point $y=+a$, in region 1. A field map is shown in Fig. 3.2 for

---

[†] The above solutions can be confirmed by showing that the tangential component of field strength is continuous across the boundary. This is easily seen, since the tangential components due to both images are equal, being given by

$$\left(\frac{\varepsilon_2-\varepsilon_1}{\varepsilon_2+\varepsilon_1}\right)\frac{q}{2\pi\varepsilon_0}\frac{x}{(x_2+a^2)}$$

and the applied field is obviously continuous.

4*

the case $\mu_2/\mu_1 = 5$, and it demonstrates the "refraction" of the flux lines at the boundary.
General remarks. Certain general points emerging from the above discussion should be emphasized. First, the fields are calculated as the resultant of the applied field and the image fields. Secondly, the images for the two regions are of equal magnitude, but whilst they are of the same sign for charge they are of opposite sign for currents. Finally, for the region where the actual charge or current is, the image is in the same position as an optical image of the applied source in the boundary; whilst, for the region not containing the influence, the image is in the same position as the influence.

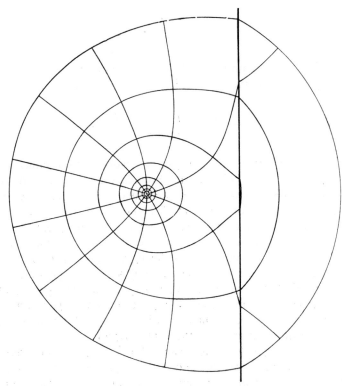

Fig. 3.2

Complex variable form. The solutions obtained above are used later in the book in complex variable form, which is generally the most suitable for calculation and is given below. The field of a line charge or a line current may be expressed by

$$w = \frac{1}{2\pi} \log(t - t_0), \qquad (3.5)$$

where $t = t_0$ gives the position of the current in the t-plane. Hence the field of a charge and its images is simply expressed by the sum of terms of the above form. Taking, for instance, the case of a charge at a distance a from an infinite plane, with the charge at the point $t = ja$ and the plane coinciding with the real axis, the field of the charge and its image is

$$w = \frac{1}{2\pi} \left[\log(t - ja) - \left(\frac{\varepsilon_2 - \varepsilon_1}{\varepsilon_1 + \varepsilon_2} \right) \log(t + ja) \right]. \qquad (3.6)$$

If the plane is conducting ($\varepsilon_2 = \infty$), this solution reduces to

$$w = \frac{1}{2\pi} \log \left(\frac{t-ja}{t+ja}\right).$$

Similarly, it can be shown that the field of a line current near an infinitely permeable plane is expressed by

$$w = \frac{1}{2\pi} \log (t-ja)(t+ja)$$

$$= \frac{1}{2\pi} \log (t^2+a^2), \tag{3.7}$$

and that of a line current near an impermeable plane by

$$w = \frac{1}{2\pi} \log \left(\frac{t-ja}{t+ja}\right). \tag{3.8}$$

3.2.2. *Parallel plane boundaries*

The solution for the field in the region between two parallel boundaries, due to a charge or current in that region, has been given by Kunz and Bayley[7] and Hague, p. 173. For either case, two solutions arise depending upon whether the boundaries are equipotentials or flux-lines.

For the case of the electric field due to a line charge between conducting (equipotential) boundaries, the images are as shown in Fig. 3.3 and the field map as shown in Fig. 3.4.

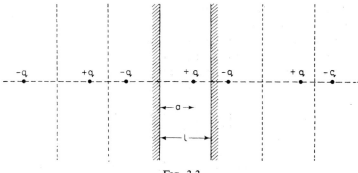

FIG. 3.3

It is seen that there is an infinite number of images, which occur as equally spaced pairs of equal and opposite charges and which are symmetrically disposed about each boundary line. It is to be noted that (because of this last feature) the flux crosses the boundary lines at right angles, indicating that the boundaries are equipotential, and so confirming that the solution is a valid one.

By summing the complex potential functions, for the positive and negative charges separately, it is simply shown, see reference (8), that the solution for the region between the boundaries is

$$w = \frac{1}{2\pi} \log \frac{\sin \pi[(t+a)/2l]}{\sin \pi[(t-a)/2l]}, \tag{3.9}$$

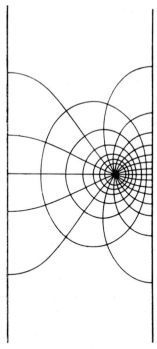 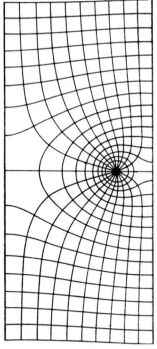

FIG. 3.4   FIG. 3.5

where the field is in the *t*-plane and the distances *a* and *l* and the origin are as shown in Fig. 3.3. This solution obviously applies also to the magnetic field of a line current between parallel impermeable boundaries.

For infinitely permeable boundaries the images of a line current have the same positions as those above, but they are all of the same sign. A map of the field is shown in Fig. 3.5, and it is seen that the images are such as to make the boundary lines magnetic equipotentials. Since the images are all of the same sign as the line current, the field is expressed by

$$w = \frac{1}{2\pi} \log \left[\sin \pi \left(\frac{t+a}{2l} \right) \right] \left[\sin \pi \left(\frac{t-a}{2l} \right) \right]. \tag{3.10}$$

These solutions have been applied to a variety of problems, e.g. by Kunz and Bayley to the calculation of the capacitance of a wire between conducting boundaries by Walker[8] to determining the characteristics of a triode valve, by Hague to the calculation of force on a conductor in a machine air gap, and by Frankel[9] to the determination of the impedance of conductors near parallel boundaries.

3.2.3. *Intersecting plane boundaries*

Solutions have been obtained, in the first instance by Lodge,[1] for a number of problems involving intersecting boundaries. The range of these solutions is, however, limited: a maximum of four boundaries can be handled, and the angles of intersection in the field region must in all cases be submultiples of π. (These limitations do not apply to conformal transformation methods, see section 10.2.) The discussion is to be restricted to the magnetic field since the analogy with the electric field is now obvious and since flux-line boundaries occur only in the magnetic case.

Two intersecting boundaries. Consider two straight boundaries of zero permeability intersecting at an angle π/n, where n is an integer, and enclosing a region containing a line current, i. The images which can be used to give the field *inside* the region are shown in Fig. 3.6(a), (b), and (c) for the values $n = 2, 3$ and 4. (With the value $n = 1$ the boundary becomes an infinite straight line, discussed earlier.) For all values of n, the image currents lie on a circle, its centre at the point of intersection of the boundaries and passing through the current, i. The currents in the complete system (the images and the actual currents) are alternately positive and negative and are symmetrically disposed about the boundaries. Consequently, this system gives rise to flux lines along the boundaries, so confirming the validity of the solution.

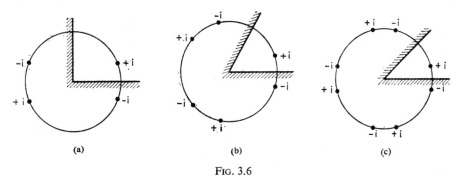

(a)   (b)   (c)

Fig. 3.6

It can be shown that in general the number of images is $(2n-1)$, and that the angles subtended by the images at the intersection of the boundaries form the series

$$\left(\frac{2\pi}{n}-\theta\right), \quad \left(\frac{2\pi}{n}+\theta\right), \quad \left(\frac{4\pi}{n}-\theta\right), \quad \left(\frac{4\pi}{n}+\theta\right), \quad \ldots$$

$$\ldots, \quad \left(\frac{2\pi}{n}(n-1)+\theta\right), \quad (2\pi-\theta),$$

where θ is the angular displacement of the current from one of the boundary lines.

When the boundaries are infinitely permeable, the images are in the same positions as the above, but they all have the same sign as the actual current.

There have been many applications of these solutions to practical problems, a recent example occurring in the calculation of force on the end-windings of turbogenerators.[10] *Three intersecting boundaries.* There are only four combinations of three intersecting boundaries which have interior angles which are submultiples of π. These combinations are:

$$\frac{\pi}{3}, \quad \frac{\pi}{3}, \quad \text{and} \quad \frac{\pi}{3};$$

$$\frac{\pi}{2}, \quad \frac{\pi}{4}, \quad \text{and} \quad \frac{\pi}{4};$$

$$\frac{\pi}{2}, \quad \frac{\pi}{3}, \quad \text{and} \quad \frac{\pi}{6};$$

and

$$\frac{\pi}{2}, \quad \frac{\pi}{2}, \quad \text{and} \quad 0.$$

This last combination may also be regarded as a special case of four intersecting boundaries or of two parallel boundaries with a line of symmetry. Taking the former view, the solution is thus given by reduced forms (m or $n = \pm 1$ only) of eqns. (3.11) and (3.12). Hague, p. 188, gives considerable attention to this case, interpreting the boundary as that of a deep slot. There is, however, little practical interest in the other cases and they are not discussed further here.

Four intersecting boundaries. Because of the limitation in the values of the angles of intersection, only one combination of four boundaries is possible, namely, that with all the interior angles equal to $\pi/2$. The positions of the images of a line current for the field with flux-line boundaries are shown in Fig. 3.7. The distribution of the images is doubly periodic, and they are symmetrical about each of the boundary lines, so that, as required, these are flux lines.

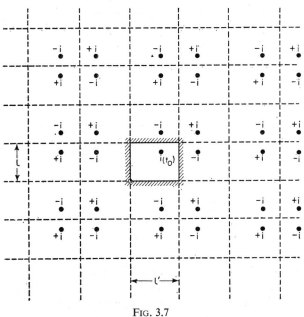

FIG. 3.7

Letting the field be in the t-plane, the origin and dimensions being as shown in Fig. 3.7, the field solution is

$$w = \frac{1}{2\pi} \log \prod_{\substack{m=-\infty \\ n=-\infty}}^{\infty} \frac{(t+\bar{t}_0+a)(t-\bar{t}_0+a)}{(t+t_0+a)(t-t_0+a)}, \tag{3.11}$$

where m and n are integers and where

$$a = 2ml' + 2jnl.$$

Equation (3.11) has been used, for example, to determine the capacitance of a wire in a rectangular cylinder.[7]

The images of a line current in the infinitely permeable boundary occupy the same positions as those for a flux-line boundary, but they all have the same sign as the current itself, and so the solution is

$$w = \frac{1}{2\pi} \log \prod_{\substack{m=-\infty \\ n=-\infty}}^{\infty} (t+\bar{t}_0+a)(t-\bar{t}_0+a)(t+t_0+a)(t-t_0+a). \qquad (3.12)$$

In section 5.6 this image distribution is considered in connection with the field of distributed currents.

3.2.4. *Inductance of parallel bus-bars near an iron surface*

To show the usefulness and simplicity of the image method, it is now applied to the calculation of the inductance of two bus-bars carrying equal and opposite currents $\pm I$ near to a highly permeable surface. The surface is assumed infinitely permeable for simplicity (though this assumption is not a necessary condition for the problem to be solved by images) and lying along the real axis of the t-plane. The positions of the currents are as

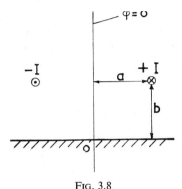

FIG. 3.8

shown in Fig. 3.8 and the radii of the conductors are r, it being assumed that r is small compared to the distances a or b. The required image currents are $+I$ at $t = a-jb$ and $-I$ at $t = -a-jb$. Hence the complete field solution for the region above the iron boundary is

$$\omega = \frac{I}{2\pi} \log \frac{[t-(a+jb)]\,[t-(a-jb)]}{[t-(-a+jb)]\,[t-(-a-jb)]}$$

which can be simplified to

$$\omega = \frac{I}{2\pi} \log \frac{(t-a)^2+b^2}{(t+a)^2+b^2}. \qquad (3.13)$$

The imaginary axis is a line of symmetry and along it lies the flux line $\varphi = 0$ as may be confirmed by showing that the real part of eqn. (3.13) is zero for all purely imaginary values of t. If the small quantity of flux which passes within the conductors is neglected, the total flux passing between the two bus-bars is twice that passing between the line $\varphi = 0$ and a point $t = (a+jb+r)$ on the surface of the conductor at $t = a+jb$. (Since r is small the surface of the conductor is very nearly a flux line.)

Hence the total flux linking the bus-bars is given by

$$2Rl \left[\frac{\mu_0 I}{2\pi} \log \frac{|(r+jb)^2+b^2|}{(2a+jb+r)^2+b^2} \right]$$

and so the inductance per unit length of the bus-bar circuit is

$$\frac{\mu_0}{\pi} Rl \log \left| \frac{(r+jb)+b}{(2a+jb+r)+b} \right|. \tag{3.14}$$

In a similar way it is possible to calculate from the electric field the capacitance of parallel transmission lines (see Bewley, p. 47, and reference 11).

3.3. Circular Boundaries

Whereas for the single straight-line boundary there are only two images to be considered, for the single circular boundary there are four sets of images—the two for the charge inside the boundary and the two for the charge outside. Solutions are developed in detail for the latter case using the concept of equivalent surface charge. The results for the case of a charge interior to the boundary are given for both regions. Consideration is also given to the field of a circular cylinder in a uniform applied field. The applied field is due to charges at infinity, but the boundary effects can still be accounted for by images, which in this case form a doublet.

3.3.1. *Charge or current near a circular boundary*

Line charge near an isolated cylinder. Consider first the field of a line charge in a medium of permittivity $\varepsilon_1 \varepsilon_0$ near a circular cylinder of permittivity $\varepsilon_2 \varepsilon_0$ (Fig. 3.9). The normal component of the applied field strength E_n at any point P on the circular boundary is given by

$$E_n = \frac{q \cos \alpha}{2\pi \varepsilon_0 \varepsilon_1 b}, \tag{3.15}$$

the angle α and distance b being as shown in Fig. 3.9. The normal component of field due to the equivalent surface charge distribution is then, from eqn. (3.3),

$$E'_n = \frac{q}{2\pi \varepsilon_0 \varepsilon_1} \left(\frac{\varepsilon_2 - \varepsilon_1}{\varepsilon_2 + \varepsilon_1} \right) \frac{\cos \alpha}{b}. \tag{3.16}$$

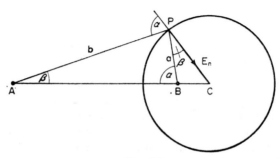

FIG. 3.9

However, by consideration of triangles PAB and PBC, where B is the inverse of A in the circle, it can be shown that

$$a \cos \alpha + b \cos \beta = \frac{b \sin (\alpha + \beta)}{\sin \alpha},$$

and

$$a \sin \alpha = r \sin (\alpha - \beta),$$

which may be combined to give

$$\frac{\cos \alpha}{b} = \frac{\cos \beta}{a} - \frac{1}{r}. \tag{3.17}$$

Hence eqn. (3.16) can be rewritten as

$$E'_n = \frac{q}{2\pi\varepsilon_0} \left(\frac{\varepsilon_2 - \varepsilon_1}{\varepsilon_2 + \varepsilon_1}\right) \frac{\cos \beta}{a} - \frac{q}{2\pi\varepsilon_0} \left(\frac{\varepsilon_2 - \varepsilon_1}{\varepsilon_2 + \varepsilon_1}\right) \frac{1}{r}, \tag{3.18}$$

and this gives the normal component of field strength along the perimeter of the circle due to the distributed charge. But this is the normal component of field due to a charge $-q(\varepsilon_2 - \varepsilon_1)/(\varepsilon_2 + \varepsilon_1)$ at inverse point B together with a charge $+q(\varepsilon_2 - \varepsilon_1)/(\varepsilon_2 + \varepsilon_1)$ at the centre of the circle. Therefore, since these charges are inside the boundary they are the required image charges which, together with the applied charge, give the field outside the circular boundary.

For the region inside the boundary, the images giving rise to E'_n must be outside the boundary, and must give rise to a normal component of field

$$E'_n = -\frac{q}{2\pi\varepsilon_0} \left(\frac{\varepsilon_2 - \varepsilon_1}{\varepsilon_2 + \varepsilon_1}\right) \frac{\cos \alpha}{b}. \tag{3.19}$$

This obviously requires the image charge to be of magnitude $-q(\varepsilon_2 - \varepsilon_1)/(\varepsilon_2 + \varepsilon_1)$ and to lie at the same point as the charge q.

In a similar way to the above it is possible to establish the magnitudes and positions of the images of a charge inside the circular boundary. The field inside the circle is given by an image of magnitude $q(\varepsilon_2 - \varepsilon_1)/(\varepsilon_2 + \varepsilon_1)$ at the exterior inverse point, and the field outside is given by an image of magnitude $-q(\varepsilon_2 - \varepsilon_1)/(\varepsilon_2 + \varepsilon_1)$ at the same point as the charge q together with an image of magnitude $q(\varepsilon_2 - \varepsilon_1)/(\varepsilon_2 + \varepsilon_1)$ at the centre. Figure 3.10 shows the field of a line charge outside a boundary of circular section for the case $\varepsilon_2/\varepsilon_1 = 5$. Images solutions for a charge near a circular boundary have been used in the determination of the field in an electronic valve (see Bewley and references 3 and 7).

Line current near a permeable cylinder. The images of a line current near a circular boundary are, of course, analogous to the above image charges. When the current i is outside the boundary, the field outside is given by an image current $i(\mu_2 - \mu_1)/(\mu_2 + \mu_1)$ at the inverse point, together with a current $-i(\mu_2 - \mu_1)/(\mu_2 + \mu_1)$ at the centre, and, of course, the actual current i; the field inside is given by a current i and an image $-i(\mu_2 - \mu_1)/(\mu_2 + \mu_1)$ at the same position as the current i. When the current is inside the boundary, the field outside requires an image current $-i(\mu_2 - \mu_1)/(\mu_2 + \mu_1)$ at the same point as the current i together with an image of magnitude $i(\mu_2 - \mu_1)/(\mu_2 + \mu_1)$ at the centre; and the field inside is that of the actual current and image current $-i(\mu_2 - \mu_1)/(\mu_2 + \mu_1)$ at the inverse point.

Line charge near an earthed conducting cylinder. When an isolated conducting cylinder is near to a line charge, the field is obtained from the general solution for a circular boundary

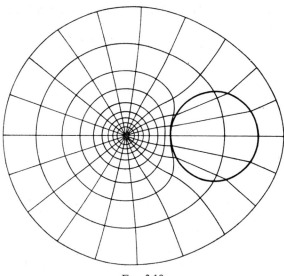

Fig. 3.10

by making the substitution $\varepsilon_2 \to \infty$. Hence, the images are of magnitude $-q$ at the inverse point and $+q$ at the centre. The conducting cylinder, being isolated, acquires a potential but has no net charge on it, since all the flux from the charge at A passes to infinity (some of it through the cylinder).

If the cylinder is earthed its potential changes to zero, which means that the image charge at its centre must change. Also, all of the flux from the charge at A now passes into the cylinder, which acquires a net charge of $-q$. For this to be possible, the image charge at the centre must change to zero, leaving the other image charge of $-q$ at B, and the potential of the cylinder changes to zero as a result of the removal of image charge from the centre.

Finally, it is interesting to note that the image solutions for charges or currents near circular boundaries reduce, as the radius tends to infinity, to those for an infinite straight-line boundary. In the next chapter, section 4.2.2, the field of a line current exterior to a permeable cylinder is analysed by the direct solution of Laplace's equation. Also, in section 7.2.3, the solution for infinite permeability is obtained by conformal transformation. In both cases, the equivalence of the solutions with that obtained by images is demonstrated.

3.3.2. Doublets: circular cylinder in a uniform field

Doublets. The influence of a circular cylinder on an applied uniform field can be accounted for by a doublet, i.e. by a combination of an infinite positive charge (or current) and an infinite negative charge (or current), an infinitesimal distance apart, so arranged that the product of charge and distance is finite. Before examining its influence on the uniform field, the field of a doublet is first developed from a consideration of that of line charges, and then the images of doublets for plane and circular boundaries are obtained. All field functions are expressed in complex variable form because of its convenience in calculation.

The complex potential function of a line charge of strength q at the point $t = a$ of the complex t-plane is given by

$$w = \frac{q}{2\pi} \log (t - a). \tag{3.20}$$

Therefore, the field due to equal positive and negative line charges at $t = a+jb$ and $t = -a-jb$ respectively is expressed by

$$w = \frac{q}{2\pi} \log\ [t-(a+jb)] - \frac{q}{2\pi} \log\ [t+(a+jb)]. \tag{3.21}$$

This can be written as

$$w = \frac{q}{2\pi} \log\left(1 - \frac{a+jb}{t}\right) - \frac{q}{2\pi} \log\left(1 + \frac{a+jb}{t}\right),$$

or, expanding in series form,

$$w = -\frac{2q}{2\pi}\left[\frac{1}{t}(a+jb) + \frac{1}{3t^3}(a+jb)^3 + \frac{1}{5t^5}(a+jb)^5 + \ \cdots\ \right]. \tag{3.22}$$

Consider now the formation of a doublet from the two discrete charges; the distance $2\,|a+jb|$ between the charges tends to zero and the magnitude of the charge q tends to infinity in such a way that $(q/\pi)\,|a+jb|$, known as the *strength*, d, of the doublet, remains finite. Under these conditions all terms but the first in eqn. (3.22) disappear, and the equation of the doublet (at the origin) becomes

$$w = -\frac{q}{\pi t}(a+jb).$$

Writing

$$d = -\frac{q}{\pi}\,|a+jb|,$$

this reduces to

$$w = \frac{d}{t}\,e^{j\alpha}, \tag{3.23}$$

where α, which equals $\tan^{-1}(b/a)$, is the inclination of the axis of the doublet to the real axis. The field of a doublet, for which $\alpha = 0$, is shown in Fig. 3.11; the flux and potential lines form orthogonal families of circles having as common tangent the axes of the doublet.

The equation of a doublet situated not at the origin but at $t = t_1$, and with its axis parallel with the x-axis, is

$$w = \frac{d}{t-t_1}, \tag{3.24}$$

and this form is used in section 10.2.2 to obtain solutions by transformation for fields exterior to boundaries of complicated shape. The relationship between w and t is that of complex inversion which is discussed in section 7.1.1.

Doublet at infinity. In a similar way to the above, it is also possible to develop the field of a doublet at the point at infinity. Consider again the two unlike charges, the field of which is given by eqn. (3.21). That equation may be rewritten as

$$w = \frac{q}{2\pi} \log\left(1 - \frac{t}{a+jb}\right) - \frac{q}{2\pi} \log\left(1 + \frac{t}{a+jb}\right) + \frac{q}{2\pi} \log\ (-1),$$

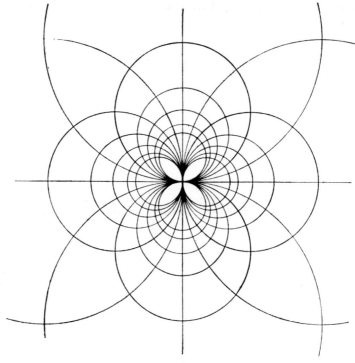

Fig. 3.11

or, expanding the first two terms on the right in a series, as

$$w = -\frac{q}{\pi}\left[\frac{t}{a+jb}+\frac{t^3}{3(a+jb)^3}+ \cdots \right]. \tag{3.25}$$

Now the doublet at infinity is formed by letting both charges move away to infinity (where they converge on the single point, $t = \infty$) whilst, at the same time, $q \to \infty$ in such a way that d, the strength of the doublet, given by

$$-\frac{2q}{(a+jb)} = de^{j\alpha},$$

remains finite. As $|a+jb|$ and q both approach infinity, eqn. (3.25) reduces to

$$w = de^{j\alpha}t, \tag{3.26}$$

which is thus the equation of the field of a doublet at infinity. It is seen that this equation describes a uniform field of strength d inclined at an angle α to the real axis, and such a field, therefore, is produced by a doublet charge at infinity. Equation (3.26) could be derived directly from eqn. (3.23) by inversion and this point is discussed in section 7.1.

Images of a doublet. It is apparent from the above developments of doublets that the images of a doublet in a straight boundary or in a circular one can be derived from the combination of the separate images of the pair of charges forming the doublet. The result for the plane surface is particularly easy to see; the image of each charge has the position of the optical image in the surface and $-(\varepsilon_2-\varepsilon_1)/(\varepsilon_2+\varepsilon_1)$ times the strength of the charge, where $\varepsilon_1\varepsilon_0$ is the permittivity in the field region, and $\varepsilon_2\varepsilon_0$ is that behind the boundary. The doublet image, therefore, has the position of the optical image of the doublet and $-(\varepsilon_2-\varepsilon_1)/(\varepsilon_2+\varepsilon_1)$ times its strength. For example, the field in the region containing a doublet at the point t_1 and influenced by a boundary coinciding with the real axis is

$$w = \frac{de^{j\alpha}}{(t-t_1)} - \left(\frac{\varepsilon_2-\varepsilon_1}{\varepsilon_2+\varepsilon_1}\right)\frac{de^{j\alpha}}{(t-\bar{t}_1)}, \tag{3.27}$$

where $\varepsilon_1\varepsilon_0$ is the permittivity in the field region, and $\varepsilon_2\varepsilon_0$ is that inside the second region, and where t_1 is the position of the doublet and $\bar{t}_1$ is the complex conjugate of t_1. The image for the field in the region of permittivity $\varepsilon_2\varepsilon_0$ can easily be found, but it is not given here.

The images of a doublet near a circular boundary can be found similarly, though rather more care is required. The case of the doublet exterior to the boundary and the solution for the region containing the doublet is examined. Consider first the images of two equal finite charges of opposite sign, at A and B, outside the boundary as shown in Fig. 3.12. The positive charge, $+q$, at A has an image $-q(\varepsilon_2-\varepsilon_1)/(\varepsilon_2+\varepsilon_1)$ at its inverse point C with respect to the circle, and an image $q(\varepsilon_2-\varepsilon_1)/(\varepsilon_2+\varepsilon_1)$ at the centre O of the circle. The images of the negative charge at B are $+q(\varepsilon_2-\varepsilon_1)/(\varepsilon_2+\varepsilon_1)$ at the corresponding inverse point D and $-q(\varepsilon_2-\varepsilon_1)/(\varepsilon_2+\varepsilon_1)$ at the centre. The two charges at the centre of the circle thus cancel one

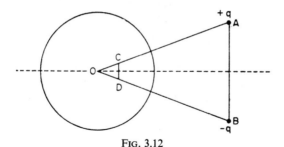

FIG. 3.12

another, leaving the pair of equal and opposite image charges at the inverse points. Figure 3.12 shows the position of these charges. As the charges at A and B approach one another, the product $q(AB)$ remaining finite, they form a doublet, the image of which is formed by the two image charges at C and D. The strength of the doublet causing the applied field is given by

$$d = q(AB),$$

and the strength of the image doublet by

$$d_i = -d \underset{A \to B}{\mathrm{Lt}} \left(\frac{CD}{AB}\right)\left(\frac{\varepsilon_2-\varepsilon_1}{\varepsilon_2+\varepsilon_1}\right). \tag{3.28}$$

But, since the triangles OCD and OBA are similar,

$$\frac{CD}{AB} = \frac{OC}{OA};$$

also if, as the pairs of charges come together,

$$OA = OB = a$$

(say), then

$$OC = OD = \frac{r^2}{a},$$ (3.29)

r being the radius of the circle. Therefore, substituting for (CD/AB) in eqn. (3.28) gives the strength of the image doublet as

$$d_i = -d\left(\frac{r^2}{a^2}\right)\left(\frac{\varepsilon_2 - \varepsilon_1}{\varepsilon_2 + \varepsilon_1}\right).$$ (3.30)

Also the position of the image is that of the inverse point of the position of the doublet with respect to the circle. The inclination of the doublet is seen, from a consideration of the similar triangles, as $AB \to 0$, to be equal and opposite to that of the doublet (with respect to the line joining them).

The image of a doublet inside the circular boundary can be found in a similar way to the above.

Cylindrical boundary in applied uniform field. The influence of a circular boundary on a uniform applied field can be accounted for simply by using a doublet at infinity to give the uniform field and then by forming images, of this doublet, in the boundary. Since the centre of the circular boundary is the inverse of the point at infinity, it is apparent that the resultant field is given by the sum of the actual doublet at infinity and the image doublet at the centre. However, it is interesting to develop this solution from a consideration of the surface charge distribution.

In a uniform electric field of strength E, the normal component at any point P on the circular boundary is $E \cos \theta$, where θ is the angle between the direction of the field, taken parallel to the real axis and a radius to the point P (Fig. 3.13). The normal field E'_n due to the

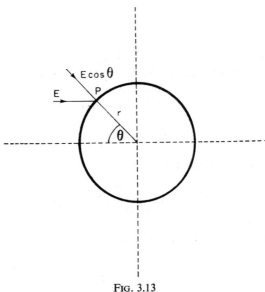

Fig. 3.13

charging of the boundary is, therefore,

$$E'_n = \left(\frac{\varepsilon_2 - \varepsilon_1}{\varepsilon_2 + \varepsilon_1}\right) E \cos \theta.$$

Consider now the field component normal to the circular boundary, due to a doublet, strength d, with its axis parallel to the field and placed at the centre of the circle. Its field is given by

$$w = \frac{d}{t},$$

and the field strength (see section 6.6.6) by

$$\frac{d}{r^2} \exp [j(\pi - 2\theta)].$$

This means that the field strength, due to the doublet, is of magnitude d/r^2, and is inclined to the real axis by an angle 2θ and so to the radius P by an angle θ. Hence, the normal component of field at the boundary due to the doublet is $(d/r^2) \cos \theta$ and so, by making

$$\frac{d}{r^2} = \left(\frac{\varepsilon_2 - \varepsilon_1}{\varepsilon_2 + \varepsilon_1}\right) E,$$

the doublet of strength d at the centre of the circle is the image required to account for the boundary effect. The expression for the field outside a circular boundary of radius r in an applied uniform field E thus becomes

$$w = E\left[t - \left(\frac{\varepsilon_2 - \varepsilon_1}{\varepsilon_2 + \varepsilon_1}\right)\frac{r^2}{t}\right]. \tag{3.31}$$

The required image for the field inside the boundary must, of course, be outside the boundary and, in fact, it is at infinity. The component E'_n opposing the applied field gives rise, inside the boundary, to a resultant normal field having along the boundary a component

$$E \cos \theta \left[1 - \left(\frac{\varepsilon_2 - \varepsilon_1}{\varepsilon_2 + \varepsilon_1}\right)\right].$$

The resultant field everywhere inside the boundary is clearly uniform and of strength

$$E\left[1 - \left(\frac{\varepsilon_2 - \varepsilon_1}{\varepsilon_2 + \varepsilon_1}\right)\right],$$

which equals

$$\frac{2\varepsilon_1 E}{\varepsilon_2 + \varepsilon_1}.$$

Figure 3.14 is a field plot for $\varepsilon_2/\varepsilon_1 = 5$.

AC 5

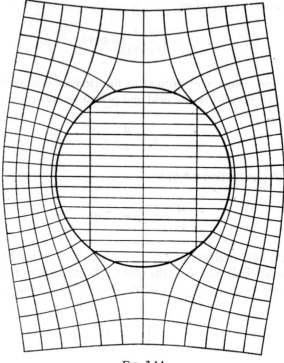

Fig. 3.14

3.4. General considerations

It has been shown that the method of images can be used to analyse fields due to line currents or charges (including those at infinity which give uniform applied fields) influenced by a variety of straight line or circular boundaries. A combination of a straight-line and a circular boundary has also been treated in several papers, e.g. by Milne-Thomson and in reference (12).

The above treatment is restricted to the fields of line influences, but it should be noted that the method also applies to influences with finite areas of cross-section. This is because any such influence can be treated as the aggregate of line influences, each of which forms its own image in the usual way. Thus, for example, the field in air of a rectangular conductor carrying current of uniform density J, and near an infinite plane of relative permeability μ, is given by introducing a rectangular image current with uniform density $J(\mu-1)/(\mu+1)$ (Fig. 3.15). (Further discussion of the images of rectangular conductors is given in Chapter 5.) However, the application of the method is restricted to straight-line boundaries, since the image in a circular boundary has an awkward shape and a non-uniform current density and this leads to excessive difficulty in the analysis.

To conclude this section and the chapter, the image method is used to demonstrate a number of general points to which its application is very suitable. The first of these is the calculation of force experienced by a boundary; the second is an indication of the error involved in the frequently made assumption that boundaries of finite permeability can be treated as being infinitely permeable; and the third is an approximate method for the treatment of field regions with boundaries in which eddy-currents are induced by an alternating field. *The total force on a boundary.* The total force acting on a boundary can be calculated very conveniently using the image method because it is equal to the total force on the images due to the applied field or to the force on the field sources due to the field of the images.

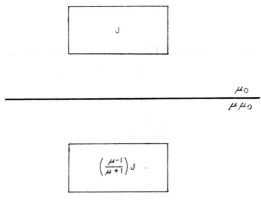

FIG. 3.15

As an example, consider the determination of the force between a line current in air and an infinite plane boundary of relative permeability μ (Fig. 3.16). The field strength H_x, at the position of the current, due to the image, is given by

$$H_x = \frac{i}{2\pi}\left(\frac{\mu-1}{\mu+1}\right)\frac{1}{2a},$$
(3.32)

there being no component in the y-direction. Therefore, the force on the current due to this field (and that on the boundary due to the current) is in the y-direction, and it is given by

$$F_y = \frac{\mu_0}{2\pi}i^2\left(\frac{\mu-1}{\mu+1}\right)\frac{1}{2a}.$$
(3.33)

This is an attractive force when the boundary is permeable, and a repulsive one when the effective permeability is less than unity.

The influence of finite permeability. The simplifying assumption, commonly made in magnetic problems, that a boundary is of infinite permeability, clearly involves some error. The magnitude of this error varies with position in a different way for each problem but, in order to demonstrate its nature, in a simple case, consideration is again given to the line current in air near an infinite plane boundary of relative permeability μ. In the air, the field is that of the current i and its image of magnitude $i(\mu-1)/(\mu+1)$ (Fig. 3.16). The component of field strength parallel to the plane is given by

$$H_x = \frac{i}{2\pi}\frac{a-y}{[(a-y)^2+x^2]} - \frac{i}{2\pi}\left(\frac{\mu-1}{\mu+1}\right)\frac{a+y}{[(a+y)^2+x^2]},$$
(3.34)

and that normal to the plane by

$$H_y = \frac{i}{2\pi}\frac{x}{[(a-y)^2+x^2]} + \frac{i}{2\pi}\left(\frac{\mu-1}{\mu+1}\right)\frac{x}{[(a+y)^2+x^2]}.$$
(3.35)

5*

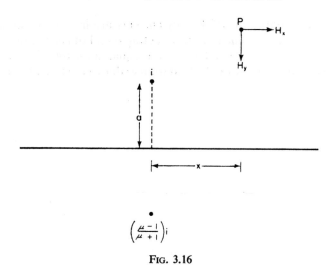

FIG. 3.16

The error involved in assuming the permeability infinite is thus given by substituting $\mu = \infty$ and subtracting the resulting values from the above equations. The error in the x-component is

$$\frac{i}{2\pi} \frac{2}{\mu+1} \frac{a+y}{[(a+y)^2+x^2]}, \tag{3.36}$$

and that in the y-component is

$$\frac{i}{2\pi} \frac{2}{\mu+1} \frac{x}{[(a+y)^2+x^2]}. \tag{3.37}$$

At any point along the boundary (given by $y = 0$) the ratio of the error to the applied field, for each component, is $2/(\mu+1)$ (which becomes less than 5 per cent for $\mu > 39$, and less than 0·5 per cent for $\mu > 399$), and this is negligibly small in practice. Further, the error decreases with movement away from the boundary, and so this value $(2/(\mu+1))$ gives the upper bound for the error at any point in the field. For most problems which can be solved for finite permeability it is found that the assumption of $\mu = \infty$ causes little error in practice, and this assumption is made for many problems considered in this book.

Eddy-current problems. All problems treated in this book are essentially steady-state ones, but an approximate method, using steady-state results, can be used for certain transient problems involving boundaries carrying eddy-currents induced by the applied field. The effect of the eddy-current is to reduce the quantity of flux crossing the boundary and, if the assumption is made that no flux penetrates (that is the boundary is a flux line), solutions may be obtained by taking the boundary permeability to be zero. The field of a line-current near a semi-infinite block of impermeable material is given by eqn. (3.8), and this equation is used later in conjunction with conformal transformation (see sections 10.2.1 and 10.2.2), for the analysis of problems involving alternating currents near complicated conducting boundaries.

References

1. O. J. Lodge, On some problems connected with the flow of electricity in a plane, *Phil. Mag.*, 5th series, **1**, 373–89 and **2**, 37–47 (1876).
2. G. F. C. Searle, On the magnetic field due to a current in a wire placed parallel to the axis of a cylinder of iron, *The Electrician* 453 and 510 (1898).
3. P. Hammond, Electric and magnetic images, *Proc. Instn. Elect. Engrs.* **107** C, 306 (1960).
4. N. Mullineux and J. R. Reed, Images of line charges and currents, *Proc. Instn. Elect. Engrs.* **111**, 1343.
5. F. Chorlton, Circle and sphere theorems in potential theory and the determination of image systems, *Birm. Coll. Adv. Tech. Technical Report*, No. 7 (1965).
6. C. J. Carpenter, The application of the method of images to machine end-winding fields, *Proc. Instn. Elect. Engrs.* **107** A, 487 (1960).
7. J. Kunz and P. L. Bayley, Some applications of the method of images, *Phys. Rev.* **17**, 147–56 (1921).
8. G. B. Walker, Electric field of a single grid radio valve, *Proc. Instn. Elect. Engrs.*, Part III, **98**, 57 (1951).
9. S. Frankel, Characteristic impedance of parallel wires in rectangular troughs, *Proc. Inst. Rad. Engs.* 182 (1942).
10. D. Mayer, Mechanicke sily pusobici na cela vinuti statoru stridavych stroju (Forces acting on the faces of stator windings in a.c. machines), *Electrotechnicky Obzor* **44** (8), 395 (1955).
11. J. H. Gridley, The shielding of overhead lines against lightning, *Proc. Instn. Elect. Engrs.* **107** A, 325 (1960).
12. B. V. Jayawant, Flux distribution in a permeable sheet with a hole near an edge, *Proc. Instn. Elect. Engrs.* **107** C, 238 (1960).
13. J. F. H. Douglas, Reactance of end-connections, *Trans. Am. Inst. Elect. Engrs.* **56**, 257 (1937).
14. P. J. Lawrenson, The magnetic field of the end-windings of turbo-generators, *Proc. Instn. Elect. Engrs.* **108** A, 538 (1961).

Additional Reference

Mack, C., The field due to an infinite dielectric cylinder between two parallel conducting plates, *Br. J. Appl. Phys.* **6**, 59 (1955).

CHAPTER 4

THE SOLUTION OF LAPLACE'S EQUATION BY SEPARATION OF THE VARIABLES

4.1. Introduction

The method of images, because of its simplicity, is of considerable value, but for many problems involving multiple boundaries or specified distributions of potential or potential gradient, it is more convenient to solve the field equations directly. In this chapter consideration is given to direct solutions of Laplace's equation, and in the next to solutions of Poisson's equation. Solutions which are available by the use of images are also available by the method discussed here, and the equivalence of solutions obtained using the two methods is demonstrated.

Essentially, the direct-solution method involves the determination of a potential function satisfying Laplace's equation and also satisfying imposed boundary and other field conditions for a particular region. This potential function is, in general, the sum of several parts (each of which separately is a solution); one part, usually in the form of a series, describes the effect of the boundary influences, and the others describe the effect of field sources such as currents and charges. (The field solution can, of course, be expressed equally in terms of the flux function, though this is not often done. The only difference is in the expression of the boundary conditions.) It must be noted, however, that the potential function cannot be determined for *any* given problem. This is not because of the difficulty of finding solutions which satisfy the equation, but because of the difficulty of choosing solutions appropriate to particular boundary and field conditions. There is an infinite number of solutions of Laplace's equation: examples are nx, $x^2 - y^2$, $e^{nx} \cdot \sin ny$, and any linear combination of these; but it is frequently impossible to find combinations of them which satisfy field conditions on boundary lines of particular shape.

However, when a coordinate system exists for which a constant value of one coordinate (or both, in certain cases) expresses the boundary shape, it is always possible to obtain a solution by a routine method. The two coordinate systems which are of most interest to the engineer or physicist are the cartesian one for the solution of fields with rectangular boundaries, and the circular-cylinder one for the solution of those with concentric circular and/or radial line boundaries. For these two coordinate systems the method of solution is discussed, and it is shown to reduce merely to the determination of constants in a general form of potential function. Discussion of other coordinate systems, e.g. parabolic and elliptic cylinder, or bipolar, may be found elsewhere. (For example, the books by Morse and Feshbach, Stratton, Weber, Bateman, and Moon and Spencer may be consulted. Particular attention is drawn to the Moon and Spencer book, the whole of which is devoted to separation-of-the-variables techniques, and which includes not only a wide variety of coordinate systems but also problems outside the scope of this book—three-dimensional fields and the diffusion and wave equations.)

When a direct solution is possible, the part of the potential function satisfying the boundary conditions can always be expressed as the product of two terms, each term being a function of one coordinate only. As a result, the partial differential equation can be converted to a pair of ordinary differential equations, related by a constant known as the "separation" constant, and so solved. A general solution is then taken which consists of a suitable linear combination of the products of the pairs of particular solutions from the ordinary equations. In practice it is not necessary to develop the general solution for each problem; instead, one begins with the known general solution for the given boundary shape and merely fits it to the boundary and field conditions of the given problem.

In the sections dealing with rectilinear boundaries, consideration is restricted to fields arising from specified flux density and potential distributions at the boundary, since the method of images gives a simpler treatment of fields due to line sources. In the sections dealing with circular boundaries, emphasis is placed on fields developed by currents and poles, but the treatment of fields arising from, or giving rise to, specified distributions of potential or potential gradient is also discussed briefly. (The fields of charges are, of course, directly analogous to those of poles.) Attention is given first to circular boundaries so that the equivalence of the solutions obtained by the method of direct solution and the method of images can be demonstrated at an early stage.

4.2. Circular boundaries

4.2.1. *The solution of Laplace's equation in circular-cylinder coordinates*

Consider first Laplace's equation in circular-cylinder coordinates in terms of the potential function ψ [eqn. (2.27)]

$$r^2 \frac{\partial^2 \psi}{\partial r^2} + r \frac{\partial \psi}{\partial r} + \frac{\partial^2 \psi}{\partial \theta^2} = 0, \tag{4.1}$$

and *assume* as a solution the potential function

$$\psi(r, \theta) = R(r) S(\theta), \tag{4.2}$$

where R is a function of r only, and S is a function of θ only. Partial differentiation of eqn. (4.2) leads to expressions for $\partial \psi / \partial r$, $\partial \psi^2 / \partial r^2$, and $\partial^2 \psi / \partial \theta^2$, and substitution of these quantities in eqn. (4.1) gives

$$\frac{1}{R} \left(r^2 \frac{d^2 R}{dr^2} + r \frac{dR}{dr} \right) = -\frac{1}{S} \frac{d^2 S}{d\theta^2}. \tag{4.3}$$

The right-hand side of this equation is independent of r and the left-hand side is independent of θ, so that both sides are equal to a constant, known as the *separation* constant, and the assumption concerning the form of the potential function is justified. For any field in which the effect of currents is ignored, the potential at (r, θ') is the same as that at $(r, \theta' + 2n\pi)$, where n is an integer, so that the solution sought for must be periodic in θ. To fulfil this condition, the constant is chosen as a positive integer, and taking it to be m^2 to avoid root signs later, gives

$$r^2 \frac{d^2 R}{dr^2} + r \frac{dR}{dr} - m^2 R = 0 \tag{4.4}$$

and

$$\frac{d^2S}{d\theta^2} = -m^2S. \tag{4.5}$$

Well-known solutions of these equations are

$$R = cr^m + dr^{-m} \tag{4.6}$$

and

$$S = g\cos m\theta + h\sin m\theta, \tag{4.7}$$

where c, d, g, and h are constants. Hence, substituting for R and S in eqn. (4.2) gives a particular solution of the original equation as

$$\psi(r, \theta) = (cr^m + dr^{-m})(g\cos m\theta + h\sin m\theta). \tag{4.8}$$

This function is known as a circular harmonic of order m.

No limitation has been imposed upon the values of the constants in this solution, so that there is an infinite number of particular solutions of the form (4.8) capable of satisfying eqn. (4.1). Since a linear combination of these particular solutions also satisfies eqn. (4.1) (see, for example, Churchill, p. 3), a more general solution can be written in the convenient form

$$\psi(r, \theta) = \sum_{m=1}^{\infty} (c_m r^m + d_m r^{-m})(g_m \cos m\theta + h_m \sin m\theta), \tag{4.9}$$

where m takes all integral values between 1 and ∞. For any fixed value of r in the range $0 \leqslant r \leqslant \infty$, the right-hand side of this equation is a Fourier series in θ (which can be used to represent any single-valued and periodic function, such as ψ) and, therefore, it is capable of describing the field due to any physically realizable boundary influences. The values of the constants (in the absence of currents and poles) depend only on the boundary conditions.

The right-hand side of eqn. (4.9) cannot account for the effect of line currents and poles. To account for these influences, additional terms are required in the potential function, and these are found to correspond to the particular solutions of the ordinary differential equations (4.4) and (4.5) for the case $m = 0$. For a line current i situated at any point, the potential function (particular solution) is $(i/2\pi)\alpha$, where the angle α, at the position of the current, measures the angular position of any point; and for a line pole p the potential function is $(p/2\pi)\log R$, where R is the distance between the pole and a point in the field (see sections 2.1 and 2.2). The complete solution, including the influence of q currents and s poles, is, therefore,

$$\psi(r, \theta) = \sum_{m=1}^{\infty} (c_m r^m + d_m r^{-m})(g_m \cos m\theta + h_m \sin m\theta)$$

$$+ \sum_q \frac{i_q}{2\pi} \alpha_q + \sum_s \frac{p_s}{2\pi} \log R_s. \tag{4.10}$$

For purposes of manipulation, the angles α_q and the distances R_s are expressed, in terms of r and θ, in series of the same form as those occurring in the first part of the function.

To express the angles α_q in series form, consider Fig. 4.1(a); a current i is situated at the point A a distance b from the origin of coordinates, and α is the angle subtended by any point P, (r, θ) at the current. From the geometry of the figure it is seen that

$$\alpha = \tan^{-1} \frac{r \sin \theta}{r \cos \theta - b}$$

$$= \text{Im} \log (r e^{j\theta} - b).$$

This function can be expanded in terms of $(b/r) e^{-j\theta}$ for $r > b$, or in terms of $(r/b) e^{j\theta}$ for $b > r$, to give two distinct convergent series, valid in the above regions, for the angle α: thus, where $r > b$,

$$\alpha = \theta + \sum_{m=1}^{\infty} \frac{1}{m} \left(\frac{b}{r} \right)^m \sin m\theta, \tag{4.11}$$

and where $r < b$,

$$\alpha = \pi - \sum_{m=1}^{\infty} \frac{1}{m} \left(\frac{r}{b} \right)^m \sin m\theta. \tag{4.12}$$

Further, where $r = b$, it is immediately obvious from the geometry that

$$\alpha = \frac{\theta + \pi}{2}. \tag{4.13}$$

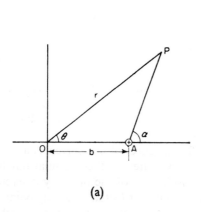 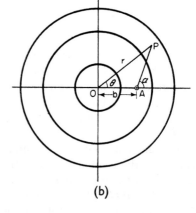

(a)   (b)

FIG. 4.1

It is to be noted that in using eqns. (4.11) and (4.12) for regions not actually containing the current, their series parts are not employed explicitly, being in effect included in the general Fourier series describing boundary effects. Thus the only term to be added to the general solutions is $(i/2\pi)\theta$ for a region enclosing that containing the current, and zero for a region enclosed by the one containing the current. This is explained with the aid of Fig. 4.1(b) in which are shown three annular regions of different permeability, the middle one containing a current i at A. For the inner region, $r < b$ everywhere, so that eqn. (4.12) is relevant for the expansion of α; whilst, for the outer region, $r > b$ everywhere, and eqn. (4.11) is relevant. Thus, combining the series parts of these equations with the general series, the terms to be added to the general solutions to account for the effect of the current in these regions can be reduced to 0 and $(i/2\pi)\theta$ respectively. In the middle region, a change of 2π in α is associated with changes in θ of 2π or 0, depending upon whether r is greater or less than b, and so eqns. (4.11) and (4.12) must be employed separately as appropriate.

The physical interpretation of the above is that the term $(i/2\pi)\theta$ accounts for the non-conservative property of the field; it is therefore required for any region in which closed paths can be drawn to contain the current but not for regions in which no path contains the current. The series terms account for the remainder of the effect of the current, the boundary magnetization, and so are naturally described by the general series of eqn. (4.10).

In a similar manner to that above (using the cosine rule) it can be shown (see Smythe, p. 65) that the expansions in terms of r and θ of the function $\log R$ for a line pole (or charge), situated at the point A (Fig. 4.1a) are as follows:

where $r > b$,

$$\log R = \log r - \sum_{m=1}^{\infty} \frac{1}{m} \left(\frac{b}{r}\right)^m \cos m\theta; \qquad (4.14)$$

where $r < b$,

$$\log R = \log b - \sum_{m=1}^{\infty} \frac{1}{m} \left(\frac{r}{b}\right)^m \cos m\theta; \qquad (4.15)$$

and where $r = b$, (θ not a multiple of π),

$$\log R = \log b - \sum_{m=1}^{\infty} \frac{1}{m} \cos m\theta. \qquad (4.16)$$

Use of all three of these equations is necessary only in the potential function for the region containing the pole. In all other regions the effect of the pole can be attributed to boundary magnetization and so included in the general series expression.

Before considering some examples it should be noted that in employing the method of separation of the variables it is simplest to use separate potential functions for each region of a field, these functions being connected by the boundary conditions between the regions. The boundary conditions in terms of flux density and field strength connect the *gradients* of the potential functions (see section 2.3) so that, in general, the resulting solution gives values of potential which are discontinuous across the boundaries. If it is required to derive a solution in which values of potential are continuous everywhere, appropriate constants, determined from additional equations expressing the equality of the potential functions at a boundary, can be added to the potential functions. However, when the field strength rather than the absolute potential values are required, this offers no advantages and, as in the examples, it is not done.

4.2.2. *Iron cylinder influenced by a current*

As a simple example of the determination of the constants from the general solution, and to show the relationship of the present method to that of images, consider the field of a current outside an iron cylinder (Fig. 4.2). The development of the solution involves the determination of two potential functions—one for the air region ψ_A and one for the iron ψ_I, the two being connected by the boundary conditions at the surface of the cylinder. The constants of ψ_A are chosen so that the boundary requirements at infinity and at the cylindrical surface are satisfied, those of ψ_I so that the requirements are satisfied at the cylindrical surface and at the the centre of the cylinder, which is taken to be the origin of coordinates.

It should first be noted that the whole field is symmetrical about the line $\theta = 0$ through the current and the centre of the circle. Thus, choosing the origin of the potential functions ψ_A and ψ_I to be the line $\theta = 0$, it is seen that, since the field due to the current is odd, the

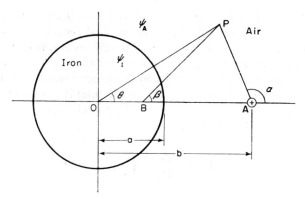

FIG. 4.2

potential functions of the whole field are odd and therefore can contain no cosine terms. Hence the general solution describing boundary influences in both regions, eqn. (4.9), can be reduced to

$$\psi(r, \theta) = \sum_{m=1}^{\infty} (c_m r^m + d_m r^{-m}) \sin m\theta. \tag{4.17}$$

In the air region, r can become infinite, but, as the potential at infinity must be finite (equipotential lines stretch from infinity to the current and $(\partial\psi/\partial r)_{r=\infty} = 0$) all values of c_{Am} must be zero. Also the air space contains a current, so it is necessary to include a term $(i/2\pi)\alpha$ in the potential function, giving

$$\psi_A = \frac{i}{2\pi}\alpha + \sum_{m=1}^{\infty} d_{Am} r^{-m} \sin m\theta. \tag{4.18}$$

In the iron, r can be zero so that, as the potential at the origin must be finite, d_{Am} must be zero. This region does not contain a current or pole, so there are no additional terms. Hence

$$\psi_I = \sum_{m=1}^{\infty} c_{Im} r^m \sin m\theta, \tag{4.19}$$

and it remains to determine d_{Am} and c_{Im} from the continuity of the two functions ψ_A and ψ_I on the cylinder surface. The boundary conditions there, eqns. (2.70) and (2.71), are, for the tangential component of field strength,

$$-\left(\frac{\partial\psi_A}{r\,\partial\theta}\right)_{r=a} = -\left(\frac{\partial\psi_I}{r\,\partial\theta}\right)_{r=a}; \tag{4.20}$$

and for the radial component of flux density,

$$-\left(\frac{\partial\psi_A}{\partial r}\right)_{r=a} = -\mu\left(\frac{\partial\psi_I}{\partial r}\right)_{r=a}, \tag{4.21}$$

where μ is the relative permeability of the iron. Substituting in ψ_A for α from eqn. (4.12), then differentiating ψ_A and ψ_I and substituting in eqns. (4.20) and (4.21), gives

$$d_{Am} - a^{2m}c_{Im} = \frac{i}{2\pi}\frac{a^{2m}}{mb^m},$$ (4.22)

and

$$d_{Am} + \mu a^{2m}c_{Im} = -\frac{i}{2\pi}\frac{a^{2m}}{mb^m},$$ (4.23)

which, when solved for c_{Im} and d_{Am}, lead to the following complete solutions:

$$\psi_A = \frac{i}{2\pi}\left(\alpha + \sum_{m=1}^{\infty}\frac{\mu-1}{\mu+1}\frac{1}{m}\frac{a^{2m}}{b^m r^m}\sin m\theta\right),$$ (4.24)

and

$$\psi_I = -\frac{i}{(\mu+1)\pi}\sum_{m=1}^{\infty}\frac{1}{m}\frac{r^m}{b^m}\sin m\theta.$$ (4.25)

The radial and tangential components of field strength are given by $-(\partial\psi/\partial r)$ and $-(\partial\psi/r\,\partial\theta)$ respectively. In forming them, α is expressed in terms of θ as described in the previous section.

The equivalence of the solution obtained by the method of images to that obtained above may be shown by summing the potential functions of the source and image currents, and by representing them in series form using eqns. (4.11) and (4.12). With current i at A the field outside the cylinder is represented by image currents of $i(\mu-1)/(\mu+1)$ at B, the inverse point of A, and $-i(\mu-1)/(\mu+1)$ at 0. The potential function for these is

$$\psi_A = \frac{i}{2\pi}\left(\alpha + \frac{\mu-1}{\mu+1}\beta - \frac{\mu-1}{\mu+1}\theta\right),$$ (4.26)

and substituting for β using eqn. (4.11) ($r_A > OB$), and remembering that $OB = a^2/b$, this expression becomes that obtained above (4.24). The equivalence of the solutions relevant to the iron region may be demonstrated similarly.

A number of similar harmonic function solutions involving a single cylindrical boundary and a single current are given by Hague and Moullin, but in practice it is quicker to solve such simple problems using the image method. Consideration is now given to some problems in which the method of separation of the variables gives a simpler, or in some cases, the only practicable method of solution.

4.2.3. The screening effect of a permeable cylinder

Consider the effect of a permeable cylinder used, for example, to screen a galvanometer from the earth's magnetic field. The cylinder, of relative permeability μ and having the dimensions shown in Fig. 4.3, is placed in a uniform magnetic field of strength H. The complete solution for the field in all the regions involves three potential functions: ψ_A for the outer air space, ψ_I for the iron, and ψ_B for the inner air space, the region now of particular interest. Since there are no currents or poles in the field and since, from symmetry, $\psi(r, \theta') = \psi(r, -\theta')$, so that there are no sine terms in the solution, the general form for each of these potential functions can be reduced to

$$\psi = \sum_{m=1}^{\infty}(c_m r^m + d_m r^{-m})\cos m\theta.$$ (4.27)

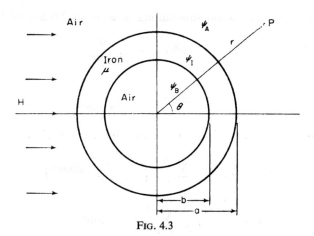

FIG. 4.3

Examining the outer air space first, it is seen that the boundary condition at infinity is not, as in the last example, one of potential value but is of potential gradient. At infinity the field strength is uniform and its radial component is

$$H \cos \theta = -\frac{\partial \psi_A}{\partial r}; \tag{4.28}$$

that is

$$H \cos \theta = -\sum_{m=1}^{\infty} (m c_{Am} r^{m-1} - m d_{Am} r^{-(m+1)}) \cos m\theta, \tag{4.29}$$

from which it is seen that m can have the value unity only and, hence, substituting $m = 1$ and $r = \infty$, that $c_{A1} = -H$. Thus, for the outer air space, eqn. (4.27) reduces to

$$\psi_A = -\left(Hr - \frac{d_{A1}}{r}\right) \cos \theta. \tag{4.30}$$

Expressing the equality of the radial components of flux density and of the tangential components of field strength on opposite sides of each iron–air interface gives

$$\left(\frac{\partial \psi_A}{\partial r}\right)_{r=a} = \mu \left(\frac{\partial \psi_I}{\partial r}\right)_{r=a}, \tag{4.31}$$

$$\frac{1}{a}\left(\frac{\partial \psi_A}{\partial \theta}\right)_{r=a} = \frac{1}{a}\left(\frac{\partial \psi_I}{\partial \theta}\right)_{r=a}, \tag{4.32}$$

$$\mu \left(\frac{\partial \psi_I}{\partial r}\right)_{r=b} = \left(\frac{\partial \psi_B}{\partial r}\right)_{r=b}, \tag{4.33}$$

and

$$\frac{1}{b}\left(\frac{\partial \psi_I}{\partial \theta}\right)_{r=b} = \frac{1}{b}\left(\frac{\partial \psi_B}{\partial \theta}\right)_{r=b}. \tag{4.34}$$

Then, since from eqn. (4.30)

$$\left(\frac{\partial \psi_A}{\partial r}\right)_{r=a} = -\left(H + \frac{d_{A1}}{a^2}\right) \cos \theta, \tag{4.35}$$

it is clear from this equation and eqns. (4.32) and (4.33) that the only value which m may take in ψ_I and ψ_B is also unity. So

$$\psi_I = -\left(c_{I1}r - \frac{d_{I1}}{r}\right)\cos\theta, \tag{4.36}$$

and

$$\psi_B = -\left(c_{B1}r - \frac{d_{B1}}{r}\right)\cos\theta. \tag{4.37}$$

The remaining boundary condition requires that the field strength remains finite, even when $r = 0$, so that $d_{B1} = 0$ and

$$\psi_B = -c_{B1}r\cos\theta. \tag{4.38}$$

Thus the field in the central air space has the direction of the line $\theta = 0$, and its strength' $-\partial\psi_B/\partial r$, is c_{B1} (a constant) and so uniform. Differentiating ψ_A, ψ_I and ψ_B, and substituting in the interface boundary conditions, (4.31) to (4.34), gives

$$\left.\begin{aligned}
H + \frac{d_{A1}}{a^2} &= \mu\left(c_{I1} + \frac{d_{I1}}{a^2}\right), \\[2mm]
H - \frac{d_{A1}}{a^2} &= \left(c_{I1} - \frac{d_{I1}}{a^2}\right), \\[2mm]
\mu\left(c_{I1} + \frac{d_{I1}}{b^2}\right) &= c_{B1}, \\[2mm]
c_{I1} - \frac{d_{I1}}{b^2} &= c_{B1}.
\end{aligned}\right\} \tag{4.39}$$

From these equations the constants are determined. The constant c_{B1}, which from eqn. (4.38) is seen to be the field strength inside the cylinder, is

$$c_{B1} = \frac{4\mu a^2 H}{a^2(\mu+1)^2 - (\mu-1)^2 b^2}, \tag{4.40}$$

and, in practice, since $\mu \gg 1$, this expression can be simplified to give

$$c_{B1} \doteqdot \frac{4}{\mu}\frac{H}{1-(b/a)^2}. \tag{4.41}$$

It is seen, therefore, that the strength of the field inside the cylinder varies inversely with the permeability for given proportions of the cylinder. For example, with mumetal at field strengths up to 1 oersted for which the permeability is 10,000 or more, it would be reduced to 0·04 per cent of the outside value. A map of the whole field is shown in Fig. 4.4.

A discussion of this problem in terms of surface polarity has been given by Hammond.[1] *Doublet representation of the cylinder.* An interesting result which can be demonstrated from the above solution is that the effect of a cylinder of finite thickness on a uniform field can be accounted for by the use of a doublet current. (It is shown in section 3.3 that a doublet can also be used to account for the effect of a *solid* cylinder on a uniform field.) By substituting $t = r\,e^{j\theta}$ in eqn. (3.23) it is readily seen that the potential function in circular-

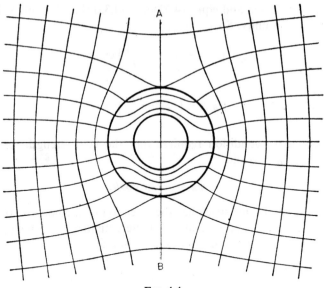

FIG. 4.4

cylinder coordinates of a doublet at the origin, with its axis on the line $\theta = 0$ and of strength d, is given by

$$\psi = R\left(\frac{d}{r} e^{-j\theta}\right)$$

$$= \frac{d}{r} \cos \theta. \tag{4.42}$$

But the potential function ψ_A for the region outside the cylinder is given by the right-hand side of eqn. (4.30), the first term of which is the potential function of the applied uniform field and the second of which is recognizable from the above as the potential function of a doublet. Thus the effect of the hollow cylinder on the uniform field can be accounted for by a doublet, placed at the centre of the cylinder, of strength $d = d_{A1}$, where it can be simply shown that

$$d_{A1} = \frac{(\mu-1)\,[1-(b/a)^2]\,a^2H}{(\mu+1)\,[1-(\mu-1)^2\,b^2/(\mu+1)^2\,a^2]}. \tag{4.43}$$

4.2.4. *The force between rotor and stator conductors in a cylindrical machine*

The magnetic circuit of a cylindrical machine is often assumed to consist essentially of a thick outer iron shell (the stator) separated by an air space (the air gap) from a solid iron cylinder (the rotor) coaxial with it (Fig. 4.5). Consideration is given here to the calculation of the tangential force between typical rotor and stator conductors, the presence of slots being neglected. This is done here by first evaluating the field at the position of the rotor conductor due to the stator current.

    Let the potential functions in the four regions of the field be: ψ_O in the outer air space, ψ_S in the stator, ψ_G in the air gap, and ψ_R in the rotor. Noting that the stator region contains a current, that in all regions the field is symmetrical and the potential is odd about the line

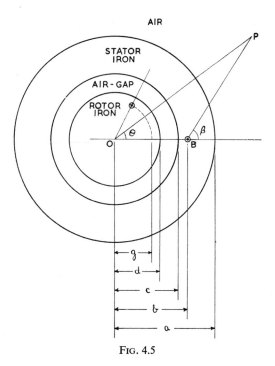

FIG. 4.5

$\theta = 0$, and that r can become infinite in the outer air region and zero in the rotor, then the potential functions for the field (due to the stator current alone) may be written:

for the outer air space,

$$\psi_O = \frac{i_S}{2\pi}\theta + \sum_{m=1}^{\infty} D_{Om}r^{-m}\sin m\theta; \tag{4.44}$$

for the stator,

$$\psi_S = \frac{i_S}{2\pi}\beta + \sum_{m=1}^{\infty} (C_{Sm}r^m + D_{Sm}r^{-m})\sin m\theta, \tag{4.45}$$

where β is the angle subtended at the stator conductor;

for the air gap,

$$\psi_G = \sum_{m=1}^{\infty} (C_{Gm}r^m + D_{Gm}r^{-m})\sin m\theta; \tag{4.46}$$

and for the rotor,

$$\psi_R = \sum_{m=1}^{\infty} C_{Rm}r^m \sin m\theta. \tag{4.47}$$

Capital letters are used for the constants to avoid confusion with the dimensions.

There are three interfaces in the field region, and substituting the above four equations in the boundary conditions at the interfaces, six equations arise from which the constants of the potential functions may be evaluated. Using the notation of Fig. 4.5, and for iron parts of relative permeability μ, the values of these constants are:

$$D_{Om} = \frac{i_S\mu}{\pi(1+\mu)} \left\{ \begin{array}{l} b^{2m}[(1+\mu)^2 c^{2m} - (\mu-1)^2 d^{2m}] \\ -c^{2m}[(\mu^2-1)(c^{2m}-d^{2m})] \end{array} \right\} \frac{1}{Q}, \tag{4.48}$$

$$C_{Sm} = \frac{i_S}{2\pi} \left(\frac{\mu-1}{\mu+1}\right) \frac{1}{a^{2m}} \left\{ \begin{array}{l} b^{2m}[(\mu+1)^2 c^{2m} - (\mu-1)^2 d^{2m}] \\ -c^{2m}[(\mu^2-1)(c^{2m}-d^{2m})] \end{array} \right\} \frac{1}{Q} \tag{4.49}$$

$$D_{Sm} = \frac{i_S}{2\pi} c^{2m} \left\{ \left[1 - \left(\frac{\mu-1}{\mu+1}\right)\left(\frac{b}{a}\right)^{2m}\right][(\mu^2-1)(c^{2m}-d^{2m})] \right\} \frac{1}{Q}, \tag{4.50}$$

$$C_{Gm} = -\frac{i_S}{\pi} \mu(\mu+1) c^{2m} \left\{1 - \left(\frac{\mu-1}{\mu+1}\right)\left(\frac{b}{a}\right)^{2m}\right\} \frac{1}{Q}, \tag{4.51}$$

$$D_{Gm} = \frac{i_S}{\pi} \mu(\mu-1) c^{2m} d^{2m} \left\{1 - \left(\frac{\mu-1}{\mu+1}\right)\left(\frac{b}{a}\right)^{2m}\right\} \frac{1}{Q}, \tag{4.52}$$

$$C_{Rm} = -\frac{2i_S}{\pi} \mu c^{2m} \left[1 - \left(\frac{\mu-1}{\mu+1}\right)\left(\frac{b}{a}\right)^{2m}\right] \frac{1}{Q}, \tag{4.53}$$

where

$$Q = \left\{ \begin{array}{l} [(\mu+1)^2 c^{2m} - (\mu-1)^2 d^{2m}] \\ -\left(\frac{\mu-1}{\mu+1}\right)\left(\frac{c}{a}\right)^{2m}[(\mu^2-1)(c^{2m}-d^{2m})] \end{array} \right\} mb^m. \tag{4.54}$$

The radial component of flux density due to the stator current at the position of the rotor conductor is

$$B_r = -\mu\left(\frac{\partial\psi_R}{\partial r}\right)_{r=g}, \tag{4.55}$$

and, if the rotor conductor carries a current i_R, the tangential force F_t causing the rotor to turn is

$$F_t = i_R B_r = -i_R\mu \sum_{m=1}^{\infty} mC_{mg}g^{m-1} \sin m\theta$$

$$= -\frac{2i_S i_R}{\pi} \sum_{m=1}^{\infty} \frac{\dfrac{g^{m-1}}{b^m} \mu^2\left[1 - \dfrac{\mu-1}{\mu+1}\left(\dfrac{b}{a}\right)^{2m}\right] \sin m\theta}{\left[(\mu+1)^2 - (\mu-1)^2\left(\dfrac{d}{c}\right)^{2m}\right] - (\mu-1)^2 \dfrac{1}{a^{2m}}[c^{2m} - d^{2m}]}, \tag{4.56}$$

its sense being dependent upon the directions of i_S and i_R.

Using values of $a = 1\cdot4$, $b = 1\cdot12$, $c = 1\cdot02$, $d = 1\cdot0$, and $g = 0\cdot9$, evaluation of this expression for $\mu = \infty$ and 100 gives the results plotted in Fig. 4.6. The influence of permeability is seen to be considerable, and this is because, in the model, the currents are embedded in the iron. In an actual machine the conductors are housed in slots so that the above representation of them is unsatisfactory. However, the representation of an actual machine, achieved by placing the currents in the model on the *surfaces* of the rotor and stator, is entirely satisfactory for the calculation of the force causing rotation. As is the case in many solutions of this type, the series is rather slowly convergent: twenty-five terms are necessary to reduce the error to less than 1 per cent. The preparation of one curve of the type shown in Fig. 4.6 takes several hours of hand calculation, but is a trivial computing problem.

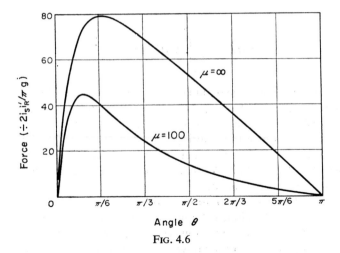

Angle θ

FIG. 4.6

4.2.5. Specified distributions of potential or potential gradient on the perimeter of a circular boundary

The preceding discussion of fields with circular boundaries is restricted mainly to cases in which the field sources are currents. The method of the section is, however, also applicable to the solution of fields (electric or magnetic) due to specified distributions of potential or potential gradient on the boundaries of circular regions. Brief consideration is now given to such problems, and the general solution—known as the Poisson integral—for the field inside a circle, due to specified values of potential on the perimeter, is developed.

The use of Fourier series. The general solution for a field, containing no line sources, in a region with a circular boundary centred on the origin, is given by the potential function of eqn. (4.9) with the addition of a constant, i. e. by

$$\psi(r, \theta) = k + \sum_{m=1}^{\infty} (c_m r^m + d_m r^{-m}) (g_m \cos m\theta + h_m \sin m\theta). \tag{4.57}$$

Taking first the case of a single region *inside* a circular boundary, $r = r'$, it is seen that, since ψ remains finite when $r = 0$, d_m is zero, and the solution becomes

$$\psi(r, \theta) = k + \sum_{m=1}^{\infty} r^m(G_m \cos m\theta + H_m \sin m\theta), \tag{4.58}$$

where $G_m = c_m g_m$ and $H_m = c_m h_m$. If the potential distribution $\psi'(r', \theta')$ is specified on the boundary, then the complete solution for the field due to this distribution is obtained by choosing the constants k, $r'^m G_m$, and $r'^m H_m$, so that the right-hand side of eqn. (4.58) is equal to $\psi'(r', \theta')$. That is, the constants must be chosen to give the Fourier series representing $\psi'(r', \theta')$ in the range 0 to 2π; hence,

$$k = \frac{1}{2\pi} \int_0^{2\pi} \psi'(r', \theta') \, d\theta', \tag{4.59}$$

$$r'^m G_m = \frac{1}{\pi} \int_0^{2\pi} \psi'(r', \theta') \cos m\theta' \, d\theta' \tag{4.60}$$

6*

and

$$r'^m H_m = \frac{1}{\pi} \int_0^{2\pi} \psi'(r', \theta') \sin m\theta' \, d\theta'. \tag{4.61}$$

For a single region *exterior* to the boundary $r = r'$, the solution is identical with the above except that r^m is replaced by $(1/r^m)$ in eqn. (4.58) and r'^m is replaced by $(1/r'^m)$ in eqns. (4.59), (4.60), and (4.61). When potential gradient is specified, the constants of eqn. (4.58) are chosen in a similar way to give the Fourier series for $\partial\psi'/\partial r$ or $\partial\psi'/r'\partial\theta$ on the boundary.

No applications to this class of problem are given here, but some are described in the papers by Rudenberg[2, 3] in which the flux distribution inside machine stators due to an impressed sinusoidal field is examined.

The Poisson integral. The general solution of the Dirichlet problem (specified boundary potentials) for the interior of a circle is known as the Poisson integral, and it is of considerable importance, particularly in connection with transformation methods (see sections 8.4 and 10.6.1). Its development from the above equations is briefly indicated here, taking, as is usual, the radius of the bounding circle to be unity. Substituting for the Fourier coefficients (with $r' = 1$) in the potential function, eqn. (4.58), gives

$$\psi(r, \theta) = \frac{1}{2\pi} \int_0^{2\pi} \psi'(\theta') \, d\theta' [1 + 2 \sum r^m \cos m(\theta - \theta')], \tag{4.62}$$

and the expression in square brackets from this equation may be reduced using the following identities:

$$1 + 2 \sum r^m \cos m(\theta - \theta') = R\left[1 + 2 \sum (r \, e^{j(\theta - \theta')})^m\right]$$
$$= R\left[\frac{1 + r \, e^{j(\theta - \theta')}}{1 - r \, e^{j(\theta - \theta')}}\right]$$
$$= \frac{1 - r^2}{1 + r^2 - 2r \cos(\theta - \theta')}. \tag{4.63}$$

Therefore eqn. (4.62) can be reduced to

$$\psi(r, \theta) = \frac{1}{2\pi} \int_0^{2\pi} \frac{1 - r^2}{1 + r^2 - 2r \cos(\theta - \theta')} \psi'(\theta') \, d\theta', \tag{4.64}$$

which is the Poisson integral, yielding directly the solution for the inside of the unit circle with any potential $\psi'(\theta')$ impressed at the circumference. (See also the discussion in section 10.6.)

4.3. Rectangular boundaries

4.3.1. *Solution of Laplace's equation in cartesian coordinates*

In cartesian coordinates Laplace's equation is

$$\frac{\partial^2 \psi}{\partial x^2} + \frac{\partial^2 \psi}{\partial y^2} = 0, \tag{4.65}$$

and a solution for it can be obtained using the same method as for the cylindrical polar form of the equation. Thus, assuming a solution

$$\psi(x, y) = X(x)\,Y(y), \tag{4.66}$$

where X is a function of x only and Y is a function of y only, the partial differential equation reduces to the two ordinary differential equations

$$\frac{d^2X}{dx^2} + m^2X = 0, \tag{4.67}$$

and

$$\frac{d^2Y}{dy^2} - m^2Y = 0, \tag{4.68}$$

in which m is a constant. Solutions to these equations lead to the particular solution

$$\psi(x, y) = (c\,\sin mx + d\,\cos mx)\,(g\,\sinh my + h\,\cosh my), \tag{4.69}$$

and the general potential function (see section 4.2.1)

$$\psi(x, y) = \sum_{m=1}^{\infty} (c_m\,\sin mx + d_m\,\cos mx)\,(g_m\,\sinh my + h_m\,\cosh my). \tag{4.70}$$

For any value of y in the range $0 \leqslant y \leqslant \infty$, this function is a Fourier series periodic in x and it is, therefore, capable of representing the field due to boundary effects. For the value of $m = 0$, the ordinary differential equations (4.67) and (4.68) give rise to the additional terms

$$k_1 + k_2x + k_3y, \tag{4.71}$$

which are also solutions of the field equation. The constant k_1 defines a reference potential, and k_2 and k_3 uniform fields in the x- and y-directions respectively. Thus, a more general solution is

$$\psi(x, y) = \sum_{m=1}^{\infty} (c_m\,\sin mx + d_m\,\cos mx)\,(g_m\,\sinh my + h_m\,\cosh my) + k_1 + k_2x + k_3y. \tag{4.72}$$

This solution does not account for the influence of line sources in the field; to enable it to do so, the potential functions of these sources have to be added to it. These functions, which are familiar in the more convenient polar forms (see section 4.2.1), are, for the current i, at (x_0, y_0),

$$\psi = \frac{i}{2\pi}\,\tan^{-1}\left(\frac{y - y_0}{x - x_0}\right), \tag{4.73}$$

and for the pole, p, at (x_0, y_0),

$$\psi = \frac{p}{2\pi}\,\log \sqrt{[(x - x_0)^2 + (y - y_0)^2]}. \tag{4.74}$$

However, a general solution is not written here since, as stated in the introduction, and as is now evident from the form of eqns. (4.73) and (4.74), the solution of problems with rectangular boundaries and line sources is more conveniently achieved using the image method.

Equation (4.70) gives, for positive integral values of m, a solution periodic in x; that is, a solution found for a region $a \leqslant x \leqslant a+\lambda$ applies to a succession of similar problems in regions $a+n\lambda \leqslant x \leqslant a+(n+1)\lambda$, where n is an integer taking values between $-\infty$ and ∞. By choosing the constant m^2 with the opposite sign, a solution periodic in y is derived.

Consideration is now given to the two (similar) boundary shapes for which Laplace's equation in cartesian coordinates may be solved by the function (4.72), and to the ways in which the solutions may be extended.

4.3.2. *The semi-infinite strip and the rectangle*

Solutions, based upon eqn. (4.72), are possible for the interior region of (i) a rectangle, and (ii) a semi-infinite strip, or rectangle with one side at infinity. They are not possible for the exterior regions for reasons discussed at the end of this subsection.

The reference potential k_1 of the field under investigation may be taken for convenience as zero, and also the presence of uniform fields parallel to the axes is disregarded, so that $k_2 = k_3 = 0$.

The semi-infinite strip. Consider first the evaluation of the constants to give a solution for the inside of the semi-infinite strip $0 \leqslant x \leqslant a$, $0 \leqslant y \leqslant \infty$ (Fig. 4.7), the edges $x = 0$,

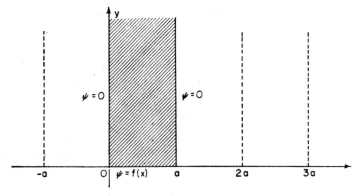

FIG. 4.7

$x = a$ and $y = \infty$ being maintained at zero potential whilst along $y = 0$ is impressed a potential distribution $f(x)$. (Physically there is only one such strip, but the mathematical treatment is the same as if there were a succession of strips side-by-side, as shown dotted in the figure.)

The boundary conditions for the potential in the sheet are:

$$\psi(0, y) = 0, \tag{4.75}$$

$$\psi(a, y) = 0, \tag{4.76}$$

$$\psi(x, \infty) = 0, \tag{4.77}$$

and
$$\psi(x, 0) = f(x). \tag{4.78}$$

The first of them requires that

$$d_m = 0, \tag{4.79}$$

and the second that

$$m = \frac{\alpha\pi}{a}, \tag{4.80}$$

where α is an integer. From the third condition, eqn. (4.77),

$$g_m \tanh \infty + h_m = 0,$$

and, therefore,

$$g_m = -h_m. \tag{4.81}$$

Substituting these values for m, d_m, and g_m in eqn. (4.70), and combining sinh my and cosh my in exponential form, gives

$$\psi(x, y) = \sum_{\alpha=1}^{\infty} C_\alpha \sin \frac{\alpha\pi}{a} x \exp\left[-(\alpha\pi/a)y\right], \tag{4.82}$$

where C_α is written for $c_\alpha h_\alpha$. The fourth condition, eqn. (4.78), requires that

$$f(x) = \sum_{\alpha=1}^{\infty} C_\alpha \sin \frac{\alpha\pi}{a} x, \tag{4.83}$$

or that C_α be chosen as the coefficients of the Fourier sine series for $f(x)$. Now it is well known (see, for example, Churchill) that the coefficients C_α of the Fourier sine series representing the function $f(x)$ in the range 0 to a are given by

$$C_m = \frac{2}{a} \int_0^a f(x) \sin mx \, \mathrm{d}x. \tag{4.84}$$

So, for example, if the potential along $y = 0$ is unity at all points, then m is odd only with

$$C_{(2\alpha-1)} = \frac{4}{\pi(2\alpha-1)}, \tag{4.85}$$

and the potential at any point in the strip is given by

$$\psi(x, y) = \frac{4}{\pi} \sum_{\alpha=1}^{\infty} \frac{1}{2\alpha-1} \exp\left\{-[(2\alpha-1)/a]\pi y\right\} \sin\left[\frac{(2\alpha-1)\pi}{a}\right]x. \tag{4.86}$$

From a family of curves of the variation of ψ with x for a range of values of y, the lines of constant potential can be found by interpolation. In the same manner, lines of flux may be derived from a family of curves for the flux function φ, this being obtained from the potential function by the Cauchy–Riemann relationships as described earlier (see sections 2.4.3 and 2.4.4). Curves of flux and potential are shown in Fig. 4.8.

This solution is probably visualized most easily in electrostatic terms, but it is instructive to note that it refers also to certain arrays of currents. Firstly, the field is identical with that which would be set up (in the positive half-plane) by an infinite array of currents of alternate sign placed on the x-axis at ..., $-a$, 0, a, $2a$, ..., all the currents being of magnitude 4 amperes. (A current of n amperes produces a potential difference $n/4$ between two lines meeting at right-angles in the current.) Secondly, the field is also that which would be set up by currents at 0 and a of magnitude 2 amperes and of opposite sign, where the boundaries

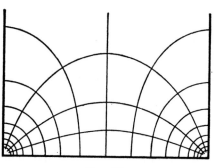

FIG. 4.8

$x = 0$ and $x = a$ are infinitely permeable. Finally, a further equivalent field array is derivable from each of the above by considering the x-axis as the edge of an infinitely permeable boundary in the negative half-plane, and the currents to have half the magnitude of those above.

The rectangle. The solution for the field inside a rectangular region is also of considerable interest. For instance, it has been used to investigate the field in a plane section of an electron multiplier, where different potentials exist on the pairs of sides diagonally opposite (see Zworykin *et al.*, p. 369); the treatment involved the superposition of the separate solutions for the different potential distributions on each of the sides of the rectangle (with the potentials on the remaining sides, in each case, assumed zero). Consideration is given below to typical basic solutions for the rectangular region.

Figure 4.9 (a) and (b) shows such a region $0 \leqslant x \leqslant a$, $0 \leqslant y \leqslant b$, and the field in this region is to be solved for two different sets of boundary conditions. The difference between

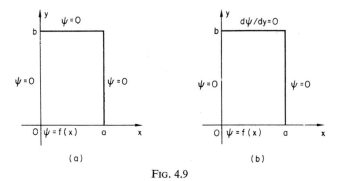

FIG. 4.9

the two sets of conditions is that in the first the edge $y = b$ is a line of zero potential, whilst in the second it is impervious to flux. The remaining conditions along $x = 0$, $x = a$, and $y = 0$, in both cases, are identical with those, eqns. (4.75), (4.76), and (4.78), in the example of the strip. Consequently they result in the same values for the constants d_m, m, and C_α, eqns. (4.79), (4.80), and (4.83). The remaining condition on $y = b$ requires in the first case ($\psi = 0$) that

$$g_m \sinh mb + h_m \cosh mb = 0$$

or

$$\frac{g_m}{h_m} = -\coth mb; \qquad (4.87)$$

and in the second case, the condition $(\partial\psi/\partial y = 0)$ on $y = b$ requires that

$$mg_m \cosh mb + mh_m \sinh mb = 0$$

or

$$\frac{g_m}{h_m} = -\tanh mb. \tag{4.88}$$

Thus, the solutions are; for the case with the line $y = b$ at zero potential,

$$\psi(x, y) = \sum_{\alpha=1}^{\infty} C_\alpha \sin\left(\frac{\alpha\pi}{a} x\right) \frac{\sinh [\alpha\pi(b-y)/a]}{\sinh [\alpha\pi b/a]}, \tag{4.89}$$

and for the case with the line $y = b$ impervious to flux,

$$\psi(x, y) = \sum_{\alpha=1}^{\infty} C_\alpha \sin\left(\frac{\alpha\pi}{a} x\right) \frac{\cosh [\alpha\pi(b-y)/a]}{\cosh [\alpha\pi b/a]}, \tag{4.90}$$

in both of which the values C_α are the coefficients of the Fourier sine series representing $f(x)$ [see eqn. (4.84)].

Interconnected regions. The results of this sub-section may be extended to interconnected rectangular regions provided that these have the same width (and so m), see section 5.5, by using the equations expressing, at the boundaries, the continuity of the normal component of flux, and the tangential component of field strength, but this technique is adequately discussed in the treatment of circular boundaries.

The reason that it is not feasible to derive solutions for the (infinite) region exterior to a rectangular boundary is coupled with the use of such inter-regional continuity equations. In order to treat the whole exterior region, this must be divided into several sub-regions (formed by extending the sides of the rectangle to infinity), and then the separate potential functions for each sub-region must be connected to those in adjacent regions by the continuity equations. But the potential functions for the regions infinite in both the x- and y- directions, being integrals (see the next section), are very difficult to handle, and, because of this and because of the large number of continuity equations, the resulting formal solution is unmanageable.

A final example demonstrates how a solution for fields with basically simple rectangular boundaries can be used to solve a physical problem having boundaries the *shape* of which is described by a complicated equation.

4.3.3. *Pole profile in the inductor alternator for a sinusoidal flux distribution*

The inductor alternator is the only machine, even in theory, capable of providing a supply having a sinusoidal wave-shape. The pole shape necessary to achieve this has been determined by F. W. Carter (published in a paper by Walker[4]) and by Hancock[5], their methods being rather different from the one employed here. The shape is represented approximately in Fig. 4.10, where the armature surface is shown as a straight line and the rotor surface is shaded. Since the pole pitch t is small, the representation of the air gap as straight involves only small approximation. Considering the iron to be infinitely permeable, the rotor profile which gives a sinusoidal flux distribution at the armature surface can be determined in the following way. The potential distribution in the typical semi-infinite region—between the dotted lines and the x-axis—which gives the required distribution of flux density along the x-axis, is first found. Then, since the rotor surface is equipotential $(\mu = \infty)$, its required

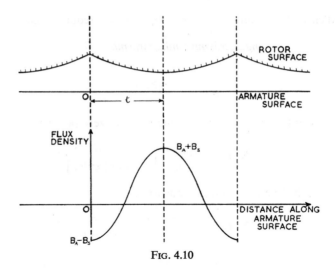

FIG. 4.10

shape is that of any convenient equipotential line of the field, determined from the solution for the semi-infinite space.[†]

The field is naturally periodic in x and may be analysed by considering the typical vertical strip bounded by the lines $x = 0$, $x = t$, and $y = 0$. Now the lines $x = 0$ and $x = t$ obviously coincide with flux lines, and it is required that the flux density on $y = 0$ should vary harmonically with period $2t$, so that the solution must satisfy the boundary conditions

$$\left(\frac{\partial \psi}{\partial x}\right)_{x=0} = 0, \tag{4.91}$$

$$\left(\frac{\partial \psi}{\partial x}\right)_{x=t} = 0, \tag{4.92}$$

and

$$\left(\frac{\partial \psi}{\partial y}\right)_{y=0} = B_A - B_S \cos\left(\frac{\pi}{t} x\right), \tag{4.93}$$

where B_A is the steady and B_S is the peak periodic component of flux density on the armature surface. In addition, the armature surface can be assumed, like the rotor, to be infinitely permeable, so that $(\partial \psi/\partial x)_{y=0} = 0$. This condition is fulfilled and the solution is given its simplest form by taking the potential at the armature surface to be zero. Then the remaining boundary condition is

$$\psi(x, 0) = 0. \tag{4.94}$$

The general solution for the region, eqn. (4.72), is

$$\psi = \sum_{m=1}^{\infty} (c_m \sin mx + d_m \cos mx)(g_m \sinh my + h_m \cosh my) + k_1 + k_2 x + k_3 y, \tag{4.95}$$

and substituting from this in the boundary conditions gives the values of the constants.

[†] This device of representing a physical boundary of non-simple shape by a field equipotential line is of considerable value in extending simpler solutions, and it has wide application. See, for example, section 8.2.3 and reference 9, chapter 8.

The first condition, eqn. (4.91), gives

$$c_m = 0$$

and

$$k_2 = 0;$$

(4.96)

and the second, eqn. (4.92),

$$m = \frac{\alpha\pi}{t},$$

(4.97)

where α is an integer. Further, from eqn. (4.94),

$$h_m = 0$$

and

$$k_1 = 0,$$

(4.98)

so that, substituting for all these constants in eqn. (4.95) and differentiating with respect to y gives

$$\frac{\partial\psi}{\partial y} = \sum_{\alpha} \frac{\alpha\pi}{t} d_m g_m \cos\frac{\alpha\pi x}{t} \cosh\frac{\alpha\pi y}{t} + k_3.$$

(4.99)

Thus, using eqn. (4.93), it is seen that

$$\alpha = 1 \quad \text{(only)},$$
$$k_3 = B_A,$$

and

$$d_m g_m = -\frac{t}{\pi} B_S,$$

(4.100)

so that finally the solution for the potential is

$$\psi = B_A y - \frac{t}{\pi} B_S \sinh\frac{\pi}{t} y \cos\frac{\pi}{t} x.$$

(4.101)

When $\psi = 0$, this reduces to the equation $y = 0$ of the armature surface, and for any other value gives an equation for a possible rotor shape.

The profile best suited to a given application depends upon the ampere turns available to set up the magnetic potential difference ψ between rotor and stator, upon the ratio of mean flux density to the amplitude of its variation, and upon a consideration of the shapes which are physically realizable and which may be manufactured conveniently. Having decided suitable values for ψ and B_A/B_S, eqn. (4.101) must be evaluated for pairs of values of x and y giving points lying on the rotor surface. Examination of the equation shows that x must be evaluated for a range of values of y between y_0 and y_t, the end points, which have to be determined graphically. Considerable accuracy is required in the arithmetic, since differences between quantities of the same order of size are involved. Figure 4.11 shows a field map of some physically realizable profiles.

It is evident that the profile for a flux wave of any shape can be found, for if the wave form is $f(x)$ it may be represented as a Fourier cosine series $\sum B_m \cos m\theta$. Then the required equipotentials have the form

$$\psi = B_A y - \sum_{m=1}^{\infty} \frac{B_m}{m} \sinh my \cos mx.$$

(4.102)

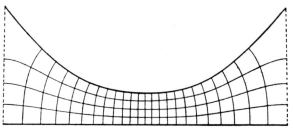

Fig. 4.11

4.4. Conclusions

Integral solutions. In the preceding sections the general solutions developed are based upon the treatment of the fields as periodic, either inherently so as when polar coordinates are used, or artificially so as when cartesian coordinates are used. Because of this it is possible to express the single-valued potential function of the field region in a general form applicable to any physically realizable boundary conditions by taking only *integral* values of the separation constant m and utilizing Fourier series. An alternative form of expression for the potential function, which is necessary when the field cannot be treated as periodic in the coordinate system used (e.g. when it is infinite in both directions and cartesian coordinates are used), is as an integral over the whole range of values of m; that is, for the coordinate system treated above, as

$$\psi(r, \theta) = \int_{-\infty}^{\infty} f(m) RS \, dm, \tag{4.103}$$

and as

$$\psi(x, y) = \int_{-\infty}^{\infty} f(m) XY \, dm, \tag{4.104}$$

where R, S, X, and Y are respectively functions of r, θ, x, and y only. These integrals, which can be regarded as the limiting forms of the Fourier series (see Byerly, p. 52), are much more difficult to handle than the series and they are of limited application: in general, an appropriate choice of coordinate system or the use of images obviates the need for them and no further discussion is given here. However, one interesting application is described by Smythe (p. 70) who analyses the field, due to a line charge, within, and exterior to, a dielectric wedge.

(Note that in section 10.6.1 an alternative form of solution for the field in a region bounded by an infinite, straight boundary on which is impressed a potential distribution, is obtained from the Schwarz complex potential function. Note, too, the relationship of this function to the Poisson integral for the unit circle.)

Scope of the method of separation of variables. For the coordinate systems considered it is seen that the scopes of the method of separation of variables and the method of images are similar. However, the types of problem to which each is best suited differ: thus, for problems involving line sources and relatively few boundaries and (with the exception of other cases treated in chapter 3) giving rise to a finite number of images (see also section 5.4), the image method is simpler; but for other problems having line sources and multiple boundaries, separation of the variables is to be preferred; and for all problems involving specified distributions of potential or gradient it is essential.

Throughout the chapter currents have been considered in the form of lines, but it is to be noted that the fields of *current sheets* can be simply derived by integration (over the range of the sheet) of the solution for a line current. See, for example, section 5.2.2 and Hague or, for a different approach, Ferraro.

Sinusoidally distributed current sheets on cylindrical surfaces have been extensively employed in the study of rotating electrical machine field problems, particularly the three-dimensional fields in the end region, and examples are included in the Additional References. Other closely related applications of the separation-of-variables technique to end-field problems are also included for completeness.

The question of the treatment of multi-region problems has been touched upon at a number of points, and in summary it may be noted that any number of regions can be treated provided that the functions for each region involve the same constant of separation (m). With concentric circular boundaries, the constant is automatically the same, and several examples are given in this chapter. Other examples in connection with circular boundaries, and many related problems of more general interest, are given by Ralph[6]. With rectangular boundaries, the constant is the same provided that all the regions are adjacent and have the same width. No examples have been included in this chapter, but the procedure should be immediately apparent by comparison with the circular-boundary cases. It is, anyway, made clear through two problems considered in sections 5.5.2 and 5.5.3—these problems differ from those of this chapter only in that the field in one of the regions satisfies the slightly more complex Poisson equation rather than the Laplace equation.

The series forms of the solutions obtained by direct methods may not be convenient for computation because the series are slowly convergent; this, however, is not a serious difficulty, particularly when computing facilities are available. The fact that the solution is in terms of either the flux function or the potential function is slightly disadvantageous as compared with solutions (obtained by transformation or images) which are expressed in complex-variable form, but, except for field maps, both functions are rarely required and, if they are, the Cauchy–Riemann equations can be used.

In the next chapter, solutions are given in terms of the vector potential function A for Poisson's equation, and it will be evident that the solutions described in this chapter could also be obtained in terms of this function. However, by so doing no advantages would accrue and, indeed, for a solution requiring values of scalar potential, the use of A is inconvenient.

References

1. P. HAMMOND, The magnetic screening effects of iron tubes, *Proc. Instn. Elect. Engrs.* **103** C, 112 (1956).
2. R. RUDENBERG, Über die Verteilung der magnetischen Induktion in Dynamoankern und die Berechnung von Hysterese und Wirbelstromverlusten, *Elektrotechn. Z.* **27** (6), 109 (1906).
3. R. RUDENBERG, Energie der Wirbelströme in elektrischen Bremsen und Dynamomaschinen, *Samml. Elektrotech. Vortr.* **10**, 269 (1907).
4. J. H. WALKER, The theory of the inductor alternator, *Proc. Instn. Elect. Engrs.* **89** II, 227 (1942).
5. N. N. HANCOCK, The production of a sinusoidal flux wave with particular reference to the inductor alternator, *Proc. Instn. Elect. Engrs.* **104** C, (new series), 167 (1957).
6. M. C. RALPH, Eddy currents in rotating electrical machines, Ph.D. thesis, University of Leeds, 1967.

Additional References

ASHWORTH, D. S., and HAMMOND, P., The calculation of the magnetic field of rotating machines: Part 2, The field of turbo-generator end-windings, *Proc. Instn. Elec. Engrs.* **108** A, 527 (1961).

HONSINGER, V. B., Theory of end-winding leakage reactance, *Trans. Am. Instn. Elect. Engrs.* **78** III, 417 (1959).

REECE, A. B. J., and PRAMANIK, A., Calculation of the end region field of a.c. machines, *Proc. Instn. Elect. Engrs.* **112,** 1355 (1965).

SMITH, R. T., End component of armature leakage reactance of round rotor generators, *Trans. Am. Instn. Elect. Engrs.* **74** III, 636 (1958).

TEGOPULOS, J. A., Determination of the magnetic field in the end zone of turbine generators, *Trans. Am. Instn. Elec. Elect. Engrs.* **82** III, 562 (1963).

CHAPTER 5

THE SOLUTION OF POISSON'S EQUATION: MAGNETIC FIELDS OF DISTRIBUTED CURRENT

5.1. Introduction

This chapter is devoted to solutions for the magnetic fields of currents distributed with uniform density[†] J over conductors with various shapes of cross-section. Within such regions of distributed current, the field is expressed in terms of vector potential (for, as is explained in section 2.2.3, scalar magnetic potential has no meaning in such regions). The magnitude of the component (in the z-direction) of the vector potential A at all points inside the current-carrying region obeys Poisson's equation,

$$\frac{\partial^2 A}{\partial x^2} + \frac{\partial^2 A}{\partial y^2} = -\mu\mu_0 J; \tag{5.1}$$

and, at all points outside it, where $J = 0$, it obeys Laplace's equation. Hence for all these fields it is necessary to consider solutions of Poisson's (and Laplace's) equation in terms of the vector potential function, the form of solution depending upon the shapes and permeabilities of both the conductor and the boundary (if present).

With regard to the boundary shapes which can be treated, the simpler cases involving a single plane or two parallel planes present no difficulty, at least in principle, and can be approached directly by way of the method of images. The most general shapes which can be treated analytically are rectangular or are synthesizable from rectangles (e.g. an L-shaped region). In cases with a single plane boundary or two parallel ones, any value of permeability can be treated but, with three or more boundary lines, a maximum of only two of these, which must be parallel to each other, can be of *finite* permeability. The remaining boundaries must be either of infinite permeability (i.e. equipotential) or of zero permeability (i.e. flux line). The special case of concentric circular boundaries with a current at their centre is not discussed, but a solution can be simply obtained because flux and boundary lines coincide (see section 2.2.3).

With regard to the kinds of conductor that can be treated, the important considerations again relate to the permeabilities and the shapes. All the conductors considered are, with the exception of infinitely permeable ones discussed in section 5.3, of permeability μ_0, i.e. of the same permeability as the immediately surrounding medium. (The solutions for these are, of course, identical with those for conductors of permeability $\mu\mu_0$ contained in a medium of permeability $\mu\mu_0$, except that the flux densities and vector potentials are μ times

[†] The assumption of uniform current density is not a serious practical limitation since, for most applications, large conductors are stranded or laminated in order to make the distribution as uniform as possible.

smaller.) The general problem for conductors of magnetic material of permeability different from that of the surrounding medium is normally analytically intractable for the reasons outlined briefly in section 5.3.1. Two special cases may be noted for which this limitation does not apply: firstly, the field of a circular conductor can be treated because flux lines coincide with the surface of the conductor; and, secondly, as described in reference 2 of Chapter 6, transformation methods can be used for non-concentric circular conductors. (Possible extensions to this situation, however, are indicated by the recent work of Pramanik,[24] who used double Fourier series in a similar manner to that discussed in section 5.6 to deal with permeable, eddy-current-carrying conductors, but with these *contained within an outer rectangular boundary of infinite permeability.*)

For conductor materials of permeability μ_0, a variety of cross-section shapes can be treated, both when boundaries are present and when they are not. It is, however, necessary that the shapes be representable by some reasonably simple mathematical expression(s). Much the most important cross-section in practice is the rectangle, and chief attention is given to this in what follows. An example of a triangular section conductor in the presence of rectangular boundaries is included in reference 25. It should be noted that the method of superposition can be used conveniently to build up more complicated shapes (or combinations) of conductors. For example, a rectangular conductor with a circular hole (for cooling purposes) can be treated by superposing the solution for a rectangular current distribution with that for a circular one having the dimensions of the hole and current density equal, but opposite, to that in the rectangle. (Any boundary influences must, of course, be treated equally with the two shapes of conductor cross-section, so that the superposition is valid.)

With boundaries and conductors of the above types two general methods of solution can be distinguished. The first is applicable to problems in which the field is not periodic —having boundaries which give rise to a finite number (including zero) of images—and it consists essentially of integrating the vector potential function of a line current over the regions of the currents (including the images)—see sections 5.2 and 5.4. The second applies when the field is periodic in one coordinate direction, or both, and may be regarded as being due to an infinite number of images. This method is based upon the use of Fourier series, single or double, discussed in sections 5.5 and 5.6. For the special class of problems involving conductors of infinite permeability, section 5.3, a particular class of solution is possible though the details vary for each conductor shape.

For problems not falling into any of the above classes, analytical solutions have not been found possible, and numerical methods must be used. These important methods are the subject of Chapter 11.

Finally, before commencing detailed discussion of the various analytical solutions, it is helpful to indicate two general features concerning the nature of the fields of distributed currents. As discussed in section 2.2.3, all such fields contain a kernel about which the flux circulates, and at which the "lines of no-work" meet. For a symmetrical conductor in air, the kernel is at the centre of the conductor, see Figs. 5.2 and 5.6. When the conductor is influenced by a permeable boundary (or current of like sign), the kernel moves towards the boundary, see Figs. 5.8 and 5.10, and when influenced by an impermeable boundary (or current of opposite sign) the kernel moves away from the boundary. A second feature is that as a flux line crosses the boundary of a current-carrying region it does not suffer a sudden change in direction unless the conductor material has a permeability different from the surrounding medium; instead, it experiences a change in curvature, a point which is discussed in reference (8), p. 104.

5.2. Non-magnetic conductors in air

5.2.1. *The method: vector potential of a line current*

The field of a distributed current, in the absence of magnetic material, can be found by the superposition of the fields of an infinite number of line current elements which together constitute the distributed current. Hence the distribution of A can be calculated by integration of the vector potential function of a line current over the section of the conductor.

The expression for the vector potential function of a line current is not given earlier in explicit form, but it can be found simply from that for a circular conductor, eqn. (2.60). For a conductor of radius a, permeability μ_0, and carrying a current of density J, the expression for A is

$$A = \tfrac{1}{2}\mu_0 J a^2 \log r + \text{constant.}$$

Since the origin of A is arbitrary and since the current i in the conductor is $\pi J a^2$, this equation can be written

$$A = \frac{\mu_0}{2\pi} i \log r. \tag{5.2}$$

But, since the field exterior to a circular conductor is identical with that of a line current at its centre, eqn. (5.2) gives also the form of the vector potential function of a line current i. (This equation could have been developed also from the flux function of a line current, since the flux and vector potential functions are equivalent.) The field of a conductor of any shape can be determined, provided that eqn. (5.2) can be integrated over the section of the conductor. Further, the principle of superposition can be used to synthesize relatively complicated conductor shapes from simple ones. [It is to be noted that as an alternative to the determination of vector potential, as described above, it may be convenient, in some cases, to obtain components of the field directly by integration (over the conductor section) of the expressions for H_x and H_y for a line current.]

Examples demonstrating the application of eqn. (5.2) to the determination of fields and forces due to distributed currents are given below; and many practically important results (developed by similar means) for the inductances of a wide range of conductor shapes are given in the book by Grover.

5.2.2. *The field of a rectangular bus-bar*

Consider a bus-bar of rectangular cross-section with sides of length $2a$ and $2b$, as shown in Fig. 5.1. For a current I in the bar, the current density is $I/4ab$, and the current i carried by a filament of cross-section $dx'\,dy'$, where x' and y' are the coordinates of any filament in the bar, is $(I/4ab)\,dx'\,dy'$. The field of all the elements of the bar is given by substituting for i in eqn. (5.2) and integrating over the section of the rectangle. The field at any point (x, y), distance r from a typical filament with coordinates (x', y'), is then

$$A = \frac{I\mu_0}{8\pi ab} \int_{-a}^{a} \int_{-b}^{b} \log r \, dx' \, dy'.$$

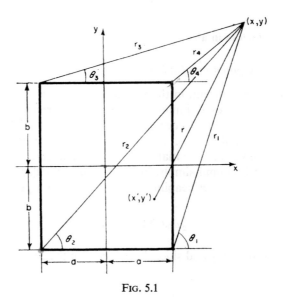

FIG. 5.1

Expressing r in terms of the coordinates gives

$$A = \frac{I\mu_0}{16\pi ab} \int_{-a}^{a} \int_{-b}^{b} \log[(x'-x)^2 + (y'-y)^2] \, dx' \, dy',$$

and this can be integrated in terms of simple functions. The result, in the form given by Strutt[1], is

$$A = \frac{I\mu_0}{16\pi ab} \left\{ \begin{array}{l} (a-x)(b-y)\log[(a-x)^2+(b-y)^2] \\ +(a+x)(b-y)\log[(a+x)^2+(b-y)^2] \\ +(a-x)(b+y)\log[(a-x)^2+(b+y)^2] \\ +(a+x)(b+y)\log[(a+x)^2+(b+y)^2] \\ +(a-x)^2\left[\tan^{-1}\frac{b-y}{a-x}+\tan^{-1}\frac{b+y}{a-x}\right] \\ +(a+x)^2\left[\tan^{-1}\frac{b-y}{a+x}+\tan^{-1}\frac{b+y}{a+x}\right] \\ +(b-y)^2\left[\tan^{-1}\frac{a-x}{b-y}+\tan^{-1}\frac{a+x}{b-y}\right] \\ +(b+y)^2\left[\tan^{-1}\frac{a-x}{b+y}+\tan^{-1}\frac{a+x}{b+y}\right] \end{array} \right\}. \tag{5.3}$$

This expression for the vector potential is, of course, the flux function of the field, so that curves corresponding to constant values of A are flux lines. A set of flux lines for a conductor having $b = 2a$ is shown in Fig. 5.2.

The field components H_x and H_y are given by

$$H_x = \frac{1}{\mu_0}\frac{\partial A}{\partial y} \quad \text{and} \quad H_y = -\frac{1}{\mu_0}\frac{\partial A}{\partial x}.$$

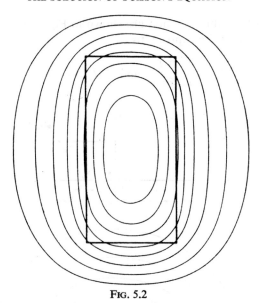

FIG. 5.2

The resulting expressions are rather long, but can be simplified by expressing them in terms of the distances r_1, r_2, r_3, and r_4 from a point in the field to the corners of the rectangle, and the angles θ_1, θ_2, θ_3, and θ_4 which the lines to the corners make with lines parallel to the x-axis. This gives

$$H_x = \frac{I}{8\pi ab}\left[\begin{array}{l}(y+b)\,(\theta_1-\theta_2)-(y-b)\,(\theta_4-\theta_3)\\[2mm]+(x+a)\log\dfrac{r_2}{r_3}-(x-a)\log\dfrac{r_1}{r_4}\end{array}\right], \tag{5.4}$$

$$H_y = -\frac{I}{8\pi ab}\left[\begin{array}{l}(x+a)\,(\theta_2-\theta_3)-(x-a)\,(\theta_1-\theta_4)\\[2mm]+(y+b)\log\dfrac{r_2}{r_1}-(y-b)\log\dfrac{r_3}{r_4}\end{array}\right]. \tag{5.5}$$

As the length of one side of the rectangle tends to zero the field approaches that of a current sheet, and it is to be noted that the current per unit area is infinite but that the line density is finite. Such sheets are useful for the representation of thin, distributed windings (see also section 4.4), and as an approximation for thin, rectangular conductors. The field of one is now determined.

Consider a sheet of width $2a$, Fig. 5.3(a) carrying a total current I. In an elemental strip of width $\mathrm{d}x'$ the current i is $(I/2a)\,\mathrm{d}x'$ and, substituting this value in eqn. (5.2), the vector potential for the sheet is expressed by

$$\begin{aligned}
A &= \frac{I\mu_0}{4\pi a}\int_{-a}^{a}\log r\,\mathrm{d}x'\\[2mm]
&= \frac{I\mu_0}{8\pi a}\int_{-a}^{a}\log\left[(x'-x)^2+y^2\right]\mathrm{d}x'\\[2mm]
&= \frac{I\mu_0}{8\pi a}\left\{\begin{array}{l}(a+x)\log\left[(a+x)^2+y^2\right]+(a-x)\log\left[(a-x)^2+y^2\right]\\[2mm]+2y\left[\tan^{-1}\dfrac{a+x}{y}+\tan^{-1}\dfrac{a-x}{y}\right]-4a\end{array}\right\}. \tag{5.6}
\end{aligned}$$

Flux lines determined from this equation are shown in Fig. 5.3(b). The components of field strength can be found from the expressions

$$H_x = \frac{1}{\mu_0} \frac{\partial A}{\partial y} \quad \text{and} \quad H_y = -\frac{1}{\mu_0} \frac{\partial A}{\partial x}.$$

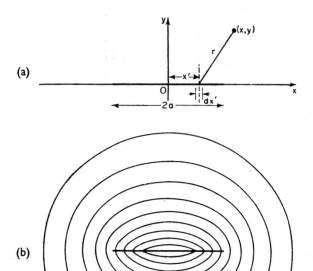

FIG. 5.3

5.2.3. The force between parallel rectangular bus-bars

The forces between adjacent conductors carrying large currents are a matter of consider-able practical importance; see, for example, references 2 and 3. The force between two parallel, rectangular bus-bars[4] is analysed here and, as in the previous section, the approach is to treat the actual currents as being the aggregates of many current filaments. The force between the bars is given by integration, over the sections of the bars, of the force between typical filaments, one from each bar. The integration has to be performed over both cross-sections, and is most simply done in three stages: first, the force between two filaments, one from each bar, is found; then this result is used to obtain the force between typical parallel current sheets from each bar; and, finally, the total force is obtained by integrating the result for the sheets over the areas of both rectangles.

Let the bars have a depth $2b$, a width $2a$, and a distance D between centres, Fig. 5.4(a), and let them carry currents i_1 and i_2. Figure 5.4(c) shows two current filaments a distance r apart, and Fig. 5.4(b) shows two current sheets, of depth $2b$ separated by a distance d. The first step is to find the force between the line currents, remembering that the force on a line current i_1 in a field of density B is Bi_1 per unit length. At radius r from a line current i_2, the flux density is $\mu_0 i_2/2\pi r$ and, hence, the force F_f between unit lengths of the filaments, Fig. 5.4(c), carrying currents i_1 and i_2, is

$$F_f = \frac{\mu_0 i_1 i_2}{2\pi r}. \tag{5.7}$$

Consider next the force between two elemental sheets, one from each rectangular conductor [Fig. 5.4(b)]. The sheets can be divided into current filaments of width dy_1 and dy_2,

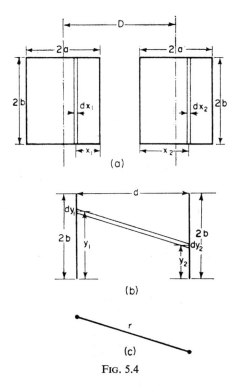

(a)

(b)

(c)

FIG. 5.4

and if the total currents in the sheets are k_1 and k_2, the currents in typical filaments are $k_1 dy_1/2b$ and $k_2 dy_2/2b$ respectively. Thus, from eqn. (5.7), the force between typical filaments is

$$F_f = \frac{\mu_0 k_1 k_2}{8\pi b^2} \cdot \frac{dy_1 dy_2}{[d^2+(y_1-y_2)^2]^{1/2}}. \tag{5.8}$$

This force has components parallel to both axes but, from considerations of symmetry, it can be seen that the resultant force on the sheets is *normal* to them. Therefore the resultant force F_s between sheets is given by integration over the two sheets of the normal component of F_f as

$$F_s = \frac{\mu_0 k_1 k_2 d}{8\pi b^2} \int_{-b}^{b} \int_{-b}^{b} \frac{dy_1 dy_2}{[d^2+(y_1-y_2)^2]}$$

$$= \frac{\mu_0 k_1 k_2}{2\pi b} \left[\tan^{-1} \frac{2b}{d} - \frac{d}{4b} \log \left(1 - \frac{4b^2}{d^2} \right) \right]. \tag{5.9}$$

This equation can now be used to find the force between the bus-bars by expressing k_1 and k_2 in terms of the actual currents I_1 and I_2, expressing d in terms of D, a, x_1, and x_2, and by integrating the resultant expression over the widths of the rectangular sections. Now it is seen that

$$k_1 = \frac{I_1 \, dx_1}{2a}, \qquad k_2 = \frac{I_2 \, dx_2}{2a},$$

and

$$d = D - 2a + x_1 + x_2;$$

and, therefore, the force, F_b, between bars is

$$
\begin{aligned}
F_b &= \frac{\mu_0 I_1 I_2}{8\pi a^2 b} \int_0^{2a} \int_0^{2a} \left\{ \tan^{-1} \left[\frac{2b}{D - 2a + x_1 + x_2} \right] \right. \\
&\quad \left. - \left[\frac{D - 2a + x_1 + x_2}{4b} \right] \log \left[1 + \frac{4b^2}{(D - 2a + x_1 + x_2)^2} \right] \right\} dx_1 \, dx_2 \\
&= \frac{\mu_0 I_1 I_2}{32 \pi a^2 b^2} \left\{ 2b \left[(D + 2a)^2 - \frac{4b^2}{3} \right] \tan^{-1} \frac{2b}{D + 2a} \right. \\
&\quad + 2b \left[(D - 2a)^2 - \frac{4b^2}{3} \right] \tan^{-1} \frac{2b}{D - 2a} \\
&\quad - 4b \left(D^2 - \frac{4b^2}{3} \right) \tan^{-1} \frac{2b}{D} - D \left(4b^2 - \frac{d^2}{3} \right) \log \frac{(D^2 + 4b^2)}{D^2} \\
&\quad + \frac{1}{2} (D + 2a) \left[4b^2 - \frac{(D + 2a)^2}{3} \log \left(\frac{(D + 2a)^2 + 4b^2}{D^2} \right) \right] \\
&\quad + \frac{1}{2} (D - 2a) \left[4b^2 - \frac{(D - 2a)^2}{3} \log \left(\frac{(D - 2a)^2 + 4b^2}{D^2} \right) \right] \\
&\quad \left. + \frac{1}{3} (D + 2a)^3 \log \left(\frac{D + 2a}{D} \right) + \frac{1}{3} (D - 2a)^3 \log \left(\frac{D - 2a}{D} \right) \right\}.
\end{aligned}
\tag{5.10}
$$

The variation of this force with the distance D for a range of values of the ratio a/b is shown in Fig. 5.5: the ratio F_b/F_f, where F_f is calculated as the force between line currents I_1 and I_2 a distance D apart, is plotted against the ratio $(D - 2a)/(2a + 2b)$, i.e. against the ratio of the distance between adjacent conductor faces to the conductor perimeter.

A useful general conclusion which can be drawn from these results is that the force between two rectangular conductors having their longer sides adjacent *is always less* than the corresponding force between the central filaments, whilst the force between the rectangular conductors when their shorter sides are adjacent is *higher* than that on the central filaments. This fact has been used, for example, in determining upper bounds for the forces in turbogenerator end windings.[23]

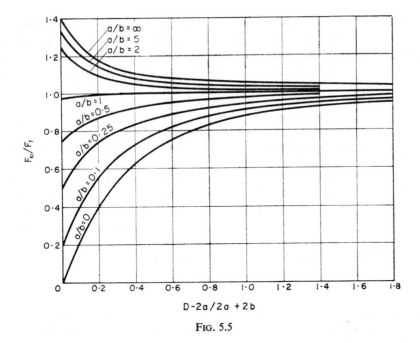

FIG. 5.5

.3. The field inside infinitely permeable conductors in air

5.3.1. *General considerations*

It is pointed out in the introduction to this chapter that, except in a small number of special cases, direct solutions cannot normally be obtained for conductors made of a material with a finite permeability (different from that of the surrounding medium). This is because in deriving a solution for these problems the region exterior to the conductor must be sub-divided into several smaller regions, and the boundary conditions connecting these lead to equations which are unmanageable (see subsection *Interconnected regions* in section 4.3.2). The same difficulty does not occur, however, when the conductors are infinitely permeable because the boundary of the conductor then coincides with a flux line. As a consequence the boundary conditions can be simply expressed, and the solutions for the field inside and outside the conductor can be found independently of one another. The fields in the exterior regions can then be found by using transformation methods (see, for example, reference 21). The form of these fields is identical with the electrostatic field exterior to charged conductors of the same shape, and the example of a conductor of rectangular cross-section is discussed in section 9.3. The solution for the field in the interior region is the subject of this section; it involves the determination of a vector potential function A which satisfies Poisson's equation and which is constant at the conductor boundary.

A function of this type can be found for only a limited number of cross-sections of conductor (and, of course, the principle of superposition cannot be used to extend the range). Further, no routine method of derivation is available, the method used in any given case being dependent upon the particular shape of cross-section of the conductor. A comprehensive bibliography of solutions mostly in terms of analogous fields, is given in reference 22.[†] The majority of these solutions are for boundary shapes of little interest to the electrical engineer, though Strutt[5] has given solutions for conductor sections with the shapes of the equilateral triangle, the ellipse, and the rectangle, the first two as finite functions, the last as an infinite series. The important practical case of the rectangle is discussed below and the solutions for the other two cases are summarized by Hague, pp. 288–90.

Before examining the method in detail, however, it is important to consider the value of solutions obtained for conductors assuming their permeability to be infinite. Firstly, the actual conductor permeability influences the boundary conditions of the field, and so the validity of the assumption, made in the analysis, that the conductor boundary coincides with a flux line. Provided, however, that μ is large, the boundary does coincide almost exactly with a flux line and the solution is entirely satisfactory. Secondly, in making any computation from the results (e.g. of the inductance of the conductor) it is important to realize that the actual numerical value of μ has a significant effect and must be known accurately. (For the theoretical case with $\mu = \infty$ it is necessary to assume, in order to retain a finite value of the term $\mu\mu_0 J$, that the current density J is zero.)

5.3.2. The field inside a highly permeable rectangular conductor

Let the boundary of the rectangle have sides $2a$ and $2b$, and let its centre be at the origin of the (x, y) plane. It is shown in Fig. 5.6 together with a plot of flux lines inside the conductor for the case $b = 2a$. The solution for the field requires the determination of a vector potential function A which satisfies Poisson's equation (5.1), and which has a constant (but unknown) value on the boundary of the rectangle, that is, satisfying the boundary conditions,

$$B_x = \frac{\partial A}{\partial y} = 0, \quad \text{when } x = -a \text{ and } a$$

and

$$B_y = -\frac{\partial A}{\partial x} = 0, \quad \text{when } y = -b \text{ and } b. \tag{5.11}$$

To form a suitable function A, which is a solution of Poisson's equation, the method is the same as that used in the solution of an inhomogeneous ordinary differential equation; it is formed as the sum of the solution to the corresponding homogeneous equation plus a suitable particular integral of the inhomogeneous one. In the present case of Poisson's equation (5.1) the homogeneous equation is that of Laplace,

$$\frac{\partial^2 A}{\partial x^2} + \frac{\partial^2 A}{\partial y^2} = 0,$$

which, from eqn. (4.72), has the general solution

$$A = \sum_{m=1}^{\infty} (c_m \sin mx + d_m \cos mx)(g_m \sinh my + h_m \cosh my) + k_1 + k_2 x + k_3 y.$$

---

† There are many important fields the analyses of which are identical with those of the magnetic fields of infinitely permeable, current-carrying conductors. These include the following: most notably, the Saint Venant problem of torsion in a linear isotropic prism, the temperature distribution in a prism in which heat is generated uniformly and the boundary temperature is everywhere constant, the motion of a non-viscous liquid of uniform vorticity circulating in a fixed prism, and the lateral displacement of a uniformly loaded homogenenous membrane, supported on a horizontal contour by a uniform edge tension.

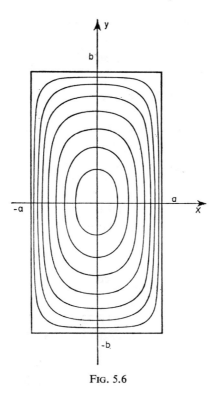

FIG. 5.6

As, however, the desired solution for A inside the rectangle must be an even function of x and y, it is necessary to take only

$$A = \sum_{m=1}^{\infty} d_m \cos mx \, \cosh my. \tag{5.12}$$

The particular integral of eqn. (5.1) must also be an even function of x and y, and the simplest is

$$A = -\tfrac{1}{2}\mu\mu_0 J x^2. \tag{5.13}$$

The complete solution of eqn. (5.1) is then given by the sum of eqns. (5.13) and (5.12) (adjusting the value of d_m) as

$$A = -\tfrac{1}{2}\mu\mu_0 J \left[x^2 + \sum_{m=1}^{\infty} d_m \cos mx \, \cosh my \right], \tag{5.14}$$

in which the constants m and d_m are to be determined from the boundary conditions, eqn. (5.11). By differentiating eqn. (5.14) with respect to y, the first boundary condition gives

$$\sum_{m=1}^{\infty} d_m \sinh my \, \cos ma = 0,$$

which requires that

$$m = \frac{n\pi}{2a},$$

where n is an odd integer. Substituting for m in eqn. (5.14) and using the second boundary condition gives

$$2x - \sum_{n=1,3,\ldots}^{\infty} \frac{n\pi}{2a} d_m \cosh \frac{n\pi b}{2a} \sin \frac{n\pi x}{2a} = 0. \tag{5.15}$$

To obtain from this equation the values of the constants d_m it is necessary to expand $2x$ in a form directly equatable with the infinite series, i.e. as a Fourier series in terms of $\sin(n\pi x/2a)$ in the range $0 < x < a$. The nth coefficient C_n of this series is

$$C_n = \frac{2}{a} \int_0^a 2x \sin \frac{n\pi x}{2a} \, dx$$

$$= \frac{(-1)^k \, 16a}{(2k+1)^2 \, \pi^2} \, ;$$

and thus, from eqn. (5.15),

$$C_n = \frac{n\pi}{2a} d_m \cosh \frac{n\pi b}{2a}$$

or

$$d_m = \frac{(-1)^k \, 32a^2}{(2k+1)^3 \, \pi^3 \cosh (2k+1)\pi b/2a} \, ,$$

where k is any integer (including zero) and $m = 2k+1$. Finally, therefore, substituting for d_m and m in eqn. (5.14), the vector potential function describing the field inside the conductor is

$$A = -\frac{1}{2} \mu \mu_0 J \left[x^2 + \sum_{k=0}^{\infty} \frac{(-1)^k \, 32a^2}{(2k+1)^3 \, \pi^3 \cosh (2k+1)\pi b/2a} \right.$$

$$\left. \cosh \frac{(2k+1)\pi y}{2a} \cos \frac{(2k+1)\pi x}{2a} \right]. \tag{5.16}$$

The term $(2k+1)^3$ in the denominator makes the series rapidly convergent (three or four terms give an accuracy much better than 1 per cent), and therefore convenient for computation. The expressions for B_x and B_y are simply found by the appropriate differentiations of eqn. (5.16). In the denominator they contain the term $(2k+1)^2$, and so they too are rapidly convergent. B_x has its maximum value at the middle of the sides $y = \pm b$ and B_y its maximum at the middle of the sides $x = \pm a$.

5.4. Simple boundaries: use of the image method

At first sight, the simplest approach to problems involving currents near boundaries would appear to be the image method; a distributed current, regardless of the particular cross-section, forms images in boundaries in a similar way to a line current (see section 3.4). However, practical use of the method is limited in two ways. Firstly, it is not generally practicable with circular boundaries because the image currents are distributed with non-uniform density over a region different in shape from that of the actual current, and so give rise to difficult or intractable mathematics. Secondly, it is not convenient to use it for plane boundaries which give rise to an *infinite* series of images because the mathematical expressions become excessively awkward. [This is apparent from a consideration of an infinite series of terms each with the form of the right-hand side of eqn. (5.3).] For this type of problem two other approaches are used and they are discussed in sections 5.5 and 5.6.

Use of the image method is, therefore, restricted in practice to straight-line boundaries giving rise to a finite number of images. No difficulty attaches to this class of problem, which involves only summation after appropriate changes in origin of the solution for the given conductor in air. No examples of this type are given here, but for a brief discussion of one of them, the field of a rectangular coil at the base of a salient pole, effectively in a right-angled corner of iron, the reader should consult the book by Hague, p. 290.

5.5. The treatment of boundaries using single Fourier series: Rogowski's method

One of the methods suitable for the treatment of problems having an infinite series of images is to make use of single Fourier series in a way developed by Rogowski. The method is essentially of the separation-of-the-variables type and is very suitable for the fields of rectangular currents surrounded by a rectangular boundary. It was developed to deal with problems[6] in transformers where, in a cross-section of the windings and core, the fields are of the above type. No solutions are given here (by this method) for the transformer problem (as such); instead, the solution for the leakage field in the slot of a rotating machine is developed. First, however, the method is explained with a simple example in the next section.

5.5.1. *Rectangular conductors in an infinite, parallel air gap*

Consider Fig. 5.7(a) which shows part of an infinitely long air gap of width g bounded by two parallel plane surfaces of permeability $\mu\mu_0$ on the lower of which lies a series of equal rectangular conductors carrying currents of magnitude I, alternatively positive and negative.

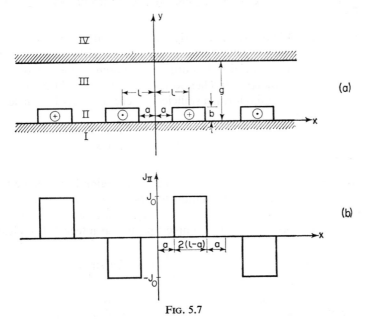

FIG. 5.7

Let the conductors have a depth b, a width $2(l-a)$, and let them be equally spaced with a distance $2l$ between their centres.

To analyse the field, the whole region is divided into several parallel subregions so chosen that, for each, a simple expression can be found to describe the current density (including $J = 0$) at all points. Separate vector potential functions are used for each region and they are chosen to satisfy, firstly, Laplace's or Poisson's equation (depending upon whether $J = 0$ or not), and, secondly, the boundary conditions between the regions.

It is convenient to subdivide the field, by lines parallel to the permeable surfaces, into four regions: I comprises the lower permeable block, II the region containing the currents between the lines $y = 0$ and b, III the region above the currents but below the upper surface, and, IV the upper permeable block.

In the regions I, III, and IV, J is everywhere zero, and so the respective vector potentials A_I, A_{II}, and A_{III} must be chosen to satisfy Laplace's equation. In the region II, A_{II} must be chosen to satisfy Poisson's equation in which J is an expression describing the current density distribution throughout the region. Since J_{II} is independent of y and varies periodically with x, in the manner shown in Fig. 5.7(b), it may be expressed in terms of the Fourier series,

$$J_{II} = \frac{4J_0}{\pi} \sum_{m=1,3,5}^{\infty} \frac{1}{m} \cos mka \cdot \sin mkx, \qquad (5.17)$$

where J_0 (the amplitude of the waveform of J) is the current density within the conductors, and $k = \pi/2l$.

The next step is the determination of the general form which the vector potentials can take. So far as the regions I, III, and IV are concerned, the form of the necessary functions can be written down immediately from the general solution of Laplace's equation (4.7.2) in which A replaces ψ.[†] The form is

$$A = \sum_{m=1}^{\infty} (c_m \sin mx + d_m \cos mx)(g_m \sinh my + h_m \cosh my) + k_1 + k_2 x + k_3 y, \qquad (5.18)$$

but this can be simplified. Firstly, since the origin of A is arbitrary, it is convenient to take $k_1 = 0$; and since the solution involves no uniform fields in either the x- or y-directions, $k_2 = k_3 = 0$. Further, with the origin of coordinates chosen as indicated in Fig. 5.7 the function of A must be odd so that $d_m = 0$. Therefore, expressing the right-hand bracketed term of eqn. (5.18) in simple exponential form, a sufficiently general form of the field valid for the three Laplacian regions is

$$A = \sum_{m=1}^{\infty} (G_m e^{mky} + H_m e^{-mky}) \sin mkx. \qquad (5.19)$$

A suitable form for the solution of Poisson's equation in region II is obtained (in the same way as in section 5.3.2) by adding the solution for the Laplace equation to a convenient particular integral for the Poisson equation, the right-hand side of which is equal to $\mu_0 J_{II}$, where J_{II} is given by eqn. (5.17). The simplest particular integral A_p is clearly

$$A_p = \frac{4J_0}{\pi} \frac{\mu_0}{k^2} \sum_{m=1,3,5,\ldots}^{\infty} \frac{1}{m^3} \cos mka \cdot \sin mka. \qquad (5.20)$$

[†] The method of developing solutions given here is not the same as was used originally by Rogowski. It is, in fact, much simpler, being based on the known solution of eqn. (4.72).

Hence the form of solution for region II may be written, combining eqns. (5.19) and (5.20), and noting that only odd values of m can arise, as

$$A_{II} = \sum_{m=1,3,5,\ldots}^{\infty} \left(G_m e^{mky} + H_m e^{-mky} + \frac{4\mu_0 J_0}{\pi m^3 k^2} \cos mka\right) \sin mkx. \qquad (5.21)$$

It remains to find the values of the constants G and H in equations having the forms of (5.19) or (5.21) which satisfy the boundary conditions between the four regions. The constants F and G are expressed in terms of C, D, E, and M (primed for G) for the functions A_I, A_{II}, A_{III}, and A_{IV} respectively. It would be possible to use all the boundary conditions immediately and to solve them simultaneously to find the constants, but the amount of work involved can be reduced by first using some of them, chosen from experience, to effect simplification of the functions A_I, A_{II}, A_{III}, and A_{IV}.

In region I at $y = -\infty$, B_x is zero; thus, $(\partial A_I/\partial y)_{y=-\infty} = 0$, and so G_m must be zero. Hence from eqn. (5.19) and noting from eqn. (5.20) that only odd values of m can arise,

$$A_I = \sum_{m=1,3,\ldots}^{\infty} C_m e^{mky} \sin mkx. \qquad (5.22)$$

No simplification is possible in A_{II} which in terms of the constants D and D' becomes

$$A_{II} = \sum_{m=1,3,\ldots}^{\infty} \left(D_m e^{mky} + D'_m e^{-mky} + \frac{4\mu_0 J_0}{\pi m^3 k^2} \cos mka\right) \sin mkx. \qquad (5.23)$$

The only simplification in A_{III} results from dropping the even values of m and so, from eqn. (5.19),

$$A_{III} = \sum_{m=1,3,\ldots}^{\infty} (E_m e^{mky} + E'_m e^{-mky}) \sin mkx. \qquad (5.24)$$

In the region IV, $B_x = 0$ at $y = \infty$ and so $F_m = 0$. Thus, from eqn. (5.19), again dropping even values of m,

$$A_{IV} = \sum_{m=1,3,\ldots}^{\infty} M'_m e^{-mky} \sin mkx. \qquad (5.25)$$

The remaining boundary conditions relating the above simplified functions, and expressing the equality of the normal components of flux and the tangential components of field at the interfaces are:

$$\left(\frac{\partial A_I}{\partial x}\right)_{y=0} = \left(\frac{\partial A_{II}}{\partial x}\right)_{y=0} \quad \text{and} \quad \frac{1}{\mu}\left(\frac{\partial A_I}{\partial y}\right)_{y=0} = \left(\frac{\partial A_{II}}{\partial y}\right)_{y=0};$$

on the surface $y = b$,

$$\left(\frac{\partial A_{II}}{\partial x}\right)_{y=b} = \left(\frac{\partial A_{III}}{\partial x}\right)_{y=b} \quad \text{and} \quad \left(\frac{\partial A_{II}}{\partial y}\right)_{y=b} = \left(\frac{\partial A_{III}}{\partial y}\right)_{y=b};$$

and, finally, on the surface $y = g$,

$$\left(\frac{\partial A_{III}}{\partial x}\right)_{y=g} = \left(\frac{\partial A_{IV}}{\partial x}\right)_{y=g} \quad \text{and} \quad \left(\frac{\partial A_{III}}{\partial y}\right)_{y=g} = \frac{1}{\mu}\left(\frac{\partial A_{III}}{\partial y}\right)_{y=g}. \qquad (5.26)$$

Substituting in these equations from eqns. (5.22)–(5.25) yields the six simultaneous equations

$$-C_m + D_m - D'_m = 0,$$
$$-\mu C_m + D_m + D'_m = -j,$$
$$\alpha^2 D_m - D'_m - \alpha^2 E_m + E'_m = 0,$$
$$\alpha^2 D_m + D'_m - \alpha^2 E_m - E'_m = -\alpha j,$$
$$\beta E_m - E'_m + M'_m = 0,$$
$$\beta E_m + E'_m - \mu M'_m = 0,$$

in which $\alpha = \exp(mkb)$, $\beta = \exp(2mkg)$ and $j = (4J_0\mu_0/\pi m^3 k^2)\cos mka$. Solving these equations gives the values of the constants as

$$C_m = 4[1 - e^{-mkb}]\left[\frac{e^{2mkg}(\mu+1) + e^{mkb}(\mu-1)}{e^{2mkg}(\mu+1)^2 + (\mu-1)^2}\right]\frac{J_0\mu_0}{\pi m^3 k^2}\cos mka, \tag{5.27}$$

$$D_m = 2e^{-mkb}\left[\frac{(\mu-1)e^{mkb}\{(\mu+1)e^{mkb} - 2\} - e^{2mkg}(\mu+1)^2}{e^{2mkg}(\mu+1)^2 - (\mu-1)^2}\right]\frac{J_0\mu_0}{\pi m^3 k^2}\cos mka, \tag{5.28}$$

$$D'_m = 2e^{-mkb}\left[\frac{(\mu-1)^2 e^{2mkb} - (\mu+1)e^{2mkg}\{(\mu-1) + 2e^{mkb}\}}{e^{2mkg}(\mu+1)^2 - (\mu-1)^2}\right]\frac{J_0\mu_0}{\pi m^3 k^2}\cos mka, \tag{5.29}$$

$$E_m = 2(\mu-1)[1 - e^{-mkb}]\left[\frac{e^{mkb}(\mu+1) - (\mu-1)}{e^{2mkg}(\mu+1)^2 - (\mu-1)^2}\right]\frac{J_0\mu_0}{\pi m^3 k^2}\cos mka, \tag{5.30}$$

$$E'_m = 2(\mu+1)e^{2mkg}[1 - e^{-mkb}]\left[\frac{e^{mkb}(\mu+1) + (\mu-1)}{e^{2mkg}(\mu+1)^2 - (\mu-1)^2}\right]\frac{J_0\mu_0}{\pi m^3 k^2}\cos mka, \tag{5.31}$$

$$M'_m = 4e^{2mkg}[1 - e^{-mkb}]\left[\frac{e^{mkb}(\mu+1) + (\mu-1)}{e^{2mkg}(\mu+1)^2 - (\mu-1)^2}\right]\frac{J_0\mu_0}{\pi m^3 k^2}\cos mka, \tag{5.32}$$

where $k = \pi/2l$, and m is an odd integer. The term m^{-3} in all of these expressions results in rapid convergence of the series and, for each value of A, two or three terms only need be taken to achieve an accuracy of 1 per cent. The map shown in Fig. 5.8 is for the case $\mu = \infty$.

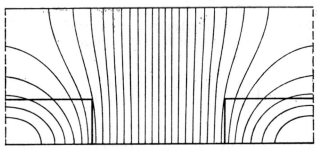

Fig. 5.8

This solution is in itself of little practical value: the example was chosen in its simplest form to demonstrate Rogowski's method of analysis. However, by making only slight modification to the value for J_{II}, as given by eqn. (5.17), to allow for more general periodic arrays of currents, it is possible to obtain solutions of a form used by several writers in the analysis of machine problems. Attention is turned now to the application of Rogowski's method to boundaries of finite size.

5.5.2. *Finite boundaries: rectangular conductor in a slot*

The method described above makes use of the fact that the current density distribution, being periodic with one coordinate in the range $-\infty$ to ∞, is expressible as a Fourier series. However, for any physically realizable problem (using cartesian coordinates), the current density does not fulfil this requirement, and, in order to make simple use of the method, it is necessary to assume that one pair of parallel boundaries are infinitely permeable (or wholly impermeable). Then, from a consideration of the images in them, it is seen that these boundaries can be replaced by an infinite, periodic array of images of the actual currents.

Consider now the determination of the field of a rectangular conductor at the bottom of a parallel-sided slot, the dimensions being as shown in Fig. 5.9(a). This field is of particular interest in the determination of the leakage inductance of conductors in a rotating machine.

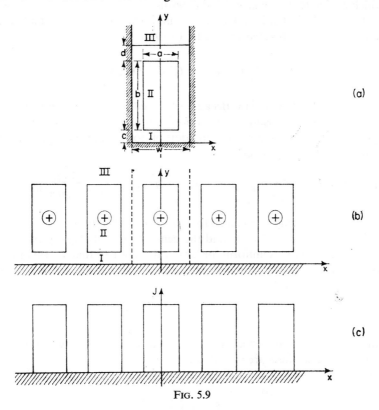

FIG. 5.9

If the sides of the slot, at $x = \pm w/2$, are assumed infinitely permeable, their effect on the field inside the slot is the same as that of an infinite array of equally spaced rectangular conductors, of the same size and carrying the same current, I, as the one in the slot, see Fig. 5.9(b). Thus the actual problem can be replaced by an equivalent one similar to that discussed in the previous section, and the methods described there can now be used to obtain its solution.

It is reasonable to assume the slot bottom also to be infinitely permeable, and doing this has the advantage that the field need only be divided into three regions. Region I is between the lines $y = 0$ and $y = c$, region II is between $y = c$ and $y = c+b$, and region III is between $y = c+b$ and $y = c+b+d$. In the first and third of these, Laplace's equation is obeyed and, in the second, Poisson's equation.

The current density distribution in region II has the form shown in Fig. 5.9(c). Because of the position of the origin, the Fourier series representing it consists only of cosine terms (as opposed to sine terms for the previous example) and because the "image" currents are of the same sign a constant term is involved. Thus

$$J_{\text{II}} = J_0 \frac{a}{w} + \frac{2}{\pi} \sum_{m=1}^{\infty} \frac{1}{m} \sin\left(\frac{mka}{2}\right) \cos mkx, \tag{5.33}$$

where J_0 is the modulus of the current density in the conductors and $k = 2\pi/w$. It is to be noted that, because the expressions for all the vector potential functions must be of the same type as that for J_{II}, they are given by the appropriate equations from (5.18), (5.19), and (5.20), in which cosine terms replace the sine terms.

In region I, the field is Laplacian (with no uniform components), and, therefore, the appropriate form of A_{I} is obtained from eqn. (5.19) as

$$A_{\text{I}} = \sum_{m=1}^{\infty} (C_m e^{mky} + C'_m e^{-mky}) \cos mkx. \tag{5.34}$$

In region II the field is Poissonian. Because of the constant term $J_0 a/w$ in the expression for J_{II}, $\partial^2 A_{\text{II}}/\partial y^2$ must be a constant (only periodic terms in x are permissible) and the particular integral must contain a term in y^2 in addition to those given in eqn. (5.20). Thus (remembering the change to cosine terms)

$$A_{\text{II}} = D_0 + D'_0 y + D''_0 y^2 + \sum_{m=1}^{\infty} (D_m e^{mky} + D'_m e^{-mky}) \cos mkx + \frac{2\mu_0 J_0}{\pi m^3 k^2} \sin\left(\frac{mka}{2}\right), \tag{5.35}$$

the constants D_0 and D'_0 being introduced to allow for continuity of A.

In region III the field is Laplacian, and the general form of A_{III} is given by eqn. (5.18). In this region the field becomes uniform for large values of y so that a term in y is necessary. Also, for continuity, E_0 must be retained. However, y can become infinite, so there can be no terms involving e^{mkx} and, hence,

$$A_{\text{III}} = E_0 + E'_0 y + \sum_{m=1}^{\infty} E'_m e^{-mky} \cos mkx. \tag{5.36}$$

Finally, the constants of the functions A_{I}, A_{II}, and A_{III} are evaluated from the remaining boundary conditions and these are as follows:

(a) The condition that $\partial A_1/\partial y = 0$, when $y = 0$ gives, by equating coefficients, not only $C'_0 = 0$ (chosen above), but also

$$C_m = C'_m. \tag{5.37}$$

(b) At the boundary between the regions I and II, that is, on $y = c$,

$$A_{\text{I}} = A_{\text{II}}, \tag{5.38}$$

$$\frac{\partial A_{\text{I}}}{\partial x} = \frac{\partial A_{\text{II}}}{\partial x}, \tag{5.39}$$

and

$$\frac{\partial A_1}{\partial y} = \frac{\partial A_{II}}{\partial y}.$$ (5.40)

Substituting, from eqns. (5.34) and (5.35) and equating coefficients gives, for the non-periodic terms of eqn. (5.38),

$$0 = D_0 + D_0'c + D_0''c^2;$$ (5.41)

for the periodic terms of eqn. (5.39),

$$C_m e^{mkc} + C_m' e^{-mkc} = D_m e^{mkc} + D_m' e^{-mkc} + \frac{2\mu_0 J_0}{\pi m^3 k^2} \sin\left(\frac{mka}{2}\right);$$ (5.42)

and for the periodic and non-periodic terms of eqn. (5.40),

$$C_m e^{mkc} - C_m' e^{-mkc} = D_m e^{mkc} - D_m' e^{-mkc},$$ (5.43)

and

$$0 = D_0' + 2D_0''c$$ (5.44)

respectively.

(c) At the boundary $y = c+b$ there are, in terms of A_{II} and A_{III}, conditions corresponding directly with eqns. (5.38), (5.39), and (5.40) giving relations between the constants corresponding directly with eqns. (5.41)–(5.44).

(d) In region III, for large values of y the flux density is uniform with the value I/w; thus

$$\left(\frac{\partial A_{III}}{\partial y}\right)_{y \to \infty} = \frac{I}{w},$$ (5.45)

giving, from eqn. (5.36),

$$E_0' = \frac{I}{w}.$$ (5.46)

Hence, there are ten equations, one from each of sections (a) and (d), and four from each of sections (b) and (c). These can be solved simultaneously to give the ten remaining values of the constants C, D, E, and so the complete solutions for A_I, A_{II}, and A_{III}. A map of the flux lines derived from these expressions is shown in Fig. 5.10: in terms of one unit of flux

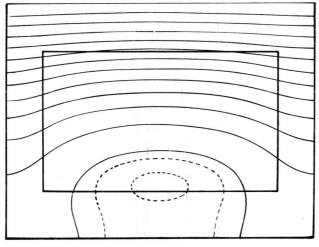

Fig. 5.10

between continuous lines, the quantity of flux between the kernel and the first dotted line is 0·08, that between the first and second dotted line is 0·24, and that between the second dotted and the first continuous line is 0·2.

Slot of finite depth: leakage inductance. The solution for a slot of finite depth can be obtained by making simple modifications to the above analysis. This is done making the assumption, usual in most calculations of slot leakage fields, that the slot mouth, say at $y = c+b+d$, is closed by a straight flux line. Then the boundary conditions described in paragraph (d) are replaced by those at $y = c+b+d$; that is,

$$\frac{\partial A_{III}}{\partial x} = 0, \quad \text{and} \quad A_{III} = 0$$

(for convenience). Also it is to be noted that, since the upper region III is finite, A_{III} can contain terms in e^{mky} as well as in $\exp^{-mky}$ and that, since the origin of A is defined at $y = c+b+d$, $C_0 \neq 0$.

This problem, and a number of similar ones, involving two conductors in a slot, have been analysed by Robertson and Terry.[7] Their object was to examine the error involved in calculating the slot leakage inductance on the assumption that all the flux lines are straight and they developed an expression which shows this error to be less than 4 per cent for all practical conductor sizes. However, this does not mean that the above simplifying assumption is, in general, valid for problems with distributed currents: it is not valid, for example, when the cross-sections of the conductors are irregular or are small compared with the boundary dimensions as, for example, in a transformer.

5.5.3. *Scope of the method*

Rogowski's method is valuable in the solution of fields of distributions of current which are rectangular or can be synthesized from rectangles interior to an open (three-sided) or closed rectangular boundary: the restrictions as to current and boundary shape both follow from the need to express the current densities, in all regions, as singly periodic functions. One side, or two (opposite parallel) sides, of the boundary can be of finite permeability, but, for simplicity, the other two opposite sides must be:

   (i) infinitely permeable [as in Fig. 5.9(a)]; or
  (ii) flux lines [e.g. the section of Fig. 5.7(a) bounded by the lines $x = 0$ and $x = 2l$]; or
 (iii) one infinitely permeable and the other a flux line [e.g. Fig. 5.10 or the section of Fig. 5.7(a) bounded by the lines $x = 0$ and $x = l$].

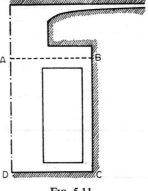

FIG. 5.11

The method, as described, could be used with many or (with the above restriction) complicated conductors. However, the large number of regions and separate vector potentials which would be required make this a complicated matter. Instead, in such cases, the solution for a single conductor should be obtained in general terms and this then used, with the principle of superposition, to synthesize the required current distribution.

The practical application of Rogowski's method to problems in which all the boundaries have finite (other than zero) permeabilities is not possible because of the large number of regions into which the field would have to be divided. Thus, for example, if the field consisted of the rectangular region $0 \leqslant x \leqslant 2l$, $0 \leqslant y \leqslant g$, Fig. 5.7(a), surrounded by a medium extending to infinity in all directions and having the same permeability everywhere (the simplest generalization), it would be necessary to divide the field into twelve regions (by the lines $y = 0$, $y = b$, $y = g$, and $x = 0$, $x = l$) and to evaluate twelve potential functions. Four of these functions would be non-periodic, and the resulting complexity of the boundary condition equations would render the problem intractable.

It should, however, be noted that solutions can be obtained if the boundary conditions on the sides of the rectangle are expressed in the form of a Fourier series (see section 4.3). This property was used by Stevenson and Park[8] to obtain an approximate solution for the field distribution in and near the exciting winding of a salient pole with a tip (Fig. 5.11). First, the field in the air gap and the inter-polar space was obtained approximately using graphical methods (described in reference 8 and also discussed in a paper by Stevenson[9]). From this field, the component of flux density normal to, and the component of field strength tangential to, the line AB, were obtained. Then, finally, these were used as one set of boundary conditions on the rectangle $ABCD$ (BC and CD being equipotentials and AC a symmetrical flux line) within which the field due to the winding was obtained using Rogowski's method. (Equally the initial boundary conditions could be obtained numerically.)

5.6. The treatment of boundaries using double Fourier series: Roth's method

5.6.1. *The method*

The method of solution which is now to be discussed, and which was originally described and developed in an electrical engineering context by Roth in a long series of papers,[10–16] is applicable to a similar class of problems as Rogowski's method. However, it is much easier to use algebraically than Rogowski's method, the solution being obtained as a single function, a double Fourier series, which defines the field in the whole region of the conductors and the air. In addition it has recently been developed to deal with a number of problems for which Rogowski's method is not suited.

To demonstrate the general method, consider an infinitely permeable, rectangular boundary containing an arrangement of p rectangular conductors, as shown by the continuous lines in Fig. 5.12. Let the lower left-hand corner of the boundary be the origin of coordinates, and let the dimensions be as shown. The order of procedure in establishing the solution is the opposite of that used in Chapter 4 or earlier in this chapter: here a function is first found which satisfies the boundary conditions of the problem, and the constants of the function are then chosen to satisfy the field equations.

The necessary form of the vector potential function can be seen from a consideration of the method of images (see section 3.2.3) and, as discussed below, this will also serve to reveal certain limitations of the method. By this method the boundaries (Fig. 5.12) could be replaced by a doubly infinite set of images, shown dotted, and so the desired function defining the resultant field inside the rectangle must be periodic with both coordinates. Its general form is given, therefore, by the product of two single Fourier series, one in x and the other in y, as

$$A = \sum_m \sum_n B_1 \cos mx \cos ny + \sum_m \sum_n B_2 \cos mx \sin ny$$
$$+ \sum_m \sum_n B_3 \sin mx \cos ny + \sum_m \sum_n B_4 \sin mx \sin ny, \qquad (5.47)$$

in which B, m, and n are constants dependent upon the field and boundary conditions.

As the boundaries are infinitely permeable, there are no tangential components of flux at their surfaces, and so the boundary conditions are

$$\left(\frac{\partial A}{\partial x}\right)_{x=0} = 0, \qquad (5.48)$$

$$\left(\frac{\partial A}{\partial y}\right)_{y=0} = 0, \qquad (5.49)$$

$$\left(\frac{\partial A}{\partial x}\right)_{x=a} = 0, \qquad (5.50)$$

and

$$\left(\frac{\partial A}{\partial y}\right)_{y=b} = 0. \qquad (5.51)$$

Differentiating A from eqn. (5.47) and substituting in eqns. (5.48) and (5.49) gives $B_2 = B_3 = B_4 = 0$ (there can be no sine terms). Hence, the necessary form of the vector potential is

$$A = \sum_m \sum_n B_{m,n} \cos mx \cos ny. \qquad (5.52)$$

The constants m and n of this function are dependent upon the two remaining boundary conditions, eqns. (5.50) and (5.51). Differentiating A and substituting in these equations gives, respectively,

$$B_{m,n}m \sin ma \cos ny = 0, \qquad (5.53)$$

for all values of y, and

$$B_{m,n}n \cos mx \sin nb = 0, \qquad (5.54)$$

for all values of x. Equation (5.53) is satisfied when ma is an even multiple of $\pi/2$, and so

$$m_h = 2(h-1)\frac{\pi}{2a}, \qquad (5.55)$$

where h is an integer taking all values from 1 to ∞. Similarly, eqn. (5.54) is satisfied when nb is an even multiple of $\pi/2$, and so

$$n_k = 2(k-1)\frac{\pi}{2b}, \qquad (5.56)$$

where k is an integer taking all values from 1 to ∞. Hence, substituting for m_h and n_k in eqn. (5.52), the final form of the vector potential function is

$$A = \sum_{h=1}^{\infty} \sum_{k=1}^{\infty} B_{h,k} \cos(h-1)\frac{\pi x}{a} \cos(k-1)\frac{\pi y}{b}.$$ (5.57)

For this function to be the required field solution it must also obey Poisson's equation over the cross-section of the conductors, and Laplace's equation elsewhere in the air region. Hence, differentiating eqn. (5.52) and substituting in the field eqn. (5.1), this requires that

$$\sum_h \sum_k (m_h^2 + n_k^2) B_{h,k} \cos m_h x \cos n_k y = \mu_0 J,$$ (5.58)

where the current density, J, is equal to J_j (a constant) over the jth conductor, and to 0 over all the air regions. Equation (5.58) is satisfied provided that $B_{h,k}$ is chosen as the general coefficient of the double Fourier series of cosine terms, the sum of which is equal to $\mu_0 J_j$

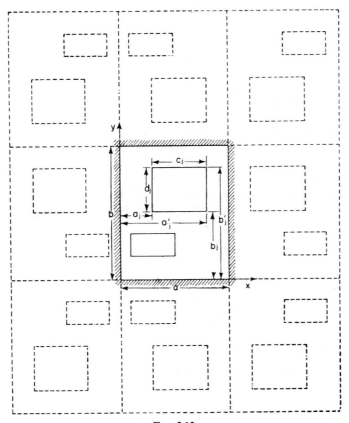

FIG. 5.12

over the area of each of the conductors, and 0 elsewhere. The values of $B_{h,k}$ can be determined by a method exactly analogous to that used in finding the coefficients of a single Fourier series, namely by multiplying both sides of eqn. (5.58) by $\cos m_h x \cos n_k y$ and by integrating over the area of the field space (one period of the doubly infinite array, Fig. 5.12).

Since all other integrated terms vanish, this gives, for the left-hand side of the equation,

$$\int_0^a \int_0^b (m_h^2 + n_k^2)\, B_{h,k} \cos^2 m_h x \cos^2 n_k y \, dx\, dy = \frac{ab}{4}(m_h^2 + n_h^2)\, B_{h,k};$$ (5.59)

and for the right-hand side,

$$\int_0^a \int_0^b \mu_0 J \cos m_h x \cos n_k y \, dx\, dy,$$

which, since J is zero except over the area of the p conductors, is equal to

$$\mu_0 \sum_{j=1}^{p} J_j \int_{a_j}^{a_j'} \int_{b_j}^{b_j'} \cos m_h x \cos n_k y \, dx\, dy$$

$$= \mu_0 \sum_{j=1}^{p} J_j \left[\frac{(\sin m_h a_j' - \sin m_h a_j)}{m_h} \cdot \frac{(\sin n_k b_j' - \sin n_k b_j)}{n_k} \right].$$ (5.60)

Hence, equating the right-hand sides of eqns. (5.59) and (5.60) gives

$$B_{h,k} = \frac{4\mu_0}{ab}\,\frac{1}{(m_h^2 + n_k^2)} \sum_{j=1}^{p} J_j \left[\frac{(\sin m_h a_j' - \sin m_h a_j)}{m_k} \cdot \frac{(\sin n_k b_j' - \sin n_k b_j)}{n_k} \right].$$ (5.61)

This equation gives the value of $B_{h,k}$ provided that h or k are not equal to 1, but, when either h or k, or both, are equal to 1 (i.e., when m_h or n_k, or both, are zero) eqn. (5.61) takes special forms and it is necessary to return to equation (5.58) to evaluate $B_{1,k}$, $B_{h,1}$, and $B_{1,1}$. The three cases are now taken in order.

(a) For $h = 1$ and $k > 1$ (i.e. $m_h = 0$ and $n_k > 0$) the left-hand side of eqn. (5.58) reduces (after integration) to $B_{1,k}\, n_k^2\, ab/2$. On the right-hand side the term

$$\frac{\sin m_1 a_j' - \sin m_1 a_j}{m_1}$$

is indeterminate, as $m_1 = 0$, but differentiating top and bottom with respect to m_1 and again putting $m_1 = 0$ gives, for its limiting value,

$$a_j' - a_j = c_j$$

(see Fig. 5.12). Therefore

$$B_{1,k} = \frac{2\mu_0}{n_k^2 ab} \sum_{j=1}^{p} J_j c_j \left[\frac{\sin n_k b_j' - \sin n_k b_j}{n_k} \right].$$ (5.62)

(b) For $h > 1$ and $k = 1$ it is easily shown in a similar manner that

$$B_{h,1} = \frac{2\mu_0}{m_h^2 ab} \sum_{j=1}^{p} J_j d_j \left[\frac{\sin m_h a_j' - \sin m_h a_j}{m_h} \right].$$ (5.63)

(c) Similarly for $h = 1$ and $k = 1$ it is easily shown that $B_{1,1}$ is a constant which, because the origin of A is arbitrary, can be ignored.

Finally, therefore, the solution for the field, inside and outside the conductors, is given by the single vector potential function A, eqn. (5.57), in which the value of $B_{h,k}$ is obtained from eqn. (5.61) in the general case, but from eqns. (5.62) and (5.63) when h or k equals 1.

This general solution is employed directly in the next section to obtain the field in a transformer winding. It applies, of course, to any number and arrangement of conductors within the boundary and, in practice, to adapt it to suit other problems only a change in the value of m or n is involved (see section 5.6.3).

5.6.2. *The forces on, and the inductance of, a transformer winding*

The forces experienced by the windings of transformers are very large and, to ensure that no damage results from them, it is important to be able to estimate their magnitudes. It is also important to be able to calculate the inductances of the windings. Analysis of forces and inductances has been made by Roth in papers referred to earlier—reference 12 is devoted to a problem similar to that examined below, and reference 15 to the case of a transformer with cylindrical windings—and, more recently, by Billig[17] (who has summarized some of Roth's work in English), Rabins,[18] and Vein.[19]

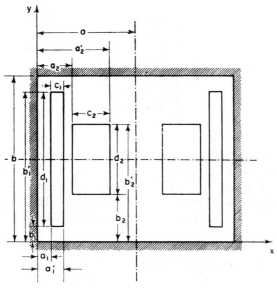

FIG. 5.13

Consider the arrangement of the windings in the "window" of a simple core-type transformer, the cross-section of which is shown in Fig. 5.13; it is a special case of that represented in Fig. 5.12. Also, it is symmetrical about the vertical line through the centre of the window and so, in performing the analysis, it is sufficient to consider one half of the window area. The line of symmetry is an equipotential line of the field and so, by choosing the coordinate axes as shown, the left-hand half corresponds directly with the field space in Fig. 5.12. Consequently, the solution given in eqns. (5.57) and (5.61)–(5.63) is, with the number of conductors p, equal to 2, the solution for the field in the present problem. It has been used to prepare the field map given in Fig. 5.14.

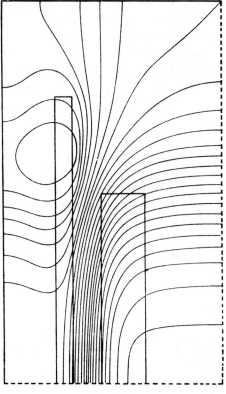

Fig. 5.14

The force F on a conductor is calculated from the components of flux density acting within the conductor. These are simply obtained in terms of the gradients of the vector potential and are used to obtain expressions for the forces on a filament of cross-section $dx\,dy$ which are then integrated over the conductor sections. Thus the x- and y-components of F, per unit length in the z-direction, acting on the jth conductor, are, respectively,

$$F_{xj} = \int_{a_j}^{a_j'} \int_{b_j}^{b_j'} J_j B_y \, dx\,dy = -J_j \int_{a_j}^{a_j'} \int_{b_j}^{b_j'} \frac{\partial A}{\partial x} \, dx\,dy$$

and

$$F_{yj} = \int_{a_j}^{a_j'} \int_{b_j}^{b_j'} J_j B_x \, dx\,dy = J_j \int_{a_j}^{a_j'} \int_{b_j}^{b_j'} \frac{\partial A}{\partial y} \, dx\,dy.$$

Substituting the values of $\partial A/\partial x$ and $\partial A/\partial y$ from eqn. (5.57) gives, respectively,

$$F_{xj} = -J_j \sum_{h=1}^{\infty} \sum_{k=1}^{\infty} B_{h,k} \left[\frac{(\cos m_h a_j' - \cos m_h a_j)(\sin n_k b_j' - \sin n_k b_j)}{n_k} \right] \quad (5.64)$$

and

$$F_{yj} = J_j \sum_{h=1}^{\infty} \sum_{k=1}^{\infty} B_{h,k} \left[\frac{(\cos n_k b_j' - \cos n_k b_j)(\sin m_h a_j' - \sin m_h a_j)}{m_h} \right], \quad (5.65)$$

where $B_{h,k}$ is, in general, given by eqn. (5.61) but has special values when h or $k = 1$ [see eqns. (5.62) and (5.63)]. It is to be noted that in the present case, because of the symmetry about the x-axis, $F_{yj} = 0$ for both conductors.

5.6.3. *Conductor in slot: calculation of inductance*

To show the great simplicity of the analysis in Roth's method as compared with that of Rogowski, consider again the problem of a conductor in a parallel-sided slot in a block of infinite permeability, shown with all dimensions in Fig. 5.15. Further, consider the case in which the slot mouth is assumed to be closed by a straight flux line at $y = b$.

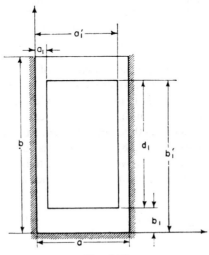

FIG. 5.15

The boundary conditions on the slot sides and bottom, those of zero tangential component of flux density on the lines $x = 0$, $y = 0$, and $x = a$, are identical with those expressed, for the detailed analysis, in eqns. (5.48)–(5.50). Therefore the form of the vector potential for the field in the slot is expressed by eqn. (5.52) and the value of m_h is given by eqn. (5.55). The remaining boundary condition differs from those considered previously in this section. At the line $y = b$ the field is bounded by a flux line, i.e. by a line of constant vector potential to which it is most convenient to attribute the value zero. Thus

$$(A)_{y=b} = 0, \tag{5.66}$$

and, substituting for A from eqn. (5.52), this requires that

$$\cos mx \cos nb = 0$$

for all values of x. nb must, therefore, be an odd multiple of $\pi/2$, so that

$$n_k = (2k-1)\frac{\pi}{2b}. \tag{5.67}$$

Substituting for m_h from eqn. (5.55) and for n_k from eqn. (5.67) in eqn. (5.52) then gives the vector potential function

$$A = \sum_{h=1}^{\infty} \sum_{k=1}^{\infty} B_{h,k} \cos\left[(h-1)\frac{\pi}{a}\right] \cos\left[(2k-1)\frac{\pi}{2b}\right]. \tag{5.68}$$

This function must also satisfy the field equations, and this is achieved, exactly as in section 5.6.1, by choosing the coefficients $B_{h,k}$ to be the coefficients of a double Fourier series the values of which are given by eqn. (5.61) with $p = 1$. Note, from eqn. (5.67), that when $k = 1$, $n_k \neq 0$, and so it is only necessary to consider the special values of $B_{h,k}$ covered by eqn. (5.62).

This example is a particularly suitable one with which to demonstrate the analytical determination of the inductance L of a conductor, and the use of eqn. (2.64),

$$L = \frac{1}{SI} \int \int_S (A_0 - A)\, dx\, dy,$$

to this end is described below. The physical quantity of most interest is the slot leakage inductance, i.e. the inductance due to flux crossing the slot from tooth to tooth and circulating within the conductor itself. It is usual to consider this flux to be bounded by the straight flux line crossing the slot mouth, and in this case the value of vector potential function A_0 associated with this line is 0 [see eqn. (5.66)]. Further, the form of the function A describing the field is given by eqn. (5.68), where $B_{h,k}$ is calculated for the total current I in the conductor. Therefore, using the above notation for the conductor and slot dimensions and substituting for A, gives the expression for inductance as

$$L = \frac{-1}{c_j d_j I} \int_{a_j}^{a_j'} \int_{b_j}^{b_j'} \sum_h \sum_k B_{h,k} \cos(h-1)\frac{\pi x}{a} \cos(2k-1)\frac{\pi y}{2b}\, dx\, dy$$

$$= \frac{-2ab}{\pi^2 c_j d_j I} \sum_h \sum_k \frac{B_{h,k}}{(h-1)(2k-1)} \left\{ \left[\sin(h-1)\frac{\pi a_j'}{a} - \sin(h-1)\frac{\pi a_j}{a} \right] \right.$$

$$\left. \times \left[\sin(2k-1)\frac{\pi b_j'}{2b} - \sin(2k-1)\frac{\pi b_j}{2b} \right] \right\}. \tag{5.69}$$

5.6.4. Scope of the method

Roth's method, its scope and extensions, have recently been extensively discussed by Hammond[26] and by Pramanik[24, 25]. The general scope of the method may be deduced from the form of the solution, eqn. (5.47), and is basically defined by the requirement that the boundaries must effectively be equivalent to lines of symmetry within some doubly periodic field pattern. As an example, Fig. 5.12 shows a small section of such a pattern (in terms of actual and image currents), and it is immediately apparent that a basically rectangular shape of boundary is implied. Also, it can be seen, in this particular case, that the permeability of *all* of the "dividing" lines of the field can be infinite (when actual and image currents have the same sign) or zero (when the signs of the currents alternate between adjacent rectangular regions). Further consideration reveals, however, that, more generally, the boundaries can involve any combination of $\mu = \infty$ or 0. Thus, for example, Fig. 5.15 applies to a combination of three boundaries having $\mu = \infty$ and one having $\mu = 0$; and Fig. 5.16 could apply for a combination with two boundaries (*HA* and *AE*) having $\mu = \infty$, and two opposite boundaries (*EI* and *IH*) having $\mu = 0$, this being regarded, if convenient, as one quarter of a symmetrical field with four boundaries (AB, BC, CD, and DA) having $\mu = \infty$.

It is to be emphasized, however, that the word boundary in the sense just used need only be taken to be an "outer" boundary. It is possible, as Pramanik has shown, to subdivide a field bounded by a rectangle such as $ABCD$ in Fig. 5.16 into separate regions which may have different *and finite* permeabilities. For example, provided $ABCD$ has $\mu = \infty$ or 0, then the medium of $ABFH$ could be μ and that of $HFCD$ could be μ_2. It has also been shown by Pramanik that, with similar restrictions on the "outer" boundary, L-shaped regions (e.g. $ABCGIHA$) can be treated. However, with both of these extensions it is necessary to work with the complete form of eqn. (5.47) for each separate region, and this aggravates difficulties of computation which, as discussed below, can exist even with solutions involving only one double series.

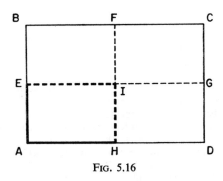

FIG. 5.16

In Roth's original work, solutions to problems involving regions of finite permeability inside an "outer" boundary were developed by introducing current sheets, of appropriate strength and distribution, along the surface of the region.[13] This approach is of considerable theoretical interest but adds to the conceptual difficulties of the solution. Additional comments on it have been given by Hammond.[26]

With regard to conductor cross-sectional shapes, although most applications of Roth's method have been concerned with rectangles placed parallel to the boundaries, there is little real restriction on the shapes or their disposition. In the first place, any shape synthesizable from rectangular elements can be treated, and there is no requirement (as there is with Rogowski's method) for the elements to be parallel to the boundaries. In fact, any shape of conductor bounded by straight lines or mathematically definable curves can be treated simply by integrating the right-hand side of eqn. (5.58) appropriately over the actual area of the conductors. An example of this procedure for the case of conductors of triangular section is given elsewhere.[25]

Perhaps the principal attraction of Roth's method, apart from its elegance, is its concise algebraic form, with a single function, which can be virtually written down by inspection, providing the solution for relatively complex combinations of Poissonian and Laplacian field regions. Unfortunately, a price has to be paid for this convenience and it is, not surprisingly, that the computational effort involved in evaluating solutions can be considerable. This follows from rather slow convergence of some of the double series. To combat this difficulty, Roth and Kouskoff[20] developed methods for the reduction of the double series to single series, and these methods have been appraised very fully by De Kuijper.[27, 28] However, as pointed out by Hammond, when such reductions are possible they must yield solutions that could also have been developed by some form of the method of separation-of-the-variables (including Rogowski's method).

In any given situation when solutions are possible by single and double series, the decision as to which to choose depends upon a variety of factors, but, generally, if the solution is to be carried out by hand, then the double series approach will be likely to involve least effort overall and, in the case of highly complex field arrangements, this may also be true if a computer is to be used.

Finally, it should be noted that, as with Rogowski's method, it may sometimes be useful to combine the double-series approach with the use of graphical or numerical methods to provide boundary conditions.

References

1. M. Strutt, Das magnetische Feld eines rechteckigen, von Gleichstrom durchflossenen Leiters, *Arch. Elektrotech.* **17**, 533, and **18**, 282 (1927).
2. O. R. Schurig and M. F. Sayre, Mechanical stresses in busbar supports during short-circits, *J. Amer. Instn. Elect. Engrs.* **44**, 365 (1925).
3. T. J. Higgins, Formulas for calculating short-circuit forces between conductors of structural shape, *Trans. Am. Instn. Elect. Engrs.* **62** III, 10, 659 (1943).
4. H. B. Dwight, Repulsion between strap conductors, *Elect. World* **70**, 522 (1917).
5. M. Strutt, Das magnetische Feld im Innern ferromagnetischer Leiter von rechteckigem, dreieckigem und elliptischem Querschnitt, *Arch. Elektrotech.* **18**, 190 (1927).
6. W. Rogowski, Über das Streufeld und der Streuinduktionskoeffizierten eines Transformators mit Schiebenwicklung und geteilten Endspulen, *Mitt. Forsch Arb. VDI* **71** (1909).
7. B. L. Robertson and J. A. Terry, Analytical determination of magnetic fields, *Trans. Am. Instn. Elect. Engrs.* **48**, 4, 1242–62 (1929).
8. A. R. Stevenson and R. H. Park, Graphical determination of magnetic fields, *Gen. Elect. Rev.* **31**, 99–109 and 153–64 (1928).
9. A. R. Stevenson, Fundamental theory of flux-plotting, *Gen. Elect. Rev.* **24**, 797 (1926).
10. E. Roth, Introduction à l'étude analytique de l'échauffement des machines électriques, *Bull. Soc. Franc. Élect.* **7**, 840 (1927).
11. E. Roth, Étude analytique du champ propre d'une encoche, *Rev. Gén. Elect.* **25**, 417 (1927).
12. E. Roth, Étude analytique du champ de fuites des transformateurs et des efforts mécaniques exercés sur les enroulements, *Rev. Gén. Elect.* **23**, 773 (1928).
13. E. Roth, Étude analytique des champs thermique et magnétique lorsque la conductibilité thermique ou la perméabilité n'est pas la même dans toute l'étendue du domaine considéré, *Rev. Gén., Elect.* **24**, 137–48, and 179–87 (1928).
14. E. Roth, Étude analytique du champ résultant d'une encoche de machine électrique, *Rev. Gén. Elect.*, **32**, 761 (1932).
15. E. Roth, Inductance due aux fuites magnétiques dans les transformateurs à bobines cylindriques et efforts exercés sur les enroulements, *Rev. Gén. Elect.* **40**, 259, 291 and 323 (1936).
16. E. Roth, Champ magnétique et inductance d'un système des barres rectangulaires parallèles, *Rev. Gén. Elect.* **44**, 275 (1938).
17. E. Billig, The calculation of the magnetic fields of rectangular conductors in a closed slot and its application to the reactance of transformer windings, *Proc. Instn. Elect. Engrs.* **98** (4), 55 (1951).
18. L. Rabins, Transformer reactance calculations with a digital computer, *Trans. Am. Instn. Elect. Engrs.* **25** I (1956).
19. P. R. Vein, A method based on Maxwell's equations for calculating the short-circuit forces on the concentric windings of an idealized transformer, *Elec. Res. Assoc.* Report Q/T151 (1960).
20. E. Roth and G. Kouskoff, Sur une méthode de sommation de certaines séries de Fourier, *Rev. Gén. Elect.* **23**, 1061 (1928).
21. A. M. Vinitskii and L. Gelbukh, Rastchet unesknei induktivnoski tonkieh pryamougol'nich stalnich skin, *Elektrichestvo*, **8**, 37 (1950).
22. T. J. Higgins, A comprehensive review of Saint-Venant's torsion problem, *Am. J. Phys.* **10**, 248 (1942).

23. P. J. LAWRENSON, Forces on turbogenerator end windings. *Proc. Instn. Elect. Engrs.* **112** (6), 1144 (1965).
24. A. PRAMANIK, Magnetic field and eddy current distributions in the core-end region of a.c. machines, Ph.D. thesis, University of Birmingham, 1967.
25. A. PRAMANIK, Extension of Roth's method to two-dimensional rectangular regions containing conductors of any cross-section, *Proc. Instn. Elect. Engrs.* **116** (7), 1286 (1969).
26. P. HAMMOND, Roth's method for the solution of boundary-value problems in electrical engineering, *Proc. Instn. Elect. Engrs.* **114** (12), 1969 (1967).
27. C. E. M. DE KUIJPER, Bijdrage tot de berekening van de spreidings-reactantie van transformatoren en van de krachten, welke op de wikkelingen van transformatoren werken, Delftsche Uitgevers Meatschappij, based on Ph.D. thesis at Technische Hogeschool, Delft, 1949.
28. N. MULLINEUX, J. R. REED, and C. E. M. DE KUIJPER, Roth's method for the solution of boundary-value problems, *Proc. Instn. Elect. Engrs.* **116** (2), 291 (1969).

Additional References

HIGGINS, T. J., Inductance of hollow rectangular conductors, *J. Franklin Inst.* **230** (3), 375 (1940).
HIGGINS, T. J., New formulas for calculating short-circuit stresses in bus supports for rectangular tubular conductors, *J. Math. Phys.* **14** (3), 151 (1943).
HIGGINS, T. J., Formula for the geometrical mean distances of rectangular areas and of line segments, *J. Math. Phys.* **14** (4), 188 (1943).
HIGGINS, T. J. and MESSINGER, H. P., Equations for the inductance of three-phase coaxial busses comprised of square tubular conductors, *J. Appl. Phys.* **18**, 1009 (1947).
LAWRENSON, P. J., Calculation of machine end-winding inductances with special reference to turbogenerators, *Proc. Instn. Elect. Engrs.* **117** (6), 1129–34 (1970).

PART III

TRANSFORMATION METHODS

CHAPTER 6

INTRODUCTION TO CONFORMAL TRANSFORMATION

CONFORMAL transformation is by far the most powerful method for the analytical solution of Laplacian fields, being capable of handling boundaries of much more complicated shape than other analytical methods. It can be used to analyse the fields, for instance, between non-concentric circular cables, in waveguides or high-frequency transmission lines of many different cross-sections, exterior to charged conductors of polygonal section, and in the slotted air gap of a rotating electrical machine. Also, in general, the solutions take very simple forms and yield, readily, expressions for flux density and permeance in magnetic fields (or potential gradient and capacitance in electrostatic fields) and, in many cases, allow the direct calculation of field maps. The chief limitation in applying transformation techniques, is that, for most problems, boundaries have to be assumed to be equipotentials (infinitely permeable or infinitely conducting), or coincident with flux lines, or combinations of these two types.

Because of the great value of the transformation method, it is discussed in considerable detail. This chapter, and the two immediately following, treat the basis of the method and discuss its simpler applications; the last two chapters of this part deal with more difficult applications and should not be attempted by the reader unfamiliar with the earlier ones.

6.1. Conformal transformation and conjugate functions

6.1.1. *Conformal transformation*

Consider the properties of a regular[†] function of the type

$$z = f(t) = x(u, v) + jy(u, v), \tag{6.1}$$

which defines a complex variable $z = x + jy$ as some function of another complex variable $t = u + jv$. A particular value, t', can be represented by a point in the Argand diagram of t, Fig. 6.1(a). Through eqn. (6.1) some particular value (or values) z' corresponds to t', and it may be represented by a point in the Argand diagram of z, Fig. 6.1(b). Further, there is a similar correspondence for a succession of pairs of points in the t- and z-planes so that to some curve $t't''$ there corresponds a curve $z'z''$ which is said to have been *transformed or mapped* from $t't''$ by eqn. (6.1).

---

† See the footnote on p. 26.

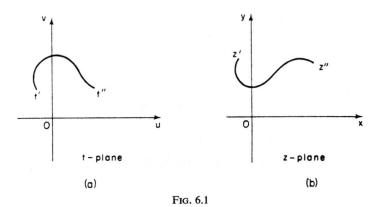

FIG. 6.1

As an example, consider the curves in the z-plane which correspond to straight lines in the t-plane through the equation

$$z = \sin t. \tag{6.2}$$

To facilitate this, it is necessary to form equations, connecting x and y, from which either u or v is absent. Expanding $\sin(u+jv)$ gives

$$z = \sin u \cos jv + \cos u \sin jv$$
$$= \sin u \cosh v + j \cos u \sinh v,$$

so that, equating the real and imaginary parts,

$$x = \sin u \cosh v, \tag{6.3}$$

and

$$y = \cos u \sinh v. \tag{6.4}$$

Squaring and adding these equations eliminates u to give

$$\frac{x^2}{\cosh^2 v} + \frac{y^2}{\sinh^2 v} = 1, \tag{6.5}$$

and squaring and subtracting them eliminates v, so that

$$\frac{x^2}{\sin^2 u} - \frac{y^2}{\cos^2 u} = 1. \tag{6.6}$$

A straight line parallel to the u-axis in Fig. 6.2(a) is described by $v = $ constant. However, for a constant value of v, eqn. (6.5) represents an ellipse in the z-plane, so that any straight line parallel to the u-axis is transformed by the equation $z = \sin t$ into an ellipse in the z-plane, Fig. 6.2(b). Any straight line parallel to the v-axis has the equation $u = $ constant, and from eqn. (6.6) it is seen that such a line is transformed into a hyperbola in the z-plane.

In this simple example it is possible to obtain recognizable equations for the transformed curves. However, this is not usually possible, and to obtain the curves it is often necessary that a series of particular values of t be substituted in the equation. Doing this in the example for the real axis between 0 and ∞ (i.e. $v = 0$), it is seen, from eqns. (6.3) and (6.4), that

(a) $y = 0$, for all values of u,

(b) as u varies from 0 through $\pi/2, \pi$ and $3\pi/2$ to 2π, x varies from 0 through 1, 0 and -1 to 0.

This shows [as is apparent from eqn. (6.3)] that x is periodic and that consequently an infinite number of values of u between $-\infty$ and ∞ can give rise to one value of x. Similarly, y also is periodic with u and, in addition, it may be noted that in an equivalent

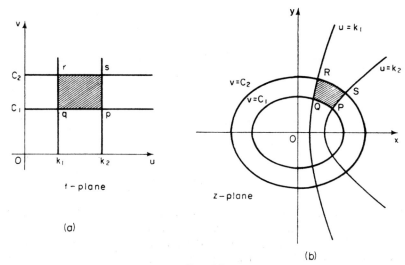

FIG. 6.2

way one point (u, v) in the t-plane can give rise to several in the z-plane. This correspondence of a single point in one plane with more than one point in another plane occurs frequently and, in some cases, requires that care be exercised when interpreting results.

The correspondence between *curves* in the two planes has been indicated, and it is evident also that particular *regions* in the two planes correspond. Thus, for example, the region between the lines $v = 0$ and $v = c_1$ corresponds in the z-plane to the interior of the ellipse for which $v = c_1$ (the ellipse for $v = 0$ degenerates, as shown above, into the straight line between $x = 1$ and -1); and the rectangle *pqrs* between the lines $v = c_1$, $v = c_2$, $u = k_1$, and $u = k_2$ corresponds to the area *PQRS* (and to images of this in the axes) in the z-plane.[†]

Transformations of the general type of eqn. (6.1) are termed *conformal*. This means they are such that if two curves cross at a given angle in one plane, the transformed curves in another plane cross at the same angle and in such a way that the senses of the two angles are the same.[‡] So, in Fig. 6.2, $p\hat{q}r = P\hat{Q}R$. In particular, if two curves cross at right-angles (as in the example), the transformed curves also cross at right-angles. This orthogonal property is, of course, characteristic of all conjugate functions (the real and imaginary parts of a regular function of a complex variable), see section 2.4.

---

† Note that any strip of width 2π in the upper half t-plane corresponds with the whole of the z-plane.

‡ The general proof of the property is simple and can be found in any elementary discussion of functions of a complex variable.

9*

6.1.2. *The solution of Laplace's equation*

In addition to possessing the above orthogonal property, conjugate functions (as shown in section 2.4) are solutions of Laplace's equation. Thus it can be seen that any conformal transformation provides a simple relationship between two Laplacian fields. The use of transformation techniques in deriving field solutions consists in the determination of a suitable equation relating a given field to another field for which a solution is known or is easily found.

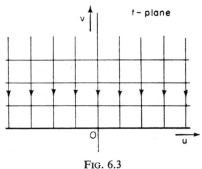

FIG. 6.3

Consider first the complex plane of *t*, Fig. 6.3, in which is represented a uniform field parallel to the axes, so that the lines $u = $ constant are flux lines and the lines $v = $ constant are equipotential lines. Let the gradient of the potential function (proportional to flux density) be k at all points.[†] Then, arbitrarily taking the *u*-axis as the zero potential line, the potential function, ψ (though not necessarily the absolute value of potential), at any point (u, v) is given by $\psi = kv$; and the flux function φ, taking the line $u = 0$ as the line $\varphi = 0$, is given by $\varphi = ku$. Thus, combining these two equations, the expression for the complex potential w of any point in the *t*-plane is

$$w = \varphi + j\psi = k(u + jv)$$

or

$$w = kt. \tag{6.7}$$

The uniform field (represented here in the *t*-plane) is the most simple of all Laplacian fields and is the basic one to which all other solutions are related. It is convenient to regard the complex potential, $w = \varphi + j\psi$, as being represented in the plane of w, so that eqn. (6.7) defines a simple form of transformation, involving only a change in scale proportional to k. Hence for any given field, the process of solution by transformation reduces ultimately to the derivation of an equation,

$$w = f(z), \tag{6.8}$$

which relates points in the given field in the *z*-plane to points in the *w*-plane. More specifically this means the boundary shape and conditions of the two planes are connected through eqn. (6.8). It will be seen that, in general, derivation of eqn. (6.8) involves the introduction of intermediate planes and variables.

[†] It will be recalled (sections 2.4.3 and 2.4.4) that the relative numerical values of φ and ψ are chosen for convenience to give the simplest form to w.

As a simple example, consider the influence of the transformation

$$z = t^{1/2} \tag{6.9}$$

on the uniform field with lines parallel to the axes in the t-plane. It is seen that the positive real axis in the t-plane transforms into the positive real axis in the z-plane, but that the negative half of the real axis, because of the root of a negative sign, becomes the imaginary axis between 0 and $j\infty$ in the z-plane. Thus, the equation transforms the upper half of the

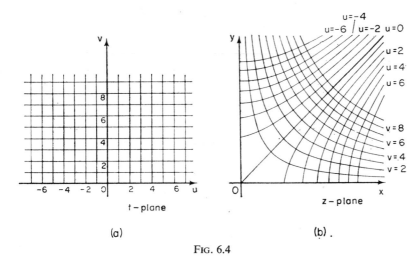

(a)   (b).

FIG. 6.4

t-plane into the first quadrant of the z-plane. The shapes of the parallel lines when transformed can be found by eliminating u or v from eqn. (6.9). Squaring this equation and equating the real and imaginary parts gives

$$x^2 - y^2 = u \tag{6.10}$$

and

$$2xy = v, \tag{6.11}$$

from which it is seen that lines parallel to the v-axis transform into rectangular hyperbolae, as also do those parallel to the u-axis, Fig. 6.4.

The physical picture of the influence of this transformation is that straight flux lines which enter an equipotential boundary coinciding with the real axis of the t-plane, become hyperbolic in shape in the z-plane and enter equipotential boundaries coinciding with the positive real and imaginary axes. The boundary conditions in the two planes are identical because of the conformal property of the transformation. Equally, the boundary in the z-plane is a limiting line ($\psi = 0$) in a set of hyperbolically shaped equipotential lines.

The equations of the flux and potential functions in the z-plane are given by elimination of t from eqns. (6.7) and (6.9) as

$$kz^2 = w, \tag{6.12}$$

or, for the flux function,

$$\varphi = k(x^2 - y^2), \tag{6.13}$$

and, for the potential function,

$$\psi = 2kxy. \tag{6.14}$$

For a given value of φ or ψ these equations map out in the z-plane the shape of the corresponding flux or equipotential line (including the boundary). Equally, substitution of any point (x, y) in eqn. (6.12) gives the complex potential w with respect to an appropriate origin, of the point (x, y).

6.1.3. The logarithmic function

The logarithmic transformation is of great importance, being used in the great majority of all solutions by transformation techniques. Its mapping properties and the three basic fields which it can be used to describe are considered below.

Field of a line current. Consider again the field of a line current. This is discussed in section 2.2 where it is shown that for a line current, i, placed at the origin of the z-plane, the complex potential function of the field is given by the function $(i/2\pi) \log z$; that is,

$$\varphi + j\psi = \frac{i}{2\pi} \log z. \tag{6.15}$$

The same equation can be developed from a consideration of the transformation

$$t = \log z. \tag{6.16}$$

Inverting this equation gives

$$e^t = z \quad \text{or} \quad e^u \cos v = x \quad \text{and} \quad e^u \sin v = y.$$

Thus

$$\sqrt{x^2 + y^2} = e^u \tag{6.17}$$

and

$$\tan^{-1}\left(\frac{y}{x}\right) = v. \tag{6.18}$$

Equation (6.17) shows that a line parallel to the v-axis in the t-plane transforms into a circle, centre the origin, in the z-plane; and eqn. (6.18) shows that a line parallel to the u-axis in the t-plane transforms into a straight line, passing through the origin and making an angle v radians with the x-axis of the z-plane (Fig. 6.5).

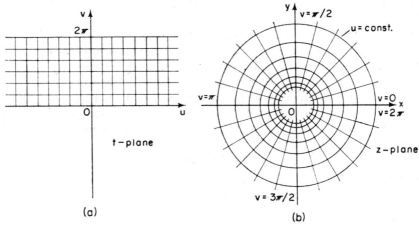

(a)

(b)

Fig. 6.5

Now if a uniform field described by

$$w = kt \tag{6.19}$$

is put in the t-plane, lines corresponding to constant values of u and v are flux and equipotential lines respectively. However, the lines in the z-plane to which these correspond are the same as the flux and equipotential lines in the field of a line current. Hence, combining eqns. (6.19) and (6.16) gives the solution for the field of a line current as

$$w = k \log z, \tag{6.20}$$

where the value of k is dependent on the magnitude of the current.

To evaluate k it is most convenient to make use of the equation for the potential function. Equating the imaginary parts of eqn. (6.20) gives

$$\psi = k \arg z. \tag{6.21}$$

Now, in tracing a path once round the current, there is a potential change of i, so that, expressing this condition in eqn. (6.21) gives $i = k2\pi$.

Thus substitution for k in eqn. (6.20) yields the complex potential function of the field due to the current, eqn. (6.15).

Field of a line charge. Equally the solution (6.20) applies, with interchange of the flux and potential functions, to the field of a line charge. But the value of the constant k is dependent in this case on the magnitude of the charge, q units per unit length of the line. The equation corresponding to (6.21) is

$$\varphi = k \arg z, \tag{6.22}$$

and, expressing the condition that a path traced round the charge cuts all the lines of flux emanating from it, this equation leads to

$$k = \frac{q}{2\pi}.$$

Field of two semi-infinite equipotential planes. The field in the upper half plane of Fig. 6.5(b) is identical with that due to two semi-infinite equipotential planes, one lying between 0 and ∞, the other between 0 and $-\infty$, with a difference in potential of $i/2$ between them. In general, with a difference in potential of ψ_1 maintained between the two halves (dividing at the origin) of the real axis, the field in the positive half plane is described by the complex potential function

$$\varphi + j\psi = \frac{\psi_1}{\pi} \log z. \tag{6.23}$$

This equation is frequently used in the analysis of fields with two (different) equipotential sections forming the boundary.

6.2. Classes of solvable problems

In order to obtain a field solution by transformation methods, two things are necessary: a transformation equation must be found relating the given field to a simpler one, and a solution must be found for this simple field. The possibility of doing the latter depends upon whether or not the transformed boundary conditions can be identified and used. In general, the problem of identifying and using these conditions can be very difficult or even impossible, but there are many important types of problem for which it is relatively simple. It is convenient to classify all problems under two headings: those in which, at the boundaries, values of φ or ψ are known or, though unknown, are constant (so that the boundary conditions for the fields on each side can be specified independently), and those in which gradient boundary conditions (excepting those with zero normal or tangential components which fall into the first group) apply between regions.

The first of these classes is by far the more important, and almost all known transformation solutions apply to it. It includes all problems involving equipotential or flux-line boundaries, or combinations of these (ψ or φ constant), and also those problems for which the potential distribution along the boundary is specified (Dirichlet problem, see section 10.6.1). Numerical solutions to the Dirichlet problem are always possible provided that the transformation of the given boundary to the unit circle or real axis can be accomplished, but analytical solutions for all but the simplest cases are extremely difficult. One of the few analytical solutions is given by Nakamura[1] who describes the solution for the field of a salient pole having a varying potential on its flank. See also section 10.6 for the general solution for a boundary having many different equipotential sections.

The second class of problem having boundaries of finite permeability or permittivity has been little explored. [The transformation of fields involving gradient boundary conditions, connecting solutions on each side of a boundary, is seen to be valid, because the transformed field is still Laplacian and the boundary conditions are preserved—the angles at which flux lines cross the interfaces are preserved, so that eqn. (2.68) is satisfied, and potential ψ must be continuous, so that eqn. (2.26) is satisfied.] In general, problems in this class are very difficult to solve by transformation methods and are appropriately treated by the methods of direct solution (Chapters 3, 4, and 5) or by numerical methods (Chapter 11). However, for any such problem for which there exists a single equation which transforms *all* regions of the field together, solutions by conformal transformation are relatively simple. An example of this sort is provided by the transformation from a circular boundary to an elliptical one in such a way that the interiors and exteriors of the two boundaries correspond. The use of this transformation to obtain the solution for a dielectric cylinder of elliptical section in an applied uniform field is given by Smythe, p. 94.

In conclusion, it should be noted that some Poissonian field solutions can also be transformed; see for example, Weber, p. 304. However the very limited exploitation of this technique does not justify its inclusion in this book. The only solution in electrical terms of which the authors are aware is that given by Higgins[2] for eccentric circular conductors, but for a complete review of the analogous problems which have been treated (see the footnote on p. 92), the reader should consult reference 22 of the previous chapter.

6.3. General considerations

The elements of the transformation method have been set out in the earlier parts of this chapter but, before proceeding to a study of the practical applications of the method, it will be helpful to elaborate on a number of general points many of which have to be considered each time the method is applied.

6.3.1. *Choice of origin*

The position of the origin in any plane is quite arbitrary. For example, the position of the corner in Fig. 6.4(b), or the position of the line current in Fig. 6.5(b) can be changed to any point $z = x_1 + jy_1$ by replacement of z by $[z - (x_1 + jy_1)]$ in eqn. (6.9) or (6.16) respectively. Similarly, replacement of t by $[t - (u_1 + jv_1)]$ makes the vertex in the z-plane correspond with the point $t = u_1 + jv_1$, Fig. 6.5(a).

6.3.2. *Multiple transformations*

In applying transformation methods it is necessary to find equations which convert a simple boundary into the more complicated boundary containing the field which it is required to solve. It is not generally possible to find, directly, one equation which does this and, usually, the w- and z-planes have to be connected through intermediate variables and boundary shapes. So, for example, the field pattern of Fig. 6.6 can be derived by com-

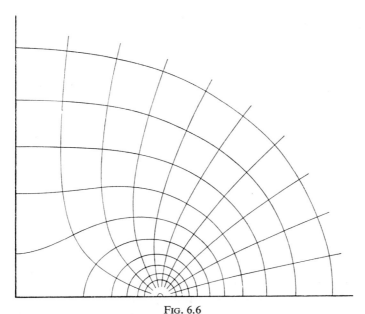

FIG. 6.6

bining transformations of the types (6.9) and (6.15). The pattern shown in Fig. 6.5(b) is mapped into the p-plane from the w-plane by the equation

$$w = \frac{i}{2\pi} \log p, \qquad (6.24)$$

and, in turn, the upper half of the p-plane is mapped into the z-plane, Fig. 6.6, by

$$z = k(p - a)^{1/2}, \qquad (6.25)$$

where a is a real constant defining the point, on the real axis of the p-plane, which corresponds to the vertex in the z-plane. Eliminating p from eqns. (6.24) and (6.25) gives the solution in the z-plane as

$$\varphi + j\psi = \frac{i}{2\pi} \log \left(\frac{z^2}{k^2} + a \right). \tag{6.26}$$

This may be regarded as describing either the field of a line current placed on the surface of an infinitely permeable corner, or one half of the field of a line current parallel to an infinite, infinitely permeable plane.

6.3.3. *Field maps*

The solution for a field in the z-plane is often derived as an equation of the form

$$w = f(z). \tag{6.27}$$

If this equation is such that it can be written

$$\begin{aligned} z &= g(w) \\ &= g(\varphi + j\psi), \end{aligned} \tag{6.28}$$

as, for instance, eqn. (6.26) can be, then a field map can be calculated directly. Substituting a value ψ_1 for ψ gives the equation connecting points (x, y) lying on the equipotential line of value ψ_1, and this line may be traced by further substitution of a succession of values of φ. Similarly, a flux line may be traced by selecting a fixed value of φ and substituting a series of values of ψ. To produce a map of curvilinear *squares* the increments in φ and ψ must be equal. The squares can be chosen to have a convenient size by using eqn. (6.27) to find the desired increment in flux or potential.

6.3.4. *Scale relationship between planes*

The equation

$$z = kf(t)$$

directly relates one *number* in the t-plane to another *number* in the z-plane. Consequently it relates a distance between two points (numbers) in the t-plane to a distance between the corresponding points in the z-plane; for example, eqn. (6.9) makes a length of 4 units measured from the origin of the t-plane correspond with 2 units of length measured from the origin in the z-plane. When plotting a field map, any convenient unit and scale may be selected, because the *shape* of a map is not dependent upon its scale. However, when numerical information, such as flux density, is required, the unit chosen must be the same for both planes; it is clear, for example, that the flux density at a point, for given boundary shapes and field excitation, is inversely proportional to the size of the field. The scale constant k is then evaluated so as to give direct correspondence between points and field strengths in the z- and t-planes (see section 6.1.3). In certain cases, to allow for differences in the angular positions of two boundaries, k may be complex.

6.3.5. *Conservation of flux and potential*

From section 6.1.2 it is evident that the complex potential associated with a particular point is identical with that of the corresponding transformed point. Therefore the flux crossing a line joining two points, or the potential difference between the points, is unchanged by transformation of the line and points. Hence, for example, the calculation of total flux—for the evaluation of capacitance or permeance—merely involves taking the difference between the flux functions for the two points.

6.3.6. *Field strength*

Since flux density and field strength are simply related by a multiplicative constant (see section 2.4.3 and 6.1.2), it is necessary to consider only one of them. The general solution for a field in the z-plane is

$$w = \varphi + j\psi = f(z),$$

and the field strength in the x-direction E_x is

$$E_x = -\frac{\partial \psi}{\partial x},$$

and the field strength in the y-direction E_y is

$$E_y = -\frac{\partial \psi}{\partial y}.$$

Consider the quantity $E_x - jE_y$. This can be written

$$E_x - jE_y = -\frac{\partial \psi}{\partial x} + j\frac{\partial \psi}{\partial y},$$

which, substituting for $\partial \psi/\partial y$ from the Cauchy–Riemann equation (2.87), gives

$$E_x - jE_y = -\frac{\partial \psi}{\partial x} + j\frac{\partial \varphi}{\partial x}$$

$$= j\left(\frac{\partial \varphi}{\partial x} + j\frac{\partial \psi}{\partial x}\right)$$

$$= j\frac{\partial w}{\partial x}$$

$$= j\frac{dw}{dz}. \tag{6.29}$$

Now the quantity $E_x - jE_y$ has the same modulus but minus the argument of the quantity $E_x + jE_y$, which is equal to the field strength **E**. Therefore from eqn. (6.29) the magnitude of the field strength is given simply by

$$|\mathbf{E}| = \left|\frac{dw}{dz}\right| \tag{6.30}$$

and the argument of the field strength by

$$\arg \mathbf{E} = -j \arg\left(\frac{dw}{dz}\right). \tag{6.31}$$

These are important results, and eqn. (6.30) particularly is frequently required.

6.4. The determination of transformation equations

It has been shown that, when the appropriate equation of transformation is known, a simple field problem can be related to a more complicated one in such a way as to give a solution for the complicated one. In applying this method of solution to a given problem it is therefore essential to be able to determine the transformation equation which relates the given boundary configuration to some simpler form for which a solution is known or easily derivable. In general, this may be extremely difficult or even impossible, but there are a number of routine methods which yield equations connecting many important classes of boundaries. These are discussed in detail in the next four chapters.

There are also many special transformation equations which are, however, of limited utility. These are not discussed but the reader who wishes to acquaint himself with some of those which are available should consult a book such as that by Köber. The book by Koppenfels and Stallman is also useful in this respect, particularly for boundaries comprised of distinct curves.

References

1. T. NAKAMURA, Application of conformal representation to flux distribution in the iron cores of electrical machines. *Elektrotechn. J.* (*Japan*) **3,** 6 (1939).
2. T. J. HIGGINS, The vector potential and inductance of a circuit comprising line conductors of different permeabilities, *J. Appl. Phys.* **13,** 390 (1942).

CHAPTER 7

CURVED BOUNDARIES

THERE are many well-known equations for the transformation of curved boundaries, but the majority of these are of limited interest. In this chapter attention is concentrated on two important transformations of value in connection with particular types of curve and also on a more general transformation applicable to a wider ranges of curves. The first of those applicable to particular curves is the *bilinear* or *Möbius* transformation which is used for the treatment of non-concentric or intersecting circular boundaries, and the second is a simple *Joukowski* transformation which can be used for elliptical (including circular) boundaries. The more general transformation applies to all curves which can be described by *parametric* equations, and, in conjunction with methods of curve fitting, it provides a technique by which approximate transformations can be found for a wide range of curves.

A number of more specialized transformations for boundaries consisting entirely of curves are given in the books by Köber and Koppenfels and Stallman. For the treatment of boundaries consisting of a combination of curves and straight lines, the reader is referred to section 10.3 and the references to Chapter 10.

7.1. The bilinear transformation

The method of direct solution (using cylindrical polar coordinates) described in Chapter 4 can be used for Laplacian fields having circular boundaries provided that these are concentric. But in practice it is often important to be able to analyse fields having non-concentric circular (including intersecting) or circular combined with linear (infinite radius) boundaries. These classes of boundary occur in the determination of, for instance, the unbalanced magnetic pull exerted on a machine rotor mounted eccentrically in the stator, the capacitance between two cables or waveguides of circular cross-section, and the reluctance of a permeable sheet with a circular hole near to its (straight) edge. All such problems may be solved by using the *bilinear transformation*. This transformation is also of value in relating a single, circular boundary to an infinite, straight-line boundary, and, in the next chapter, it is used to develop general equations which transform a circular boundary into a polygonal one. It has also been very useful in overcoming difficulties associated with the need to consider infinite regions in certain problems: Boothroyd et al.[2] have developed a circular electrolytic tank for the representation of the whole complex plane in connection with network problems, and Silvester[3] has studied eddy-current effects in isolated conductors in conjunction with both analogue and direct methods. Before going on to use the transformation to solve particular problems, its general mapping properties are first explained.

7.1.1. *Mapping properties*

The bilinear transformation is of the form

$$z = \frac{at+b}{ct+d}, \tag{7.1}$$

where a, b, c, and d are constants which may be real or complex. It is the general transformation for which one, and only one, value of one variable corresponds to one, and only one, value of the other. It can be regarded as being built up from three successive, simple transformations:

$$t_1 = ct+d, \tag{7.2}$$

$$t_2 = \frac{1}{t_1}, \tag{7.3}$$

and

$$z = \frac{a}{c} + (bc-ad)\frac{t_2}{c}. \tag{7.4}$$

The first and third of these introduce magnification, translation, and rotation of the field map depending upon the values of the constants, but do not change its *shape*. The expression $(bc-ad)$ is called the *determinant* of the transformation (7.1) and clearly, from eqn. (7.4), must not become zero. Only eqn. (7.3), which is called the *inversion* transformation (since one variable is the inverse of the other), changes the shape of the map. Therefore the complete bilinear transformation produces, for a given map, the same change of shape as does the inversion transformation. The bilinear and inversion transformations are most useful when applied to circles or straight lines, and the influence of the inversion transformation on these curves is now described.

The transformation of straight lines, and the doublet. Consider the inversion transformation

$$t = \frac{1}{z}. \tag{7.5}$$

Rationalizing this equation and equating real and imaginary parts gives

$$u = \frac{x}{x^2+y^2} \quad \text{and} \quad v = -\frac{y}{x^2+y^2},$$

which may be rewritten as

$$\left(x - \frac{1}{2u}\right)^2 + y^2 = \left(\frac{1}{2u}\right)^2 \tag{7.6}$$

and

$$x^2 + \left(y + \frac{1}{2v}\right)^2 = \left(\frac{1}{2v}\right)^2. \tag{7.7}$$

For constant values of u and v these are the equations of circles in the z-plane; that is, straight lines parallel to the axes of the t-plane are mapped into circles in the z-plane. The circles described by eqn. (7.6) have radii $(1/2u)$ and centres $(1/2u, 0)$, and those described by eqn. (7.7) have radii $(1/2v)$ and centres $(0, -1/2v)$, so that circles from each set are tangential to the axes at the origin (Fig. 7.1).

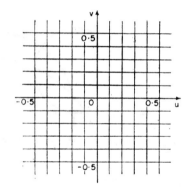

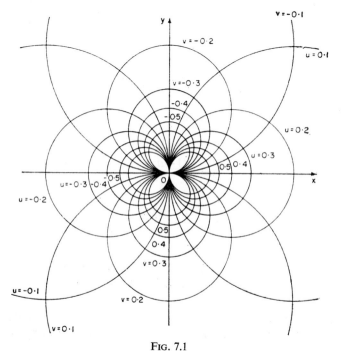

FIG. 7.1

It is important to realize that this pattern is the same as that of the flux and equipotential lines in the field of a doublet (see section 3.3.2). Thus it can be seen that the inversion transformation relates the uniform field (in the w-plane) to the field of a doublet, and the similarity between eqn. (7.5) with $z = w/d$ and eqn. (3.23) or (3.24) should be noted.

In a similar manner to the above, any straight line in the t-plane is transformed by eqn. (7.5) into a circle in the z-plane. Thus any straight-sided figure in the t-plane is transformed into a similar curvilinear figure in the z-plane. This is demonstrated in Fig. 7.2 for the case of a triangle. Note not only that ao', ob', oc' have magnitudes inversely proportional to oa, ob, oc, but that the arguments of the z-plane vectors are the negative of those in the t-plane.

The transformation of circles. The equation of a circle, centre $(a, 0)$, and radius r, in the t-plane, is

$$r = |t - a|, \tag{7.8}$$

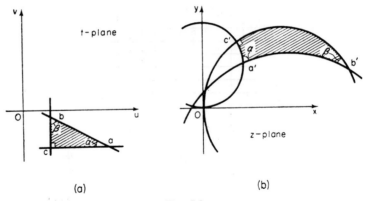

(a)   (b)

FIG. 7.2

but, since

$$t-a = \frac{1}{x+jy} - a,$$

the expression for the radius can be written

$$r = |t-a|$$

$$= \left| \frac{(1-ax)-jay}{x+jy} \right|.$$

Therefore

$$r^2 = \frac{(1-ax)^2 + a^2y^2}{x^2+y^2},$$

which may be rearranged as

$$\left(x - \frac{a}{a^2-r^2} \right)^2 + y^2 = \left(\frac{r}{a^2-r^2} \right)^2. \tag{7.9}$$

This is the equation of a circle, radius $r/(a^2-r^2)$, centre $(a/(a^2-r^2), 0)$ in the z-plane, mapped by eqn. (7.5) from the circle, radius r, centre a, in the t-plane. For $a > r$ it follows that

$$\frac{a}{a^2-r^2} > \frac{r}{a^2-r^2} > 0, \tag{7.10}$$

so that the transformed circle lies entirely in the positive real half of the z-plane. For $a = r$, i.e. for a circle passing through the origin in the t-plane, the transformed curve is a straight line,

$$2ax = 1, \tag{7.11}$$

and, for $a < r$,

$$\frac{a}{a^2-r^2} < \frac{r}{a^2-r^2} < 0, \tag{7.12}$$

so that the transformed circle extends into the positive real half of the plane, though its centre is on the negative real axis. These three cases are shown in Fig. 7.3 where a set of concentric circles, and the corresponding transformed curves, are shown.

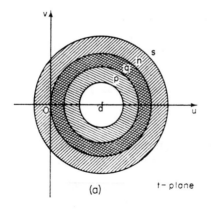

(a)   t−plane

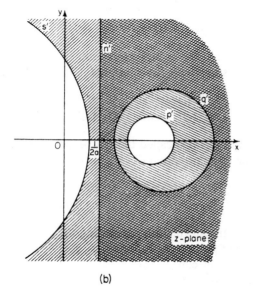

(b)

FIG. 7.3

Alternatively, by suitable positioning of the origin in the z-plane, it is possible to transform any two circles (the straight line is a circle of infinite radius) which are non-concentric into two concentric circles in the t-plane. There are three different arrays in the z-plane:

(i) a circle within another circle;
(ii) a circle near a straight line; and
(iii) a circle exterior to another circle;

and a field having boundaries falling into any one of these classes can be analysed by first determining the solution of the field for the corresponding concentric boundaries with appropriate boundary conditions.

Geometrical and complex inversion. Geometrical and complex inversion are not identical: a point having an argument θ inverts, in the geometrical case, into a point with the same argument whilst, in the complex case, the inverted point has argument $-\theta$. This is evident

AC 10

when z [eqn. (7.5)] is written in polar form,

$$t = \frac{1}{r e^{j\theta}} = \frac{1}{r} e^{-j\theta}. \tag{7.13}$$

As mentioned above, this feature is demonstrated in Fig. 7.2.

7.1.2. *The cross-ratio*

Consider again the general transformation (7.1) and let z_1, z_2, z_3, z_4 and t_1, t_2, t_3, t_4, be two sets of corresponding points. Then, by substitution in eqn. (7.1),

$$z_1 - z_4 = \frac{at_1 + b}{ct_1 + d} - \frac{at_4 + b}{ct_4 + d}$$

$$= \frac{ad - bc}{(ct_1 + d)(ct_4 + d)}(t_1 - t_4). \tag{7.14}$$

Similar expressions for $z_3 - z_2$, $z_1 - z_2$ and $z_3 - z_4$ can be derived and combined with the above to eliminate all terms involving the constants a, b, c, and d to give

$$\frac{(z_1 - z_4)(z_3 - z_2)}{(z_1 - z_2)(z_3 - z_4)} = \frac{(t_1 - t_\ast)(t_3 - t_2)}{(t_1 - t_2)(t_3 - t_4)}. \tag{7.15}$$

The right-hand side of eqn. (7.15) is called the *cross-ratio* of the four points t_1, t_2, t_3, and t_4. It is a constant for the transformation (7.1), and the relationship

$$\frac{(z_1 - z)(z_3 - z_2)}{(z_1 - z_2)(z_3 - z)} = \frac{(t_1 - t)(t_3 - t_2)}{(t_1 - t_2)(t_3 - t)} \tag{7.16}$$

defines the unique bilinear transformation connecting the curves passing through the particular corresponding sets of point t_1, t_2, t_3 and z_1, z_2, z_3.

As an example of the use of eqn. (7.16), the transformation, used in the next chapter, which converts the real axis of one plane into the unit circle in another, is developed here. The points $t_1 = 0$, $t_2 = 1$, and $t_3 = \infty$ define the real axis of the t-plane. The unit circle in the z-plane may be defined by the corresponding points $z_1 = 1$, $z_2 = j$, $z_3 = -1$. Substitution of all these values in eqn. (7.16) yields the equation transforming the real axis of the t-plane into the unit circle in the z-plane in such a way that the perimeter of the circle in the upper half of the plane corresponds with the positive real axis. This equation is

$$t = j\frac{(1-z)}{(1+z)}. \tag{7.17}$$

It is easiest to see which regions in the two planes correspond by expressing z in polar form, $z = \varrho e^{j\theta}$. This gives for eqn. (7.17)

$$t = j\frac{1 - \varrho e^{j\theta}}{1 + \varrho e^{j\theta}}$$

$$= \frac{2\varrho \sin\theta}{1 + 2\varrho \cos\theta + \varrho^2} + j\frac{(1 - \varrho^2)}{1 + 2\varrho \cos\theta + \varrho^2}. \tag{7.18}$$

Since the quantity $(1+2\varrho\cos\theta+\varrho^2)$ cannot be negative, this shows that the interior of the unit circle $(\varrho < 1)$ corresponds to the upper half of the t-plane, and that the exterior of the unit circle $(\varrho > 1)$ corresponds to the lower half of the t-plane.

7.1.3. *The magnetic field of currents inside an infinitely permeable tube*

Equation (7.17) can be used, for instance, in the analysis of the magnetic field of currents flowing inside an infinitely permeable tube of circular cross-section, taken for convenience to have unit inside radius. (The equivalent current-flow and electrostatic fields will be apparent.) First consider the field of a single current, of strength i, placed at the origin on the

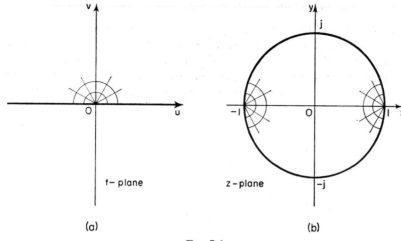

FIG. 7.4

edge of an infinitely permeable sheet in the lower half of the t-plane, see Fig. 7.4(a), and described by

$$w = \psi + j\varphi = \frac{i}{\pi}\log t. \qquad (7.19)$$

Magnetic equipotential lines from the point $t = 0$ terminate at infinity, so that at $t = \infty$ there is effectively a "return" current $-i$. When the field in the upper half plane is transformed by eqn. (7.17) into the interior of the circular tube represented in the z-plane by the unit circle, centre $z = 0$, the point $z = 1$ is made to correspond with $t = 0$, and the point $z = -1$ with $t = \infty$. Thus the field interior to the tube is that of two currents of opposite signs at the opposite ends of a diameter, Fig. 7.4(b). Elimination of t from eqns. (7.17) and (7.19) gives the solution for this field as

$$w = \psi + j\varphi = \frac{i}{\pi}\log j\frac{(1-z)}{(1+z)}. \qquad (7.20)$$

To calculate the field map in the z-plane, z is expressed in terms of $\psi + j\varphi$ [by inverting eqn. (7.20)] as

$$z = \frac{1+je^{w\pi/i}}{1-je^{w\pi/i}}, \qquad (7.21)$$

and values of w are substituted as described earlier (see section 6.3.3).

10*

The solution for the field in the w-plane with the currents at the points $z = -1$ and $z = z_s$ (on the surface of the tube) is obtained by noting that t in eqn. (7.19) must be replaced by $(t - t_s)$ where t_s is the point corresponding to z_s. The result is

$$w = \frac{i}{\pi} \log j \left\{ \frac{(1-z)}{(1+z)} - \frac{(1-z_s)}{(1+z_s)} \right\}. \tag{7.22}$$

Treatment of the more general problem in which the two (or more) currents are inside the tube away from the boundary involves the use of field solutions [in place of eqn. (7.19)] which are discussed in section 10.2.

7.1.4. *The capacitance of and the voltage gradient between two cylindrical conductors*

As a demonstration of the use of the special form of the bilinear transformation, inversion, the field between two charged cylindrical conductors, situated as shown in Fig. 7.5, is considered. The solution of this field is of value, for example, in determining the breakdown voltage of air gaps and the necessary insulation for and the capacitance of parallel cables: in the former case the space between the cylinders is filled with air, and in the latter

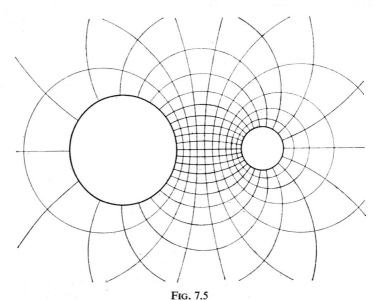

FIG. 7.5

with air or other insulating material. It should be noted that for any problem the analysis given requires the assumption that a single uniform dielectric be present between the cylinders.

Let the non-concentric boundaries of Fig. 7.5 be placed in the z-plane, see Fig. 7.6(a), and let the origin of inversion be placed d units to the left of the smaller circle, radius R_2. d must be so chosen that these circles are transformed into concentric circles in the t-plane, see Fig. 7.6(b). From eqn. (7.9) it is seen that the smaller circle, centre $(d, 0)$, in the z-plane, transforms into a circle in the t-plane with centre coordinates

$$\left(\frac{'d}{d^2 - R_2^2}, \; 0 \right)$$

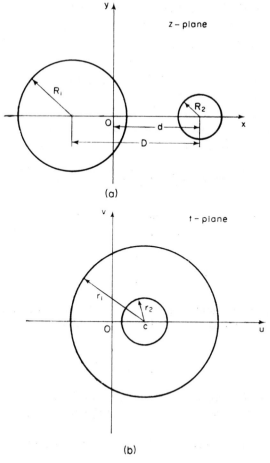

y

z – plane

R_1

R_2

O

d

D

x

(a)

v

t – plane

r_1

r_2

O

c

u

(b)

FIG. 7.6

and radius

$$r_2 = \frac{R_2}{d^2 - R_2^2}.$$

Similarly, the larger circle, with radius R_1 and centre $(d-D, 0)$ transforms into a circle, centre

$$\left(\frac{d-D}{(d-D)^2 - R_1^2}, \ 0 \right)$$

and radius

$$r_1 = \frac{R_1}{(d-D)^2 - R_1^2},$$

where D is the distance between the centres of the circles in the z-plane. For the circles in the t-plane to have the same centre, $(c, 0)$, it is necessary that

$$\frac{d}{d^2 - R_2^2} = \frac{d-D}{(d-D)^2 - R_1^2}.$$

This equation fixes the necessary position of the origin of inversion in the z-plane by defining $d^{\dagger}$ as

$$d = \frac{D^2+R_2^2-R_1^2}{2D} \pm \sqrt{\left[\left(\frac{D^2+R_2^2-R_1^2}{2D}\right)^2 - R_2^2\right]}. \tag{7.23}$$

The cylinders in the z-plane have equipotential boundaries at different potentials and so, in the t-plane, the concentric cylindrical boundaries must be equipotential lines with the same difference in potential between then. The field between two concentric cylindrical conductors, centred at the point $(c, 0)$, separated by a medium of relative permittivity ε and carrying a charge of q units per unit length is, see section 2.1.3, given by

$$w = \psi + j\varphi = \frac{q}{2\pi\varepsilon\varepsilon_0} \log(t-c). \tag{7.24}$$

By considering two points, one on each of these cylinders, most simply, $t_1 = c+r_1$ and $t_2 = c+r_2$, with the same flux function, the potential difference between them may thus be expressed as

$$\psi_1-\psi_2 = \frac{q}{2\pi\varepsilon\varepsilon_0} \log\frac{r_1}{r_2}. \tag{7.25}$$

Therefore the capacitance per unit length C between these cyliinders, i.e. the total flux per unit length (which is equal to q) divided by the potential difference, is

$$C = \frac{2\pi\varepsilon\varepsilon_0}{\log(r_1/r_2)}. \tag{7.26}$$

Further, since corresponding points and boundaries in the t- and z-planes have the same flux and potential functions, eqn. (7.26) also gives the capacitance of the non-concentric cylinders.

The potential gradient $\mathbf{E}$, at any point in the z-plane has, as is described earlier (see section 6.3.6), a magnitude

$$|\mathbf{E}| = \left|\frac{dw}{dz}\right|$$

$$= \left|\frac{dw}{dt}\right| \times \left|\frac{dt}{dz}\right|. \tag{7.27}$$

Differentiating eqn. (7.24) gives

$$\frac{dw}{dt} = \frac{q}{2\pi\varepsilon\varepsilon_0}\frac{1}{t-c},$$

and substituting for q from eqn. (7.25) gives

$$\frac{dw}{dt} = \frac{\psi_1-\psi_2}{\log(r_1/r_2)}\frac{1}{t-c}.$$

Also, from equation (7.5)

$$\frac{dt}{dz} = -\frac{1}{z^2},$$

† The value of d making r positive must be taken.

so equation (7.27) can be written

$$|\mathbf{E}| = \left| \frac{\psi_1 - \psi_2}{\log(r_1/r_2)} \frac{1}{(t-c)} \frac{1}{z^2} \right|$$

$$= \left| \frac{\psi_1 - \psi_2}{\log(r_1/r_2)} \frac{1}{(1-cz)z} \right|. \tag{7.28}$$

The maximum value of potential gradient is important in the consideration of the breakdown voltage between the cylinders. It occurs at the point on the shortest line between the cylinders at the boundary with the greater curvature. This point is $z = d - R_2$ to which corresponds the point $t = c + r_2$ and, therefore, substitution of these values in eqn. (7.28) gives the expression for maximum gradient as

$$|\mathbf{E}|_{\max} = \left| \frac{\psi_1 - \psi_2}{\log(r_1/r_2)} \frac{1}{r_2(d-R_2)^2} \right|. \tag{7.29}$$

The field map in the z-plane (Fig. 7.5) is transformed from that in the t-plane, which consists of concentric circles, and radial lines centre $(c, 0)$. It is calculated using the equation

$$z = \frac{1}{e^W + c}, \tag{7.30}$$

where

$$W = \frac{w \log(r_1/r_2)}{\psi_1 - \psi_2}, \tag{7.31}$$

derived by eliminating t from eqns. (7.5) and (7.24) and using eqn. (7.25). As will be clear from the subsection *The transformation of circles* of section 7.1.1 the same equations can also

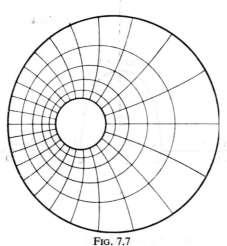

FIG. 7.7

be used to plot the field map for two cylindrical boundaries one within the other (Fig. 7.7), or, after first determining the appropriate values of d, r_1, and r_2, for one cylinder near a plane surface.

For the maps, Figs. 7.5 and 7.7, the boundary potentials and dimensions are such as to give the same quantity of total flux in each.

7.2. The simple Joukowski transformation

7.2.1. The transformation

The equation

$$Kt = z + \frac{a^2}{z} \tag{7.32}$$

where K and a are constants, and which may also be written,

$$z = \frac{K}{2}\left[t \pm \sqrt{t^2 - \left(\frac{2a}{K}\right)^2}\right], \tag{7.33}$$

transforms a circle in the z-plane into an ellipse in the t-plane, in such a way that the regions exterior to both curves, and the regions interior to them, correspond. It also has a second mapping property which is of rather more interest to electrical engineers, and this is that it can be used to transform the real axis in the t-plane into the real axis of the z-plane for $|t| \geqslant 2a/K$, and into a semicircle, radius a, centre $z = 0$, for $|t| \leqslant 2a/K$. More generally, with the equation,

$$z = \frac{K}{2}\left[t \pm \lambda \sqrt{t^2 - \left(\frac{2a}{K}\right)^2}\right], \tag{7.34}$$

where λ is a real constant ($\neq 1$), the curve in the z-plane for $|t| < 2a/K$ is an ellipse (Fig. 7.8).

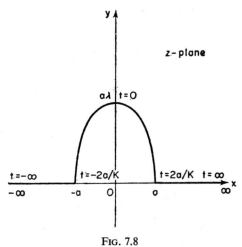

Fig. 7.8

This may be seen from the following considerations. When t is real and $|t| \geqslant 2a/K$, the expression under the root sign is positive, so that, provided the root is given the sign of t, z is wholly real with the sign of t. When t is real and $|t| < 2a/K$, the quantity under the root sign is negative, and separation of the real and imaginary parts of eqn. (7.34) gives

$$x = \frac{K}{2}u \quad \text{and} \quad y = \frac{K\lambda}{2}\sqrt{\left(\frac{2a}{K}\right)^2 - u^2},$$

from which u can be eliminated to give

$$\frac{x^2}{a^2} + \frac{y^2}{(a\lambda)^2} = 1. \tag{7.35}$$

This is the equation of an ellipse describing the shape of the curve in the z-plane corresponding with the real axis of the t-plane between $\pm 2a/K$. The ellipse cuts the y-axis at $y = \pm a\lambda$. For $\lambda = 1$ the curve degenerates into the circle of radius a. It should be noted that to transform the upper half t-plane into the region above the real axis (and curve) in the z-plane, the imaginary part of the square root must be taken as positive throughout, whilst the real part takes the same sign as the real part of t.

(This transformation is also considered in section 10.3 as a particular example of a transformation for curved boundaries with vertices.)

7.2.2. *Flow round a circular hole*

As an example of the use of eqn. (7.32) [eqn. (7.34) in which $\lambda = 1$], the very simple example of the flux distribution round a circular hole in an infinite sheet is considered (Fig. 7.9). The field is symmetrical, and so one half of it can be represented in the upper

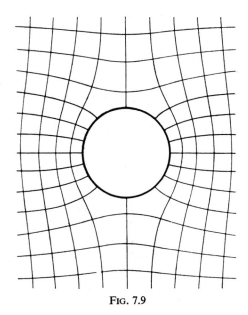

FIG. 7.9

half of the z-plane with the boundary (flux line of zero flow) shown in Fig. 7.10(a). This boundary, assuming the hole to have a radius a, is related to the real axis in the t-plane, Fig. 7.10(b), by eqn. (7.32) with $K = 2a$, i.e. by

$$t = \frac{1}{2a}\left(z + \frac{a^2}{z}\right). \tag{7.36}$$

Now the field in the z-plane is such that the whole boundary is a flux line and, therefore, the real axis in the z-plane must be a flux line. This condition is satisfied by a uniform field

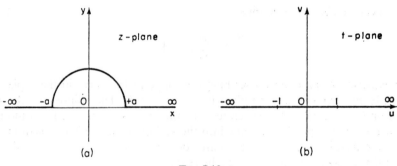

FIG. 7.10

(in the *t*-plane) parallel to the real axis described by

$$w = \psi + j\varphi = kt, \tag{7.37}$$

where k defines the density of the flow. Thus, eliminating t between eqns. (7.36) and (7.37), the solution for the field in the z-plane may be written as

$$w = \frac{k}{2a}\left(z + \frac{a^2}{z}\right). \tag{7.38}$$

Inverting this gives the form from which, by substitution of values of φ and ψ, the field map (there is only one pattern) is calculated.

It is important to note that the field described by eqn. (7.38) is formed by the superposition of two simpler fields, one uniform (z), the other that of a doublet (a^2/z). Thus, as demonstrated in sections 3.3.2 and 4.2.3, the influence of a circular hole, or cylinder, in a uniform applied field, is described by a doublet at the circle centre.

7.2.3. *Permeable cylinder influenced by a line current*

Equation (7.32) is applied now to the analysis of the field of a permeable cylinder of unit radius influenced by a line current (Fig. 7.11), a problem previously treated by images, section 3.3.1, and the direct solution of Laplace's equation, section 4.2.2. The solution of this field is useful because the circular boundary can be further transformed to yield the solution of fields, due to a current, exterior to bodies of polygonal cross-section, see section 8.4.

The boundary of one half of the field is represented in the z-plane as shown in Fig. 7.12(a), and clearly can be transformed from the real axis of the t-plane by eqn. (7.32) in which $a = 1$. Due to the current i at c, Fig. 7.12(a), a magnetic potential difference of $i/2$ exists between the two boundary lines $pqrsc$ and cl. Therefore, for it to be possible to obtain the z-plane field from the t-plane field, the point, c', corresponding to c, must divide the real axis into two equipotential portions, the values of which differ by $i/2$. That is, the field in the (whole) t-plane must be that of a line current i at c', described by

$$w = \varphi + j\psi = \frac{i}{2\pi} \log (t - c'). \tag{7.39}$$

This solution is transformed by eqn. (7.32) (remembering $a = 1$ and taking $K = 1$, so

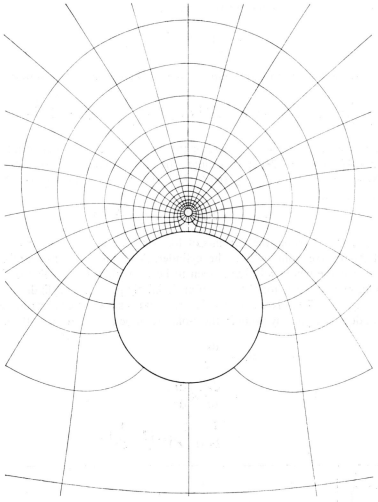

FIG. 7.11

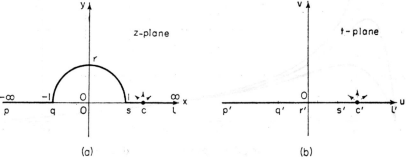

FIG. 7.12

that $s' = 1$ and $q' = -1$) into that for the z-plane,

$$w = \frac{i}{2\pi} \log \left(\frac{z^2 + 1}{z} - c' \right).$$　　　　(7.40)

By factorizing the term in brackets, eqn. (7.40) can be rewritten as follows:

$$w = \frac{i}{2\pi} \left\{ \log \left[z - \frac{1}{2} (c' + \sqrt{c'^2 - 2^2}) \right] + \log \left[z - \frac{1}{2} (c' - \sqrt{c'^2 - 2^2}) \right] - \log z \right\}.　(7.41)$$

This equation is seen to describe the field of three equal currents of magnitude i: those at the points $z = \frac{1}{2}(c' \pm \sqrt{c'^2 - 2^2})$ have similar directions, that at $z = 0$ has the opposite direction. Now, from eqn. (7.33), (with $K = a = 1$), $c = \frac{1}{2}(c' + \sqrt{c'^2 - 2^2})$, and it can easily be seen that

$$\tfrac{1}{4}(c' + \sqrt{c'^2 - 2^2})(c' - \sqrt{c'^2 - 2^2}) = 1,$$

and thus that, in the z-plane, the currents of like sign are at the positions of the actual current and its inverse point within the cylinder. Also the current of unlike sign is at the origin, so that the whole solution is seen to be equivalent to that first derived by the image method, section 3.3.1, for the case of an infinitely permeable cylinder.

The field map (Fig. 7.11) is calculated in the usual manner after inverting eqn. (7.40). Also, the flux density B_z at any point in the z-plane can be evaluated from the equation

$$B_z = \left| \frac{dw}{dz} \right|$$

$$= \left| \frac{dw}{dt} \times \frac{dt}{dz} \right|$$

$$= \left| \frac{i}{2\pi} \frac{1}{(t - c')} \times \left(1 - \frac{1}{z^2} \right) \right|.$$　　　　(7.42)

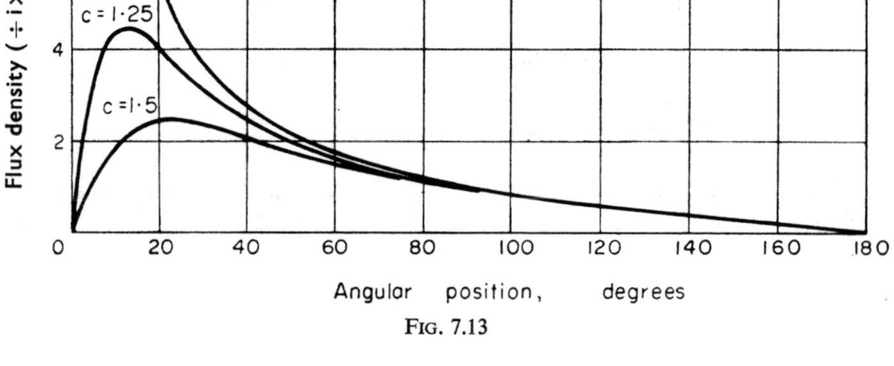

FIG. 7.13

This equation can be evaluated in two ways by expressing it wholly in terms either of z or of t. If it is expressed in terms of t it is also necessary to evaluate the points in the z-plane corresponding with the values of t for which B_z is evaluated, but, even so, this can be the simpler method when t is real. In Fig. 7.13 are shown curves of flux density over the cylinder surface. $c = 1$ gives the case in which the current touches the cylinder.

7.3. Curves expressible parametrically: general series transformations

The equation, transforming any curve, the shape of which is expressible in terms of a parameter, into a straight line, can easily be obtained. This is demonstrated in the first sub-section, and then, after consideration of a simple example, the result is used to develop general series transformations for both open and closed curved boundaries. Employing these series with curve-fitting techniques, it is possible to derive solutions for problems involving a wide range of boundary shapes.

7.3.1. The method

Consider a curve in the z-plane described by the parametric equations

and

$$\left.\begin{array}{l} x = f_1(u) \\ y = f_2(u), \end{array}\right\} \qquad (7.43)$$

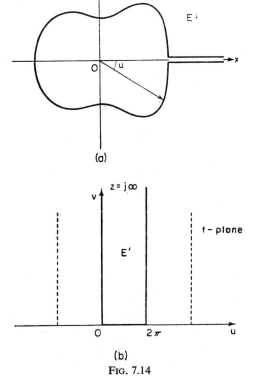

(a)

(b)

FIG. 7.14

where u is the parameter. Further, consider the equation

$$z = f_1(t) + j f_2(t), \tag{7.44}$$

where $t = u + jv$. It is apparent that when $v = 0$, t reduces to u, and eqn. (7.44) reduces to the parametric eqn. (7.43) of the curve in the z-plane. Thus the real axis of the t-plane is transformed into the curve in the z-plane by eqn. (7.44). If both of the functions, f_1 and f_2, are periodic, the curve is a closed one and the whole of it corresponds with the portion of the real axis in the t-plane between 0 and 2π. Also, the region exterior to the curve in the z-plane corresponds with the vertical strip of width 2π in the upper half of the t-plane. Figure 7.14 shows the corresponding regions (marked E and E') and the boundaries in the two planes. If the curve in the z-plane is an open one, not more than one of the functions, f_1 or f_2, is periodic, the whole of the real axis of the t-plane corresponds with the whole of the curve, and the whole of the upper half of the t-plane is transformed into the region exterior to the boundary in the z-plane.

7.3.2. *The field outside a charged, conducting boundary of elliptical shape*

As a simple example of the above method, consider the analysis of the field exterior to a charged conducting boundary with the shape of the ellipse described by

$$\frac{x^2}{a^2} + \frac{y^2}{b^2} = 1.$$

The coordinates of any point on this boundary can be expressed in terms of the parameter u by

$$x = a \cos u \quad \text{and} \quad y = b \sin u,$$

and thus, from eqn. (7.44), the equation transforming the boundary into the real axis of the t-plane, between 0 and 2π, is

$$z = a \cos t + jb \sin t. \tag{7.45}$$

The boundary in the z-plane is equipotential, and so the real axis in the t-plane must also be equipotential. Thus the required field in the t-plane is uniform, described, most simply, by the equation

$$w = t,$$

and the solution in the z-plane, substituting for t in eqn. (7.45), is

$$z = a \cos w + jb \sin w. \tag{7.46}$$

This can be simplified by substituting $a = k \cosh \alpha$ and $b = k \sinh \alpha$, where $k^2 = a^2 - b^2$, to give

$$z = k \cos (w + j\alpha). \tag{7.47}$$

Shifting the origin in the w-plane by $(\pi/2 - j\alpha)$ reduces this to

$$z = k \sin w, \tag{7.48}$$

which is the transformation equation considered earlier (section 6.1.1), generalized by the scale constant k. The field equipotential lines are ellipses described by

$$\frac{x^2}{k^2 \cosh^2 \psi} + \frac{y^2}{k^2 \sinh^2 \psi} = 1, \tag{7.49}$$

the flux lines are the hyperbolae described by

$$\frac{x^2}{k^2 \sin^2\varphi} - \frac{y^2}{k^2 \cos^2\varphi} = 1, \tag{7.50}$$

and both sets are shown in Fig. 7.15. By writing ψ for φ the solution becomes that for a field in which the elliptical boundary is a flux line.

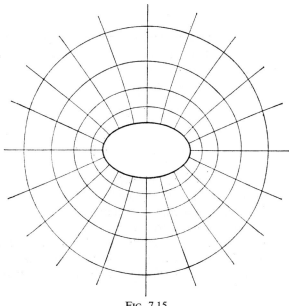

FIG. 7.15

7.3.3. General series transformations

As described in section 7.3.1, it is possible, for any curve expressible in terms of parametric equations, to derive simply a transformation equation. Parametric equations are well known for a rather limited number of curves, but a most important point is that they can be determined approximately in the form of power series for open boundaries or Fourier series for closed ones. One limitation to be noted, however, is that since a vertex in a curve requires a discontinuity in the gradient of the curve, it is not possible to treat curves having vertices by employing a power series or a finite number of terms of a Fourier series. An exception to this is the representation, with closed curves (expressed in Fourier series), of cusps (points where two parts of a curve intersect and have a common tangent).

The constants in the series can be determined by employing standard curve-fitting techniques. These techniques cannot be discussed here, but they are described in most books on numerical analysis. (See for example the book by Lanczos.)

Closed boundaries. The general equation which transforms a closed boundary in the z-plane into the real axis of the t-plane between 0 and 2π, can be written

$$z = a_0 e^{-jt} + b_0 + b_1 e^{jt} + b_2 e^{2jt} + \ldots, \tag{7.51}$$

and it is readily seen that this corresponds with the parametric equations

and
$$\left.\begin{array}{l} x = b_0 + (a_0 + b_1) \cos u + b_2 \cos 2u + \ldots \\ y = (b_1 - a_0) \sin u + b_2 \sin 2u + \ldots \end{array}\right\} \tag{7.52}$$

(which are simply Fourier series for x and y in terms of the angle u subtended at the origin, see section 7.3.1 and Fig. 7.14). For certain simple forms of eqn. (7.51) the closed boundary takes the form of well-known geometrical curves. This can be seen by substituing $v = 0$, when the transformation equation reduces to recognizable parametric equations.

(a) When $b_n = 0$ for all n,

$$z = a_0 e^{-jt}$$
$$= a_0(\cos t - j \sin t), \tag{7.53}$$

and this reduces to the parametric equation of a circle of radius a_0 centred on the origin of the z-plane. [It is easily seen that eqn. (7.53) also transforms the *interior* of the circle into the strip of width 2π in the *lower* half t-plane.]

(b) When $a_0 = (a+b)/2$, $b_1 = (a-b)/2$ and all other values of b are zero,

$$z = \frac{a+b}{2} e^{-jt} + \frac{a-b}{2} e^{jt}$$
$$= a \cos t - jb \sin t, \tag{7.54}$$

and this reduces to the equation of an allipse with semi-axes a and b and centred on the point $z = 0$. It is interesting to note that the Joukowski transformation, eqn. (7.32), can be derived by eliminating e^{jt} between eqns. (7.53) and (7.54).

(c) When terms up to and including b_2 are present, the curves range from the hypocycloid with three cusps to symmetrical aerofoils, and they have been considered in detail by Wrinch.[1] More complicated forms of eqn. (7.51) do not appear to have been studied.

Open boundaries. A general form of transformation for open boundaries without vertices in the z-plane can be written

$$z = jt + b_0 + b_1 t + b_2 t^2 + \dots. \tag{7.55}$$

This is seen to transform a curve having the equation

$$x = b_0 + b_1 y + b_2 y^2 + \dots, \tag{7.56}$$

which is equivalent to the parametric equation

and
$$\left. \begin{array}{l} x = b_0 + b_1 u + b_2 u^2 + \dots \\ y = u. \end{array} \right\} \tag{7.57}$$

As with eqn. (7.51), certain simple forms of eqn. (7.55) can be recognized immediately. For example, when all constants in eqn. (7.56) except b_0 and b_2 are zero, the points (x, y) lie on a parabola, for which the transformation equation is thus

$$z = b_0 + jt + b_2 t^2. \tag{7.58}$$

7.3.4. *Field solutions*

The solutions for fields with equipotential or flux-line boundaries are particularly simple to obtain; since the required field in the t-plane is uniform, these solutions are given, by writing $w = t$ in eqn. (7.44), as

$$z = f_1(w) + jf_2(w). \tag{7.59}$$

[See eqn. (7.46).]

It should also be noted that the fields of line sources can be derived by transformation of the appropriate image solutions: for open curved boundaries the image solution is that for an infinite plane and, for closed boundaries, it is that for three intersecting plane boundaries, two of which are parallel (see section 3.2.3). This technique could be used, for example, as an alternative to that described in section 7.3.3 for the field of a current near an infinitely permeable circular cylinder.

For a general discussion of the transformation of field sources the reader is referred to Chapter 10.

References

1. D. WRINCH, Some problems of two dimensional hydrodynamics, *Phil. Mag.* **48**, 1089 (1924).
2. A. R. BOOTHROYD, E. C. CHERRY, and R. MAKAR, An electrolytic tank for measurement of steady-state response, transient response and allied properties of networks, *Proc. Instn. Elect. Engrs.* **96** II, 176 (1949).
3. P. SILVESTER, Network analog solution of skin and proximity effect problems, *Trans. Inst. Elect. Electr. Engrs.* PAS-86, 241 (1967).

Additional References

ADAMS, E. P., Electrical distributions on circular cylinders, *Proc. Am. Phil. Soc.* **75**, 11 (1935).

ADAMS, E. P., Split cylindrical condenser, *Proc. Am. Phil. Soc.* **76**, 251 (1936).

FRY, T. C., Two problems in potential theory, *Am. Math. Mon.* **39**, 199 (1932).

HODGKINSON, J., A note on a two-dimensional problem in electrostatics, *Q. J. Math.*, Oxford Series **9**, 5 (1938).

JAYAWANT, B. V., Flux distribution in a permeable sheet with a hole near an edge, *Proc. Instn. Elect. Engrs.* **107** C, 238 (1960).

RICHMOND, W. H., Notes on the use of the Schwarz–Christoffel transformation in electrostatics (and hydrodynamics), *Proc. Lond. Math. Soc.* **22**, 483 (1923).

SNOW, C., Electric field of a charged wire and a slotted cylindrical conductor, *Bureau of Standards*, Sci. Papers **542**, 631 (1926).

SWANN, S. A., Effect of rotor eccentricity on the magnetic field in the air gap of a non-salient pole machine, *Proc. Instn. Elect. Engrs.* **110**, 903 (1963).

WRINCH, D., On the electric capacity of certain solids or revolution, *Phil. Mag.* **50**, 60 (1925).

CHAPTER 8

POLYGONAL BOUNDARIES

8.1 Introduction

In practice there are encountered very many fields with boundaries which are, or can be treated as being made up of, straight-line segments; for instance, the field of a plate condenser or microwave strip line or the field in the slotted air gap of a machine. Any polygon has a boundary consisting of straight segments and, for convenience, all boundaries of the above type are referred to as polygonal. Further, it must be noted that, whilst many of those so described are not closed in the finite regions of the plane in which they are placed, Fig. 8.1, they are closed at infinity. Whether the meeting point of two adjacent sides of a polygon occurs at a finite point, or at infinity, it is referred to as a vertex, and the angle there is referred to as a vertex angle.

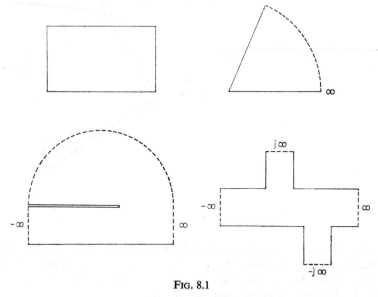

FIG. 8.1

In this chapter are described the routine methods which are available for the derivation of equations transforming a polygonal boundary into either an infinite straight line or the perimeter of a circle. These methods involve the determination, by inspection of the boundaries to be related, of a differential equation which is integrated to yield the transformation equation. This differential equation can be written down for *any shape of polygon*, but the integration of it varies from being very simple to being impossible by analytical means. The various types of integral which occur and the ways in which they are dependent upon the form of the boundary are discussed at the end of this chapter. However, so that attention is concentrated on the function of the transformation and not on the evaluation of an integral, the examples considered in this chapter involve the use of simple functions only. Examples of different boundary types are classified, in the natural way, according to the number of vertex angles which define the boundary, see section 8.2.1.

Also included in this chapter are examples of the use of transformation methods in the calculation of flux densities (e.g. sections 8.2.3 and 8.2.6), capacitance (section 8.2.3), and force on a magnetized surface (section 8.2.8). The calculation of inductance (or capacitance) of a conductor of small, circular cross-section is obtained from the field solution exactly as was demonstrated in section 3.2.4 (but see also reference 14).

8.2. Transformation of the upper half plane into the interior of a polygon

8.2.1. *The transformation*

The transformation equation which connects the real axis in one plane with the boundary of a polygon in another plane, in such a way that the upper half of the first plane transforms into the interior of the polygon, was first given, independently, by Schwarz[1] and Christoffel.[2]

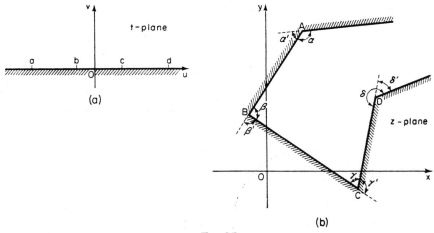

FIG. 8.2

Consider the two planes, shown in Fig. 8.2, in which the regions to be transformed lie on the unshaded sides of the boundary lines and corresponding points are similarly lettered. Then the transformation from the real axis of the t-plane to the polygonal boundary in the z-plane is obtained by integrating the equation

$$\frac{dz}{dt} = S(t-a)^{(\alpha/\pi)-1}(t-b)^{(\beta/\pi)-1}(t-c)^{(\gamma/\pi)-1}(t-d)^{(\delta/\pi)-1}\ldots. \tag{8.1}$$

This is the Schwarz–Christoffel differential equation, and in it S is a constant of scale and rotation, $a, b, c, d, \ldots$, are points on the real axis of the t-plane corresponding to the vertices of the polygon in the z-plane, and $\alpha, \beta, \lambda, \delta, \ldots$, are the interior angles of the polygon at A, B, C, D, ..., respectively. The number of products in eqn. (8.1) is equal to the number of vertices required to define the polygonal boundary.

It is most simply demonstrated that the Schwarz–Christoffel differential equation introduces a succession of vertices in the z-plane boundary, as described, by expressing eqn. (8.1) in terms of the angles α', β', γ', δ', ..., where each primed variable is equal to π minus the respective unprimed one and represents the *change in direction* of the z-plane boundary at the related vertex, Fig. 8.2(b). Equation (8.1) then takes the simpler form[†]

$$\frac{dz}{dt} = S(t-a)^{-(\alpha'/\pi)}(t-b)^{-(\beta'/\pi)}(t-c)^{-(\gamma'/\pi)}(t-d)^{-(\delta'/\pi)}\ldots,$$

and, like eqn. (8.1), describes the length and inclination of a small element dz in terms of the length and inclination of the corresponding element dt. Further, since the direction of all the elements dt constituting the real axis in the t-plane is constant, the argument of this equation defines directly, for continuously varying real values of t between $-\infty$ and ∞, the inclination of elements dz tracing out the boundary of the polygon. Now it is seen that

$$\arg\left(\frac{dz}{dt}\right) = \arg S - \frac{\alpha'}{\pi}\arg(t-a) - \frac{\beta'}{\pi}\arg(t-b) - \frac{\lambda'}{\pi}\arg(t-c) - \frac{\delta'}{\pi}\arg(t-d) - \ldots,$$

and the right-hand-side expression can be simply evaluated for real values of t. For all values of $t < a$, $\arg(dz/dt)$ remains constant because all the terms in brackets are real and negative; and, therefore, the corresponding points on the boundary in the z-plane trace out a straight line. However, as the point t passes through a, $(t-a)$ becomes positive, so that $\arg(t-a)$ changes by $-\pi$ and the argument of dz changes by α'. For values of t between a and b, $\arg(dz/dt)$, and the inclination of elements in the z-plane, remain constant at the new value, and so the corresponding portion of the z-plane boundary is a further straight line, inclined by α' radians to the first section. Again, when the point t passes through b, the term $(t-b)$ is the only one to change sign and, therefore, the slope in the z-plane changes by β' radians on passing through the point corresponding with $t = b$, thereafter remaining constant until the next vertex is reached. Similarly, as t increases further and passes through the points c, d, ..., the direction of the z-plane boundary changes by γ', δ', ..., respectively, until the whole polygon is traced out.

It must be noted that, in this way, all the angles of the polygon are completely defined by those at vertices corresponding with finite points in the t-plane; if the point $t = \pm \infty$ corresponds with a vertex, the angle at that vertex is fixed by the remaining angles since the sum of the interior angles of a polygon is $\pi(N-2)$, where N is the number of vertices. Thus the number of factors in eqn. (8.1) is $(N-1)$ when the point $t = \infty$ corresponds with a vertex, and is N when $t = \infty$ corresponds with a finite point on the boundary of the polygon.

A wise choice of corresponding points can often simplify considerably the form of solution derived from eqn. (8.1). The various possibilities for a simple boundary shape are discussed in section 8.2.3, but some general guidance is given here. For polygons which are infinite, the simplest form of solution is usually obtained by choosing the limits of the real axis to correspond with the limits of a pair of adjacent sides of the polygon which go to infinity. For polygons which are finite and also symmetrical, the point $t = \infty$ is usually chosen to lie where the line of symmetry cuts the boundary. In cases for which this line passes through a vertex and the centre of one side, the point is chosen to be at the vertex.

[†] This form of the equation is not normally used in practice because it is necessary to determine the appropriate signs to be attributed to the angles. This is avoided with eqn. (8.1).

So far consideration of eqn. (8.1) has been restricted to its argument. However, it is clear that the modulus of this equation determines the length of an element dz in terms of the length of the corresponding element dt, and thus, in transforming the real axis of the t-plane into a polygon with given dimensions, it is necessary to choose the constants $a, b, c, \ldots,$ to give the desired vertex positions in the z-plane. The ways in which this can be accomplished are demonstrated in the examples, and the general problem of scale relationship between planes is discussed in section 8.2.5.

8.2.2. *Polygons with two vertices*

Quadrant bounded by the real and imaginary axes. As a first example of the use of the Schwarz–Christoffel differential equation, consider the transformation of the real axis of the t-plane, Fig. 8.3(a), into the polygon consisting of two semi-infinite, straight lines meeting

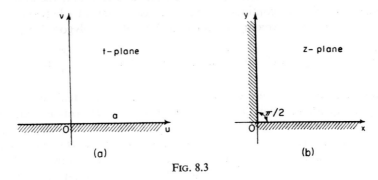

FIG. 8.3

in a right-angle at the point $z = 0$, Fig. 8.3(b), so that the region of the upper half of the t-plane becomes the region of the first quadrant in the z-plane. Let the points $t = -\infty$, $z = j\infty$, and $t = \infty, z = \infty$ correspond, and let the point $t = a$ correspond with the vertex at $z = 0$. The interior angle of the polygon at $z = 0$ is $\pi/2$, and so the Schwarz–Christoffel equation (8.1), gives

$$\frac{\mathrm{d}z}{\mathrm{d}t} = S(t-a)^{-1/2}. \tag{8.2}$$

This, when integrated, yields

$$z = S'(t-a)^{1/2}+k. \tag{8.3}$$

Now the vertex in the z-plane may be made to correspond with any real value of t since this is equivalent merely to shifting the t-plane origin (see section 6.3.1), and so, taking the simplest case, $a = 0$, eqn. (8.3) reduces to

$$z = S't^{1/2}+k.$$

The constants S' and k depend upon the points which, in the two planes, are made to correspond. The problem has been defined in such a way that the points $z = 0$ and $t = a$ correspond and, therefore, substitution of these values in eqn. (8.3) gives $k = 0$. A more general view in connection with the evaluation of k is that by leaving the origin in one plane free to take up any necessary value, the constant k can always be made equal to zero. The scale constant S' can be given any convenient value but, by defining a further pair of corresponding points in the two planes, the scale relationship is fixed. If, for example, the points

$z = 2$ and $t = 4$ are made to correspond, then $S' = 1$, and the transformation equation reduces finally to the form

$$z = t^{1/2},$$

which was used earlier (see section 6.1.2) to describe the field of flow near a perfectly conducting corner. In practice the choice of corresponding points in the two planes is not arbitrary but depends upon the nature of the field to be solved. This will become clear in the examples which follow.

Polygon with two parallel sides. As a second simple example consider the transformation of the real axis of the t-plane, Fig. 8.4(a), into the polygon formed by two infinite, parallel lines, Fig. 8.4(b). For a reason which will become evident, let the polygon be placed in the

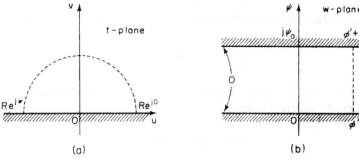

(a)   (b)

FIG. 8.4

w-plane and let the sides be a distance $j\psi_0$ apart. Let the points $t = -\infty$ and $t = \infty$ be chosen to correspond with the points $w = \infty + j\psi_0$ and $w = \infty$ respectively. Further, let the point $t = 0$ correspond with the vertex of the polygon at $w = -\infty$ where there is an interior angle of 0. Then the Schwarz–Christoffel equation gives

$$\frac{dw}{dt} = S(t-0)^{-1}, \qquad (8.4)$$

which, when integrated, is

$$w = S \log t + k. \qquad (8.5)$$

The points $w = -\infty$ and $t = 0$ have been made to correspond, so $k = 0$ and

$$w = S \log t, \qquad (8.6)$$

which will be recognized as the important equation describing, in terms of the complex potential w, the field of a line current or the field between two semi-infinite equipotential lines meeting at the origin of the t-plane (see section 6.1.3). [Note that the points $t = -\infty$ and ∞ cannot be chosen to correspond with the points $w = -\infty$ and $-\infty + j\psi_0$. This requires a vertex at $w = \infty$ to correspond with a finite value of t, which is not possible for eqn. (8.6).]

The constant S may be evaluated, as in the previous example, by the substitution of corresponding values of t and w in the transformation eqn. (8.6). To determine corresponding points between the planes, it is helpful to recall that a vertical flux line in the w-plane transforms into a semi-circular one in the t-plane.† For, considering a flux line with radius R in

---

† The shape of the flux lines is, of course, a consequence, not a cause, of the correspondence of points between the planes.

the t-plane, it is seen that its end points, $R\,\mathrm{e}^{jo}$ and $R\,\mathrm{e}^{j\pi}$, correspond with points $w = \varphi'$ and $w = \varphi' + j\psi_0$ respectively where φ' is the flux function at the radius R. Substituting these points in eqn. (8.6) gives

$$\varphi' = S \log R\,\mathrm{e}^{j\pi} = S \log R$$

and

$$\varphi' + j\psi_0 = S \log R\,\mathrm{e}^{j\pi} = S \log R + Sj\pi.$$

Subtracting these equations leads to

$$j\psi_0 = Sj\pi \quad \text{or} \quad S = \frac{\psi_0}{\pi},$$

and the equation of transformation can be written

$$w = \frac{\psi_0}{\pi} \log t.$$

A method equivalent to, but neater than, the above, can be used to evaluate the constant S. Let $t = R\,\mathrm{e}^{j\theta}$, then, by differentiation,

$$\mathrm{d}t = jR\,\mathrm{e}^{j\theta}\,\mathrm{d}\theta,$$

and eqn. (8.4) can be rewritten

$$\mathrm{d}w = \frac{SjR\,\mathrm{e}^{j\theta}\,\mathrm{d}\theta}{R\,\mathrm{e}^{j\theta}} = Sj\,\mathrm{d}\theta.$$

This equation when integrated yields

$$[w]_{w_1}^{w_2} = jS[\mathrm{d}\theta]_{\theta_1}^{\theta_2}, \tag{8.7}$$

where w_1, $R\,\mathrm{e}^{j\theta_1}$ and w_2, $R\,\mathrm{e}^{j\theta_2}$ are pairs of corresponding points. In particular, if $\theta_1 = 0$ and $\theta_2 = \pi$, then $w_2 - w_1 = j\psi_0$ and eqn. (8.7) yields $S = \psi_0/\pi$.

This evaluation of S is an example of a general method which can be used to relate the t-plane constants with the dimensions of a polygon whenever the polygon has parallel boundaries which meet at infinity. The method is used in many of the examples which follow and is discussed in section 8.2.5.

8.2.3. *Parallel plate condenser: Rogowski electrode*

As an example of the transformation for a polygon with three vertices, the effect of flux fringing on the capacitance of a parallel plate condenser, Fig. 8.5(a), is considered. The plates of the condenser, a distance $2d$ apart, are assumed, for simplicity, to be of negligible thickness; they are also assumed to be charged to potentials of ψ_1 and $-\psi_1$ respectively. The field is symmetrical and the boundaries of one half of it can be represented in the z-plane as shown in Fig. 8.5(b), where the real axis corresponds with the line of symmetry.

The Schwarz–Christoffel equation is used to transform the real axis of the t-plane, Fig. 8.5(c), into the boundary of the z-plane. The upper surface of the plate is represented by the portion of the real axis of the t-plane between $-\infty$ and a, and the lower surface of the plate

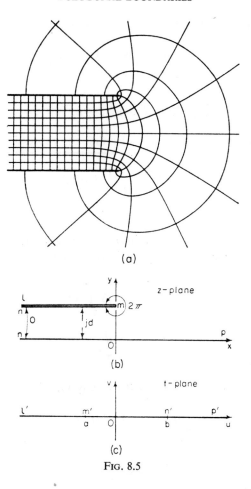

(a)

(b)

(c)

FIG. 8.5

by the portion between a and b. Corresponding points in the two planes are then:

$$t \to -\infty, \qquad z \to -\infty + jd;$$
$$t = a, \qquad z = 0 + jd;$$
$$t = b, \qquad z = -\infty$$
$$t \to +\infty, \qquad z \to +\infty + j0;$$

and, since the interior angles of the polygon are 0 at $z = -\infty$ and 2π at $z = 0 + jd$, eqn. (8.1) gives

$$\frac{dz}{dt} = S(t-a)(t-b)^{-1}. \tag{8.8}$$

The constants S, a, and b must be chosen so that the transformation equation, derived by integration of eqn. (8.8), gives the required shape and size of the z-plane boundary. However, the *shape* of the z-plane configuration is independent of the single dimension d, so that any two of these constants may be given convenient values, the third being determined to give the required value for d. By taking $a = -1$ and $b = 0$, eqn. (8.8) is simplified to the form

$$\frac{dz}{dt} = \frac{S(t+1)}{t},$$
(8.9)

which gives, when integrated,

$$z = S(t + \log t) + k.$$
(8.10)

To evaluate the constants S and k, consider first the expression of eqn. (8.9) in polar coordinates

$$dz = S\frac{(r\,e^{j\theta}+1)}{r\,e^{j\theta}} jr\,e^{j\theta}\,d\theta$$

$$= jS(r\,e^{j\theta}+1)\,d\theta.$$

Now movement round a small circle, of radius r, and centre at the point $t = 0$, corresponds, in the z-plane, with movement from the real axis to the line $z = jd$ and so, as r tends to zero, the last equation gives

$$\int_{-\infty}^{-\infty+jd} dz = jS \int_{0}^{\pi} d\theta$$
(8.11)

or

$$S = \frac{d}{\pi}.$$

Since the points $t = -1$ and $z = jd$ have been chosen to correspond, substitution of these values in eqn. (8.10) yields

$$jd = S(-1+j\pi)+k.$$

Putting $S = d/\pi$ in this equation gives

$$k = \frac{d}{\pi}$$

and

$$z = \frac{d}{\pi}(1+t+\log t)$$
(8.12)

as the equation of transformation.

The boundaries of the z-plane have a potential difference ψ_1 of half that between the capacitor plates. Hence the required field in the t-plane is expressed, from eqn. (6.23), by

$$w = \frac{\psi_1}{\pi}\log t,$$
(8.13)

and the solution for the z-plane field is, eliminating t between eqns. (8.12) and (8.13),

$$z = \frac{d}{\pi}\left(1+e^{w\pi/\psi_1}+\frac{w\pi}{\psi_1}\right).$$
(8.14)

This gives as corresponding values: $z = -\infty$, $w = -\infty$ and $z = jd$, $w = j\psi_1$.

Because the constants a and b have been chosen as -1 and 0 rather than, for instance, 0 and 1, the potential division in the t-plane is at the point $t = 0$ and the solution for the field in the t-plane has its simplest form, eqn. (8.13). Also, as pointed out in section (8.2.2), the constant k may be given the value zero and the transformation eqn. (8.12) is then

$$z = \frac{d}{\pi}(t+\log t), \tag{8.15}$$

the origin in the z-plane being displaced through a distance d/π to the right of its first position.

Capacitance. The capacitance of a parallel plate condenser is often calculated on the assumpton that the flux density everywhere between the plates has the value that wonld exist if the plates were infinite. However, the value so derived is low, because the charge density on the inner faces increases towards the edges of the plates, and also because the charge on the outer faces of the plate is ignored. At any point on the surface of the plates the charge density, ϱ is equal to the flux density there (see sections 2.1.1 and 2.5); that is,

$$\varrho = \left(\frac{\partial \psi}{\partial x}\right) \varepsilon_0$$

$$= \left|\frac{\partial w}{\partial z}\right| \varepsilon_0, \tag{8.16}$$

where ε is the relative permittivity of the medium surrounding the plates. From eqn. (8.14) by differentiation

$$\left|\frac{dw}{dz}\right| = \frac{\psi_1}{d} \frac{1}{|e^{w\pi/\psi_1}+1|},$$

and, hence, substitution in eqn. (8.16) gives the charge density at any point as

$$\varrho = \frac{\varrho_0}{|e^{w\pi/\psi_1}+1|}, \tag{8.17}$$

where $\varrho_0 = \psi_1 \varepsilon \varepsilon_0/d$ is the density of charge calculated on the assumption that the plates are infinite. Equation (8.17) shows that the charge density increases towards, and becomes infinite at, the end of the plate ($z = jd$, $t = -1$, $w = j\psi_1$).

The additional charge on the inner surface of the plates, due to the fringing, is the difference between the charge actually present and that calculated assuming the field uniform everywhere. It is equal to

$$\int_{-\infty+jd}^{jd} (\varrho-\varrho_0)\, dz,$$

which can be expressed in terms of w as

$$\int_{-\infty}^{j\psi_1} \varrho_0 \left(\frac{1}{1+e^{w\pi/\psi_1}} - 1\right) \frac{d}{\psi_1}(e^{w\pi/\psi_1}+1)\, dw,$$

and simplified and integrated, to yield

$$\frac{\varrho_0 d}{\pi}.$$

Thus this additional charge is equivalent to a lengthening of the uniform field by the amount d/π.

The charge on the outer surface of the plates is

$$\int_{jd}^{-\infty} \varrho \, dz = \frac{\varrho_0 d}{\psi_1} \int_{j\psi_1}^{-\infty} dw.$$

$$= \infty.$$

However, for a condenser of finite width, it is possible to estimate the additional charge fairly closely by assuming the charge distribution near both edges to be identical with that near the edge of the infinite plate. This is a reasonable assumption since the density of charge decreases very rapidly away from an edge—at a distance from the edge equal to $1\cdot 5d$ the charge density is less than $\varrho_0/10$—and the resulting value for the charge on the outer surface is calculable as

$$2 \int_{jd}^{-l+jd} \varrho \, dz,$$

where $2l$ is the width of the plate.

Rogowski electrode. The solution for the field of the boundary shown in Fig. 8.5(b) was used in an interesting and valuable way by Rogowski[3] to establish electrode shapes suitable for the measurement of the breakdown strength of gases and liquids. For these measurements it is important that breakdown always occurs in the uniform portion of the field. This may be ensured by choosing an electrode with the shape of an equipotential line along which the gradient is nowhere greater than ψ_1/d. For other uses of equipotential lines in representing boundary shapes, see sections 4.3.3 and 10.3.

8.2.4. *The choice of corresponding points*

For a given polygon, the choice of t-plane constants can be made in many ways, all of which lead, with varying degrees of difficulty, to the solution. To demonstrate the main considerations involved in making a choice, the various possibilities are examined for the following simple problem.

Current between two infinite, parallel permeable surfaces. A general form of this problem is treated earlier in the book by the method of images (see section 3.2.2), and it is sufficient for the present purpose to consider the particular case with the current and midway between the surfaces. Since the field is symmetrical about the line through the current normal to the parallel surfaces, it is necessary to consider only the field to one side of this line. (Equally, since this field region is also symmetrical, it would be sufficient to consider one half of it.) The boundary of the field is shown in Fig. 8.6, where the current acts at the point $z = 0$, the finite side of the polygon corresponds to the line of symmetry normal to the surfaces, and the semi-infinite lines, a distance $2d$ apart, correspond with the permeable surfaces. There are three vertices, at $l(z = j\infty)$, $m(z = -d)$, and $n(z = d)$, and the angles are respectively $0, \pi/2$, and $\pi/2$.

It is possible to choose the constants on the real axis of the t-plane in three reasonable ways and, because the polygon is defined by a single dimension, $2d$, two (or more when there is symmetry) of each set of constants can be given convenient values. The choices of pairs of corresponding points are as follows:

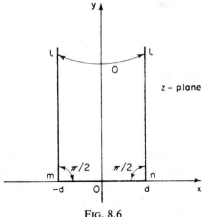

Fig. 8.6

Choice 1. The corresponding points are:

$$z = -d+j\infty, \qquad t = -\infty;$$
$$z = d+j\infty, \qquad t = \infty;$$
$$z = -d, \qquad t = -1;$$
$$z = d, \qquad t = 1.$$

Equation (8.1) gives

$$\frac{dz}{dt} = \frac{S}{\sqrt{t^2-1}},$$

which, when integrated, yields the transformation equation

$$z = S \cosh^{-1} t + k. \qquad (8.18)$$

Choice 2. The corresponding points are:

$$z = -d, \qquad t = \pm\infty;$$
$$z = \pm d+j\infty, \qquad z = 0;$$
$$z = d, \qquad t = 1.$$

Equation (8.1) gives

$$\frac{dz}{dt} = \frac{S}{t\sqrt{t-1}},$$

which yields the transformation equation

$$z = 2S \tan^{-1}\sqrt{t-1} + k. \qquad (8.19)$$

Choice 3. The corresponding points are:

$$z = 0, \qquad t = \pm\infty;$$
$$z = \pm d+j\infty, \qquad t = 0;$$
$$z = -d, \qquad t = -1;$$
$$z = d, \qquad t = +1.$$

Equations (8.1) gives

$$\frac{dz}{dt} = \frac{S}{t\sqrt{t^2-1}}.$$

which yields the transformation equation

$$z = S \log \frac{\sqrt{t^2-1}-1}{t} + k. \tag{8.20}$$

Consider now the field in the z-plane and the forms of t-plane fields appropriate to each of the above transformation equations. In the z-plane, the current I sets up a potential difference of $I/2$ between the two halves of the boundary meeting at the point, $z = 0$, and so each of the t-planes must have the same potential difference, $I/2$, between the sections of the real axis corresponding with the two halves of the z-plane boundary. In the first and third case above, this requires potential divisions at $t = 0$ and $t = \pm\infty$, and the field is given by

$$w = \frac{I}{2\pi} \log t \tag{8.21}$$

(with opposite signs in the two cases). In the second case above, potential divisions are required at $t = 0$ and at $t = t_c$ which corresponds with $z = 0$: the field is that due to currents, of equal magnitudes but opposite signs, at the points 0 and t_c, and the solution is (see section 6.1.3)

$$w = \frac{I}{2\pi} [\log t - \log(t-t_c)]$$

$$= \frac{I}{2\pi} \log \left(\frac{t}{t-t_c}\right). \tag{8.22}$$

A comparison of the three cases shows, firstly that of the three transformation equations, eqns. (8.18) and (8.19) are (marginally) simpler than (8.20) and, secondly that, of the two field equations, eqn. (8.21) is simpler. Therefore the solution of this problem is best obtained using the first choice of constants.

Substitution for the pairs of corresponding points, $z = -d$, $t = -1$, and $z = d$, $t = 1$ in eqn. (8.18) gives $k = d$ and $S = 2jd/\pi$. Then, eliminating t between eqns. (8.18) and (8.22), the solution for the field in the z-plane is

$$\left[w = \frac{I}{2\pi} \log \left[\cosh \frac{(z-d)\pi}{2jd}\right]\right]. \tag{8.23}$$

[From a consideration of the image solution to this problem it is apparent that eqn. (8.23) describes also the field due to an infinite array of equally spaced line currents.]

In the above example there is little variation in difficulty associated with the possible choices of constants, but, for more complicated problems, a wise choice can be very important. As has been seen, the two features to be considered are the difficulty of integrating and using the transformation equation, and the determination and complexity of the equation of the t-plane field. In practice, the latter is usually the more important (see section 5.2.8) and first consideration should be given to it.

8.2.5. *Scale relationship between planes*

General considerations. Use of the Schwarz–Christoffel equation automatically constructs a polygon with the required *angles*. The *dimensions* of a polygon are obtained by a suitable choice of the constants in the transformation equation. For a consideration of scale, the positions of the origins are of no concern, and so the equation can be written

$$z = Sf(t, a, b, c \ldots), \tag{8.24}$$

where f is a function not only of the complex variable t but also of the real constants a, b, $c \ldots$.

It is shown in the previous section that when only two constants a and b (in addition to the scale constant S) occur in eqn. (8.24), both can be given arbitrary values, and it is generally true that, for any number of constants a, b, c, $\ldots$, two of them can be given convenient values. This is so because, by fixing the values of two constants, a distance between the corresponding points in the z-plane is defined, and this can be given the correct value by a suitable choice of S. The remaining constants in the transformation equation are then defined by the other dimensions of the polygon. It should be noted that the number of constants to be determined is equal to the number of *ratios* of the dimensions which define the proportions of the polygon: the polygons of the above examples are defined by no more than a single *dimension* and so no constants, apart from S, require evaluation; the polygon of the succeeding section is defined by two dimensions, or one ratio, and requires the evaluation of one constant in addition to S. The two "free" constants are chosen to simplify the form of the transformation eqn. (8.24) or the solution of the field in the t-plane, and the values most commonly used are 0 and ± 1.

Evaluation of constants. Two methods are available for the evaluation of the constants which, in eqns. (8.1) and (8.24), define the proportions of the polygon. The first method merely involves substitution in the transformation equation, (8.24), for pairs of corresponding values of z and t; each substitution yields one equation connecting the constants with the polygon dimensions. In many cases the equation gives the value of a constant directly [see eqn. (8.18)], but in others a graphical method may be necessary [see eqn. (8.43)] and, in the most difficult, it is necessary to invoke an iterative numerical scheme (see section 10.5).

The second method, *the method of residues*,[†] can be applied when the polygon has parallel sides meeting at infinity and, using it, one relationship can be obtained for each such vertex. (Forms of it have already been used in the example of the field of a line current and the example of the parallel plate condenser.) At these vertices a finite change in z occurs due to an infinitely small change in t through the point corresponding with the vertex (there are poles of dz/dt at these points in the t-plane) and the related changes can be equated by integrating eqn. (8.1) between the appropriate limits.

Consider a vertex of angle 0 at the point corresponding to $t = n$, giving a term $(t-n)^{-1}$ in eqn. (8.1), and let

$$t-n = Re^{j\theta}.$$

Differentiating this equation gives

$$dt = jRe^{j\theta}\, d\theta,$$

and, substituting for t and dt, eqn. (8.1) becomes

$$dz = Sf(Re^{j\theta}, n, a, b, c, \ldots)\, d\theta. \tag{8.25}$$

[†] See footnote, p. 164.

Now if R is infinitely small, a change in θ from 0 to 2π causes t to pass through the value n and this change corresponds with one in z equal to the distance, D, between the parallel lines. Further, when $R \to 0$, $t \to n$ and eqn. (8.25) takes a simple form for, in it, θ only occurs due to the terms dt and $(t-n)$ and the factors other than $(t-n)$ remain constant as θ changes from 0 to π. The form is

$$dz = \frac{S}{R e^{j\theta}} (n-a)^{(\alpha/\pi)-1} (n-b)^{(\beta/\pi)-1} \ldots jR e^{j\theta}\, d\theta,$$

and this, when simplified and integrated between $\theta = 0$ and $\theta = \pi$, gives

$$D = j\pi S(n-a)^{(\alpha/\pi)-1} (n-b)^{(\beta/\pi)-1} \ldots, \tag{8.26}$$

in which the factor corresponding to $(t-n)$ does not appear.

When the points $t = \pm\infty$ correspond with the meeting point at infinity of two parallel sides of the polygon containing the angle 0, eqn. (8.25) can again be used, this time to relate the distance between the parallel lines directly with the scale constant. Then the term $(t-n)$ does not appear in eqn. (8.24) which, with the substitution $t = R e^{j\theta}$, becomes

$$dz = S(R e^{j\theta} - a)^{(\alpha/\pi)-1} (R e^{j\theta} - b)^{(\beta/\pi)-1} \ldots jR e^{j\theta}\, d\theta.$$

As $R \to \infty$ the constants a, b, $\ldots$, become negligible, and the equation may be rewritten

$$dz = jSR e^{j\theta [\Sigma \alpha/\pi - (N-1)+1]}\, d\theta$$

since there are $(N-1)$ terms in the equation, where N is the number of vertices. However, the sum of the interior angles of the polygon is $\pi(N-2)$ and, since the interior angle at vertex n is zero, this is equal to $\Sigma\alpha/\pi$. Thus,

$$dz = jS\, d\theta,$$

which, when integrated, gives

$$S = \frac{D}{j\pi}, \tag{8.27}$$

where D is the distance between the parallel sides. The fact that when $R \to \infty$ eqn. (8.25) reduces to a form independent of the constants a, b, c, $\ldots$, is to be expected since a circle of infinite radius corresponds, regardless of its centre, with the same path in the z-plane. Both eqns. (8.26) and (8.27), which relate the constants of the transformation equation with the dimensions of the polygon, are of considerable importance and they are used frequently.

The method just discussed of writing eqn. (8.25) in polar form and determining the value of the integral as $R \to 0$ or ∞ is equivalent to the evaluation of the integral by the method of residues.[†] The discussion so far has been restricted to an expression of the distance between parallel lines meeting at a vertex of angle of *zero*. Whilst this case is by far the

---

[†] This is discussed fully in any book dealing with the theory of functions of a complex variable but the definitions of residue, pole and zero are given here. A pole of the function $f(z)$ (with finite principal part in the Laurent expansion) is a point at which $f(z)$ is infinite; if $f(z)$ contains a term $1/(z-a)^n$ the pole at $z = a$ is said to be of order n. The residue of $f(z)$ at the pole a is the coefficient of the term $1/(z-a)$ in the Laurent expansion of $f(z)$ about $z = a$. A zero of the function $f(z)$ is a point at which $f(z)$ is zero; if $f(z)$ contains a term $(z-b)^n$ it is said to have a zero of order n at $z = b$.

most important one, it must be noted that parallel lines can also meet at an angle of $-\pi$ (this is explained in section 8.2.7) when eqn. (8.1) contains a term $(t-n)^{-2}$. This gives a pole of order 2 at n, so that the above method of integration fails.

The distance between parallel lines is, however, always given by

$$D = j\pi \times \text{residue at pole,} \tag{8.28}$$

and for a pole of order 2 the residue has to be found as the coefficient of $(t-n)^{-1}$ when the function $Sf(t, a, b, c, \ldots)$ of eqn. (8.1) is expressed as a series, or as

$$\operatorname*{Lt}_{t \to n} \frac{\mathrm{d}}{\mathrm{d}t} [(t-n)^2 \, Sf(t, a, b, c, \ldots)]. \tag{8.29}$$

8.2.6. *The field of a current in a slot*

Very many practical problems have been investigated using the boundary shape shown in Fig. 8.7(b): Jeans, p. 277, for example, described the calculation of the field at the corner of a Leyden jar; Carter[4] examined the field of a rectangular salient pole; and Stein[5] investigated the flux distribution at the corner of a transformer core. In this section, the same boundary shape is used in the calculation of flux density and of permeance factors for the field of a current in a slot of an electrical machine. The same solutions apply, though with less satisfactory approximation, to the field of a salient pole.[4] (The solution, by conformal transformation, for the field of a salient pole with a rectangular tip is possible but difficult.[6])

Since the air gap in most electrical machines is small compared with the width of an open slot, it is sufficient in the analysis to consider a single slot, of width $2p$, separated by an air gap of length g from a plane surface, see Fig. 8.7(a). Also, so far as the field in the air gap is concerned, the slot can be treated as infinitely deep (see section 9.4) with the current at the "bottom" of it. For a current $2I$ the potentials of the adjacent teeth are I and $-I$ with respect to the plane surface at zero potential, and the potential on the line of symmetry down the slot centre is also 0. Thus, one half of Fig. 8.7(a) can be represented, in the z-plane, as shown in Fig. 8.7(b) where the boundary mnq has a potential differing by I from that of the boundary mlq.

The polygon has four vertices at l, m, n, and q and, by choosing the limits of the real axis of the t-plane, Fig. 8.7(c), to correspond with q, it is defined by the angles $3\pi/2$, 0 and $\pi/2$ at l, m, and n respectively. Of the three finite points in the t-plane, l, m, and n, corresponding with these three vertices, two may be given convenient values but the third must be determined to give the required ratio of (g/p). Let the pairs of corresponding points be:

$$\text{at } l, \quad z = p+jg \quad \text{and} \quad t = -1;$$
$$\text{at } m, \quad z = j\infty \quad \text{and} \quad t = 0;$$
$$\text{and} \qquad \text{at } n, \quad z = 0 \quad \text{and} \quad t = a.$$

Then the Schwarz–Christoffel equation giving the transformation between the planes is

$$\frac{\mathrm{d}z}{\mathrm{d}t} = S(t+1)^{1/2} t^{-1}(t-a)^{-1/2}, \tag{8.30}$$

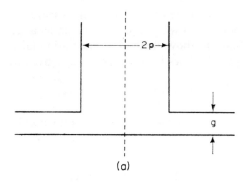

(a)

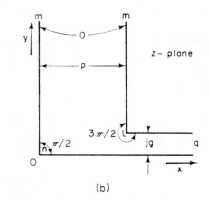

(b)

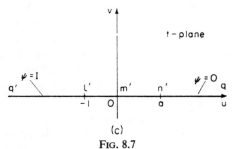

(c)

Fig. 8.7

and, making the substitution

$$u^2 = \frac{t-a}{t+1},$$

this can be simply integrated to yield

$$z = 2S\left[\frac{1}{\sqrt{a}}\tan^{-1}\frac{u}{\sqrt{a}} + \frac{1}{2}\log\left(\frac{1+u}{1-u}\right)\right]+k.$$

By choosing the origin in the z-plane to correspond with the point $t = a$ the constant k in the last equation is made zero, and the transformation equation reduces to

$$z = 2S\left[\frac{1}{\sqrt{a}}\tan^{-1}\frac{u}{\sqrt{a}} + \frac{1}{2}\log\left(\frac{1+u}{1-u}\right)\right]. \qquad (8.31)$$

The constants S and a are best evaluated by the method of residues. At the vertex q the parallel sides are a distance jg apart and the point in the t-plane corresponding to this vertex is $t = \infty$. Therefore, eqn. (8.27) gives

$$S = \frac{g}{\pi}. \tag{8.32}$$

At the vertex m the parallel lines are a distance p apart and the point in the t-plane corresponding to m is $t = 0$. Therefore eqn. (8.26) gives

$$p = \frac{j\pi S}{\sqrt{-a}},$$

and, substituting for S from eqn. (8.32), gives

$$a = \left(\frac{g}{p}\right)^2. \tag{8.33}$$

Then, putting the above values for S and a in eqn. (8.31), the complete transformation equation becomes

$$z = \frac{2g}{\pi}\left[\frac{p}{g}\tan^{-1}\frac{pu}{g} + \frac{1}{2}\log\left(\frac{1+u}{1-u}\right)\right], \tag{8.34}$$

where

$$u^2 = \frac{t-(g/p)^2}{t+1}. \tag{8.35}$$

To obtain the required field in the z-plane it is necessary to have in the t-plane a potential I along the part of the real axis between $-\infty$ and 0, and a potential 0 along the part between 0 and ∞. This gives a field in the t-plane described by

$$w = \frac{I}{\pi}\log t, \tag{8.36}$$

and elimination of t and u between eqns. (8.34)–(8.36) gives the solution for the field in the z-plane. The field map shown in Fig. 8.8 has been plotted from this solution, taking a value of 5 for the ratio p/g.

Flux density. The flux density at any point in the field is given by $|dw/dz|$. Equation (8.36) gives

$$\frac{dw}{dt} = \frac{I}{\pi}\frac{1}{t},$$

and this may be combined with eqns. (8.30) and (8.32) to give

$$\frac{dw}{dz} = \frac{I}{g}\sqrt{\frac{t-a}{t+1}}. \tag{8.37}$$

It is not possible to evaluate dw/dz directly in terms of z. Instead, a curve of flux density variation with z is obtained by (a) substitution of values of t in eqn. (8.37) to give flux density and (b) in eqns. (8.34) and (8.35) to give the corresponding values of z. Taking a range of values of t between a and ∞ yields the curve for the distribution of flux density along the plane

12*

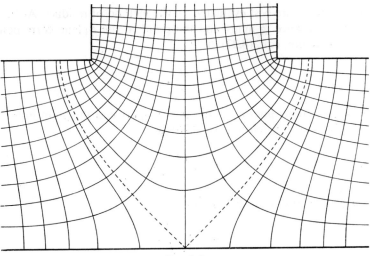

Fig. 8.8

surface, shown in Fig. 8.9. This curve is of importance for the determination of the shape of the flux wave inducing losses in the surface of poles.

Leakage flux. The main and leakage fluxes may be considered to be separated by the fictitious flux line (shown dotted in Fig. 8.8) which leaves the tooth and just touches the plane surface at $z = 0$ in Fig. 8.7(a). In design calculations of their magnitudes it is convenient, as it is in many other problems, to make use of permeance coefficients. These are used to express a quantity of flux in a non-uniform part of the field in terms of an equivalent length of a uniform field; thus, in Fig. 8.7(b) the total main flux leaving the tooth can be calculated by assuming the uniform distribution of field to exist everywhere between q and l, and by representing the quantity of flux which fringes from the tooth side (Fig. 8.8) as a certain length of the uniform field.

The value of the "fringe flux" is taken as the difference between the integral of the flux density along the plane surface from n to q and the flux which would pass between q and l,

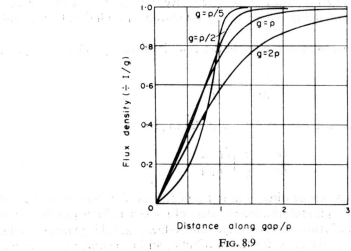

Fig. 8.9

if the density everywhere there were uniform. Both of these quantities of flux are infinite, but their difference is finite and it can be found numerically by calculating the quantities as being bounded, not at q, but by a line at some convenient point sufficiently far down the gap for the density there to be uniform. In some problems, the difference in the flux densities can be expressed in such a form that the integration yields a simple analytical result. An example of this kind occurred in connection with the capacitor fringe field, section 8.2.3., and another is given in reference 7. This is not possible in the present example, but an approximate analytical result is obtained by assuming everywhere under the tooth, a uniform field which is bounded by a straight line—from the tooth corner, normal to the plane surface—and by calculating the fringe flux as that crossing the plane surface between $x = 0$ and $x = p$. The result in the form given by Carter, expresses the fringe flux as a constant, λ, times the air-gap length, where

$$\lambda = 0{\cdot}72 \log \left[\frac{1}{4} \left(1 - \frac{p^2}{g^2} \right) + \frac{1}{90} \left(\frac{p}{g} \tan^{-1} \frac{g}{p} \right) \right],$$

and the logarithm is a common one and the angle is expressed in degrees.

8.2.7. *Negative vertex angles*

When two adjacent sides of a polygon diverge, they meet at infinity at a vertex having a negative angle. (The reader may readily confirm this by considering the sum of all the interior angles of the polygon which must add up to $\pi(N-2)$, N being the number of vertices.) Angles of this type give, in eqn. (8.1), terms which have a power less than minus one and, consequently, tend to complicate the integration. Therefore, if possible, they should be avoided by choosing the point $t = \infty$ to correspond with the vertex having the negative angle (see the following section).

Opposite parallel plates. The polygonal boundary, shown in the z-plane, Fig. 8.10(a), consists of two thin, semi-infinite, parallel plates and it has two negative angles, one of which must be accounted for explicitly in forming the transformation equation. Apart from its being a good example for the discussion of negative angles and the evaluation of the residue at a second order pole [see eqn. (8.29)], particular forms of this boundary can be used to derive solutions which are frequently transformed to obtain solutions for more complex problems.

The polygon has four vertices l, m, n, and h. Those at l and n have interior angles of 2π, whilst those at m and h have angles of $-\pi$. The angles at m and h can be visualized by considering the movement of two sides at a vertex as the angle there is *decreased* through the value 0 until the lines become parallel again.

To transform the real axis of the t-plane, Fig. 8.10(b), into the polygon, let the point $t = \infty$ correspond with the vertex m, and let the remaining pairs of corresponding points be

$$\text{at } l, \quad z = q \quad \text{and} \quad t = a;$$
$$\text{at } h, \quad z = \infty \quad \text{and} \quad t = 0;$$
$$\text{and} \qquad \text{at } n, \quad z = jp \quad \text{and} \quad t = -1.$$

Then the Schwarz–Christoffel equation is

$$\frac{dz}{dt} = S(t+1)t^{-2}(t-a), \tag{8.38}$$

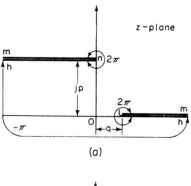

(a)

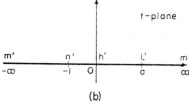

(b)

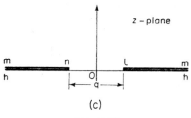

(c)

FIG. 8.10

and this integrates simply as

$$z = S\left(t+(1-a)\log t+\frac{a}{t}\right)+k. \tag{8.39}$$

The term t^{-2} in eqn. (8.38) means that there is a second order pole of dz/dt at the point $t = 0$, and so eqn. (8.28) must be used to express the distance, jp, between the parallel lines, with the constants in eqn. (8.38). The residue of dz/dt at $t = 0$ is the coefficient of $1/t$ in the expansion of the right-hand side of equation (8.38) and it is seen to be $S(1-a)$. [This result is obtained equally well using eqn. (8.29)]. Hence, eqn. (8.28) gives

$$jp = j\pi S(1-a). \tag{8.40}$$

Substitution for pairs of corresponding points in eqn. (8.39) yields two additional equations which, together with equation (8.40), are used to evaluate the constants S, a and k. Substitution for the vertex l gives

$$q = S\{(1+a)+(1-a)\log a\}+k, \tag{8.41}$$

and for the vertex n

$$jp = S\{-(1+a)+(1-a)j\pi\}+k. \tag{8.42}$$

Eliminating S and k between eqns. (8.40)–(8.42) yields the equation

$$\pi\left(\frac{q}{p}\right) = \frac{2(1+a)}{(1-a)} + \log a, \tag{8.43}$$

which, however, must be solved graphically to obtain the value of a for a given value of (q/p). When a is known, eqns. (8.41) and (8.42) are used to evaluate S and k.

Both plates on the real axis. When $p = 0$ the two plates are in line with each other and they can be represented on the real axis of the z-plane, symmetrically placed with respect to the origin, as shown in Fig. 8.10(c). Because of the symmetry, a has the value 1 and eqn. (8.39) becomes

$$z = S\left(t + \frac{1}{t}\right) + k.$$

The values of the constants S and k are given directly by substitution for the pairs of corresponding points $z = -q/2$, $t = -1$, and $z = q/2$, $t = 1$ and are

$$S = q/4 \quad \text{and} \quad k = 0.$$

Hence, the transformation equation can be written

$$z = \frac{q}{4}\left(t + \frac{1}{t}\right). \tag{8.44}$$

The solution for the field due to a potential difference ψ_1 maintained between the two plates, is often used (see, for example, section 8.2.9) and is obtained by combining eqn. (8.44) with the equation for the field in the t-plane,

$$w = \frac{\psi_1}{\pi}\log t.$$

It takes the most convenient form when $q = 2$ and, with this substitution, is

$$w = \frac{\psi_1}{\pi}\log\left(z + \sqrt{z^2 - 1}\right)$$

$$= \frac{\psi_1}{\pi}\cosh^{-1}z. \tag{8.45}$$

An alternative derivation of eqn. (8.45) is to use eqn. (8.18) and to transform directly into the w-plane.

8.2.8. *The forces between the armature and magnet of a contactor*

A number of methods of determining the forces experienced by magnetized boundaries are developed in section 2.6 and two of them are discussed here in their application to a typical problem. Consider Fig. 8.11(a) which represents the armature and magnet of a contactor or the stator and rotor of a machine. A magnetic potential difference established between the two parts gives rise to two components of force on each: firstly, there is an attractive force and, secondly, there is a force tending to align the parts symmetrically with respect to one another. An exact treatment of the boundary shape requires the use of mathematics not discussed until the next section, but completely satisfactory results for many purposes can be achieved by consideration of the portion of the boundary shown in Fig. 8.11(b). This is possible because the field over most of the length of the narrow air gap, in both figures, is uniform and uninfluenced by effects at the ends of the elements, so that the boundary of Fig. 8.11(a) can be synthesized from two boundaries (each with an appropriate length of the parallel gap) of the type shown in Fig. 8.11(b).

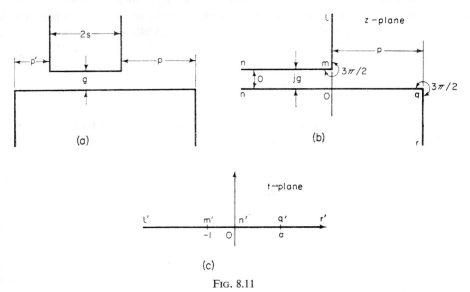

Fig. 8.11

Before the forces can be determined, the solution for the field must be obtained. Let the boundary of Fig. 8.11(b) be in the z-plane and let it be transformed into the real axis of the t-plane, Fig. 8.11(c). Further, let the pairs of corresponding points in the two planes be:

$$z = j\infty, \qquad t = -\infty;$$
$$z = jg, \qquad t = -1;$$
$$z = -\infty, \qquad t = 0;$$
$$z = p, \qquad t = a;$$
$$z = p - j\infty, \qquad t = \infty.$$

Then, since the vertex angles at the points jg, ∞, and p are, respectively, $3\pi/2$, 0, and $3\pi/2$, the Schwarz–Christoffel equation connecting the planes is

$$\frac{dz}{dt} = S \frac{\sqrt{(t+1)(t-a)}}{t}. \tag{8.46}$$

This equation, after multiplication of the numerator and the denominator of the right-hand side by $\sqrt{t-a}$, can be integrated to give

$$z = S\left[\frac{R(a+1)}{(R^2-1)} + (1-a)\tanh^{-1}R + j\sqrt{a}\log\frac{(R\sqrt{a}-j)}{(R\sqrt{a}+j)}\right] + k, \tag{8.47}$$

where

$$R = \sqrt{\frac{t+1}{t-a}}. \tag{8.48}$$

Application of eqn. (8.26) to the point $t = 0$ yields

$$S = -\frac{jg}{\pi \sqrt{a}}; \tag{8.49}$$

and in eqn. (8.47), using this value of S, substitution for the pairs of corresponding points $z = jg$, $t = -1$ and $z = p$, $t = a$ gives, respectively,

$$k = 0$$

and

$$a = 1 + \frac{2p}{g} \pm \sqrt{\left(1 + \frac{2p}{g}\right)^2 - 1}. \tag{8.50}$$

Finally, if the magnetic potential difference between the two elements in the z-plane is ψ, the solution for the field in the t-plane is seen to be

$$w = \frac{\psi}{\pi} \log t, \tag{8.51}$$

and elimination of t between this equation and eqns. (8.47) and (8.48) gives the solution for the field in the z-plane.

Consider now, by use of the method described in section 2.6.4, the determination of the "alignment" force F on the elements. Since the boundaries are infinitely permeable, this force acts wholly on the vertical faces and it may be found by using eqn. (2.116). This states, for the present problem, that F is given by the integral, over the vertical surfaces, of the square of the flux density; that is, for Fig. 8.11(b),

$$F = \tfrac{1}{2}\mu_0 \int_{jg}^{j\infty} B_z^2 \, dz.$$

This equation may be rewritten as

$$F = \frac{1}{2} \mu_0 \int_{jg}^{j\infty} \left(\left|\frac{dw}{dz}\right|\right)^2 dz$$

$$= \frac{1}{2} \mu_0 \int_{jg}^{j\infty} \left(\left|\frac{dw}{dt}\right|\right)^2 \left(\left|\frac{dt}{dz}\right|\right)^2 dz,$$

or, most conveniently, with a change of the variable of integration, as

$$F = \frac{1}{2} \mu_0 \int_{-1}^{-\infty} \left(\left|\frac{dw}{dt}\right|\right)^2 \left(\left|\frac{dt}{dz}\right|\right) dt. \tag{8.5 2}$$

Differentiation of eqn. (8.51) gives

$$\frac{dw}{dt} = \frac{\psi}{\pi} \frac{1}{t},$$

and dt/dz is given by eqns. (8.46) and (8.49) so that substituting for these derivatives in eqn. (8.52) yields

$$F = \frac{\mu_0 \psi^2 \sqrt{a}}{2\pi g} \int\limits_{-1}^{-\infty} \frac{dt}{t \sqrt{(t-a)(t+1)}}.$$

Integration of this equation leads to

$$F = \frac{\mu_0 \psi^2}{2\pi g} \left[\sin^{-1} \frac{t(1-a)-2a}{t(1+a)} \right]_{-1}^{-\infty}$$

$$= \frac{\mu_0 \psi^2}{2\pi g} \left[\sin^{-1} \left(\frac{1-a}{1+a} \right) - \frac{\pi}{2} \right]. \tag{8.53}$$

This is the force on one side of the upper element in Fig. 8.11(a) so that the resultant force on this element is

$$\frac{\mu_0 \psi^2}{2\pi g} \left[\sin^{-1} \left(\frac{1-a}{1+a} \right) - \sin^{-1} \left(\frac{1-a'}{1+a'} \right) \right], \tag{8.54}$$

a' being given by eqn. (8.50) in which p is replaced by p', the projection at the other side of the array.

As an alternative to the above method, the alignment force can be calculated by determining the rate of change, with respect to p (and p'), of the total flux entering the upper element—see section 2.6.3. The equations involved are such that this cannot be done analytically—it is not possible to form the equation $w = f(z)$—but it can be done numerically. In the computation, it is convenient to consider the actual boundary divided into two boundaries, of the type shown in Fig. 8.11(b), by the centre line of the upper element. The total flux Φ entering one half of the upper element is taken to be that crossing the boundary in the z-plane between $-s+jg$, where $2s$ is the width of the upper element, and some point jY, where Y is sufficiently large for the flux between jY and $j\infty$ to be negligible. The total flux is calculated for, and is plotted against, a range of values of p. Then, applying eqn. (2.114), the difference between the slopes of the graph for the values p and p', multiplied by $\psi/2$, gives the resultant alignment force on the upper element; thus,

$$F = \frac{\psi}{2} \left(\frac{d\Phi}{dp} - \frac{d\Phi}{dp'} \right).$$

The forces of attraction can also be found using either of these methods. It is again convenient to consider the actual boundary divided into two portions by the centre line of the upper element. With the first method, the force on the upper element, for example, is given by

$$\frac{\mu_0}{2} \left\{ \int\limits_{-s+jg}^{jg} B^2 \, dz + \int\limits_{-s+jg}^{jg} B'^2 \, dz \right\},$$

where B is the flux density on the surface $z = jg$ for a projection p, and B' is the flux density on the same surface for a projection p'. Using the second method, the force is given by

$$\frac{\psi}{2}\left(\frac{d\Phi}{dg}+\frac{d\Phi'}{dg}\right),$$

where Φ and Φ' are the quantities of flux crossing the boundary, in the z-plane, between the points $-s+jg$ and jg, for projections p and p' respectively.

In general, of the two methods, the surface integral one is to be preferred. It is easier to apply and is the more accurate one when numerical techniques have to be used, for then, the rate of change of total flux method necessitates the determination of the slope of a curve.

8.2.9. *A simple electrostatic lens*

Many important problems involve the polygon (with five vertices) shown in Fig. 8.12(a): Dreyfus[8] has examined the eletric field between the low and high voltage windings of a transformer; Carter[9] and Kucera[10] have discussed the field between unequal, opposite

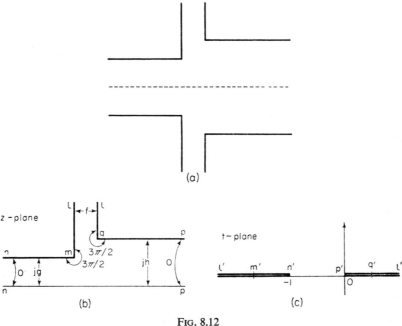

FIG. 8.12

slots; and Herzog[11] has analysed the field of an electrostatic lens. In a number of these problems the field is caused by a difference of potential maintained between the left- and right-hand sections of the polygon so that the line of symmetry is a flux line. It is shown now that the solution of such a field requires the transformation of the field between two semi-infinite plates (see section 8.2.7) and that only one choice of the t-plane constants is possible.

For a simple electrostatic lens with cylindrical symmetry the boundary of Fig. 8.12(a) represents a radial cross-section through the cathode and the anode. The anode usually consists of a thin plate (normal to the axis), but representing it as infinitely long modifies the inter-electrode field only slightly and makes the transformation a much easier one—all the constants may be evaluated by the method of residues. The field is symmetrical and to simplify the transformation it is convenient to consider one half of it as represented in the z-plane, Fig. 8.12(b); the anode with potential ψ_A is the section *lmn*, the cathode with potential 0 is the section *pql*, and *np* is the flux line along the axis of the lens.

To transform the z-plane field into a known field in the t-plane, the vertex *l* must be chosen to correspond with $t = \infty$. The t-plane field, Fig. 8.12(c), then has, on the real axis, two equipotential sections, *l'm'n'*, corresponding with *lmn*, and *p'q'l'*, corresponding with *pql*; that is, it is of the type discussed in section 8.2.7. Let the remaining pairs of corresponding points be:

$$m, \quad t = -a;$$
$$n, \quad t = -1;$$
$$p, \quad t = 0;$$
$$q, \quad t = b;$$

then the Schwarz–Christoffel equation gives

$$\frac{dz}{dt} = S\frac{\sqrt{(t+a)(t-b)}}{t(t+1)}.$$ (8.55)

The constants S, a, and b can all be evaluated simply by the method of residues; application of eqn. (8.27) to the point $t = 0$ gives

$$S = -\frac{jf}{\pi};$$ (8.56)

and application of eqn. (8.26) to the points 0 and -1 gives respectively

$$\frac{h}{f} = \sqrt{ab}$$ (8.57)

and

$$\frac{g}{f} = \sqrt{(b+1)(a-1)}.$$ (8.58)

Then, integrating eqn. (8.55), and ignoring the position of the origin in the z-plane, the transformation equation becomes

$$z = -\frac{jf}{\pi}\left\{\cosh^{-1}\left[\frac{2t+a-b}{a+b}\right] + \frac{h}{f}\cos^{-1}\left[\frac{(a-b)t-2ab}{(a+b)t}\right]\right.$$
$$\left. + \frac{g}{f}\cos^{-1}\left[\frac{2(a-1)(b+1)+(b-a+2)(t+1)}{(a+b)(t+1)}\right]\right\}.$$ (8.59)

The solution for the field in the t-plane is derived from eqn. (8.45) by halving the scale in the t-plane (so that the plates are unit distance apart) and shifting the origin by 1. The result is

$$w = \frac{\psi_A}{\pi}\cosh^{-1}(2t+1),$$ (8.60)

and with eqn. (8.59) this describes the field in the z-plane. The field strength in the z-plane, $|dw/dz|$, is evaluated from eqns. (8.55)–(8.58) and from the expression for dw/dt, found by differentiating eqn. (8.60). Its value along the axis of the lens leads to the determination of the forces acting on electrons emitted from the cathode.

8.3. Transformation of the upper half plane into the region exterior to a polygon

8.3.1. *The transformation*

It is seen from the preceding section that the Schwarz–Christoffel equation (8.1), can be used to obtain solutions for fields in regions "exterior" to certain physical boundaries. However, these exterior regions must be assumed to have boundaries extending to infinity and, in fact, they correspond, mathematically, with the interiors of polygons closed at infinity. When it is necessary to analyse fields exterior to *finite*, polygonal boundaries a different transformation equation, now to be developed, is required.

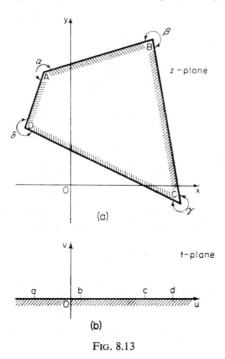

FIG. 8.13

This equation transforms the real axis of one plane, into the boundary of a polygon in another plane, in such a way that the upper half of the first plane becomes the region exterior to the polygonal boundary in the second plane. Consider a polygon in the z-plane, Fig. 8.13(a), which is to be transformed from the real axis of the t-plane, Fig. 8.13(b). The changes in direction at the vertices of the polygon are introduced by a function with the form of the right-hand side of eqn. (8.1) and, therefore, it can be assumed that the desired transformation equation has the form

$$\frac{dz}{dt} = Sf(t)\left[(t-a)^{(\alpha/\pi)-1}(t-b)^{(\beta/\pi)-1}\ldots\right], \tag{8.61}$$

where the function $f(t)$ is to be chosen to satisfy the additional conditions of this transformation, and α, β, γ, ..., are the *exterior* angles of the polygon (but the interior angles of the field).

To determine the form of $f(t)$ consider first the boundaries in the two planes. Because, in eq. (8.61), the part in square brackets completely defines the changes in direction of the z-plane boundary, $f(t)$ must cause no change in argument and, therefore, must be real, non-zero and finite, for all real values of t. Secondly, some point in the t-plane, not on the real axis, must correspond with $z = \infty$. Any non-real point in the t-plane may be chosen, but the simplest form of $f(t)$ is obtained by taking it to be $t = j$ and then, since there must be a pole (see footnote to p. 164) at the point, $f(t)$ must be a functon of $1/(t-j)$. But $f(t)$ must be real for all real values of t and this can be achieved only by introducing the product term $1/(t+j)^\dagger$ which makes

$$f(t) = g\left(\frac{1}{t^2+1}\right).$$

Again, because $t = \infty$ corresponds with a finite value of z, $f(t)$ must have a zero at $t = \infty$ and, since $\Sigma(\alpha/\pi-1) = 2$ (from the sum of the exterior angles of the polygon), this zero must be of order three or more. Thus, it is necessary that

$$f(t) = h\left[\frac{1}{(t^2+1)^n}\right],$$

where n may take all integral values between 2 and ∞. In its most general form this may be expressed

$$f(t) = \frac{L}{(t^2+1)^2} + \frac{M}{(t^2+1)^3} + \frac{N}{(t^2+1)^4} + \cdots, \tag{8.62}$$

where $L, M, N, \ldots$ are all real constants. The values of these constants are derived from the requirement that an infinitely small circle surrounding $t = j$ corresponds with an infinite circle, *traced out once only*, in the z-plane. Now the small circle round the point $t = j$ is described by

$$t-j = Re^{j\theta},$$

and substituting this value in eqn. (8.62), noting that as $R \to 0$, $t+j \to 2j$, gives

$$f(t) = -\frac{L}{4R^2 e^{2j\theta}} - \frac{M}{4R^3 e^{3j\theta}} - \frac{N}{4R^4 e^{4j\theta}} - \cdots.$$

Hence, as $dt = jRe^{j\theta}\,d\theta$, eqn. (8.61) becomes

$$\frac{dz}{dt} = -\frac{jK}{4Re^{j\theta}}\left[L + \frac{M}{Re^{j\theta}} + \frac{N}{R^2 e^{2j\theta}} + \cdots\right],$$

where K is the value, as $t \to j$, of the terms in square brackets in eqn. (8.61). In this last equation, as $R \to 0$, each term in the brackets gives an infinite circle in the z-plane corresponding with the circle in the t-plane but, as θ varies from 0 to 2π, all terms except the first cause the

---

† The pole at $t = -j$ does not disturb the behaviour of the function in the upper half plane.

infinite circle to be traced out more than once. Thus, M, N, ..., must all be zero so that, finally,

$$f(t) = \frac{L}{(t^2+1)^2},$$

and the desired transformation equation is

$$\frac{dz}{dt} = S(t^2+1)^{-2}(t-a)^{(\alpha/\pi)-1}(t-b)^{(\beta/\pi)-1} \cdots . \tag{8.63}$$

This equation differs from the Schwarz–Christoffel equation (8.1), only by the additional factor $(t^2+1)^{-2}$. However, this term causes the integration, for given vertex angles, to be more difficult than that for eqn. (8.1) with the same angles. Because of this, and because it cannot be used to derive a solution directly from the w-plane (see the example below), eqn. (8.63) is never applied to exterior regions with infinite boundaries but only to those with finite boundaries. It is used, for instance, to calculate the field distribution round a charged bus-bar, and the case of a bar with rectangular section is considered in the next chapter.

Treatment of the above boundary involves the use of non-simple functions, as indeed do all transformations using eqn. (8.63) except those for a single, thin plate and various combinations of intersecting thin plates. These are discussed briefly by Bickley[12] and the case of a single, charged plate is treated fully here.

8.3.2. The field of a charged, conducting plate

As an example of the use of eqn. (8.63) consider the field of a conducting plate carrying a charge q. Let the plate lie along the real axis of the z-plane between the points c and $-c$, Fig. 8.14(a); and let the origin of the t-plane, Fig. 8.14(b), correspond with the point $z = c$ and the limits of the real axis with $z = -c$. The exterior angle of the polygon at $z = c$ is 2π and, therefore, equation (8.63) gives for the transformation from the real axis of the

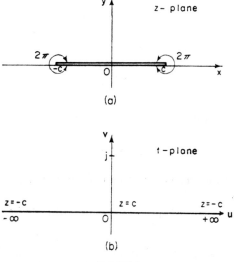

Fig. 8.14

t-plane into the z-plane boundary as

$$\frac{dz}{dt} = S\frac{t}{(t^2+1)^2}.$$

When integrated this yields

$$z = -\frac{S}{2}\frac{1}{(t^2+1)} + k,$$

from which the values of S and k are derived by substitution of the corresponding values of z and t. Hence, $k = -c$ and $S = -c$, and the equation of transformation is

$$z = c\frac{1-t^2}{1+t^2}. \tag{8.64}$$

The required field in the t-plane is different from any used previously in the book, but its form can be simply determined. First, the upper and lower surfaces of the plate correspond with the real axis in the t-plane and so a total charge q must be distributed over this real axis. And, further, since the plate is conducting, the real axis must be equipotential. Secondly, all the flux leaving the plate goes to infinity and so, as the point $t = j$ corresponds with infinity in the z-plane, there must be a charge $-q$ at $t = j$. Thus, the t-plane field is as shown in Fig. 8.15(a) and the reader will appreciate that it is one half of the field between equal unlike charges at the points $t = j$ and $t = -j$. The complex potential function describing it is, therefore (see section 3.2.1)

$$w = \frac{q}{2\pi}\log\frac{t-j}{t+j}.$$

Putting $jW = 2w\pi/q$, this equation may be rewritten

$$t = j\frac{1+e^{jW}}{1-e^{jW}}$$

$$= -\cot\left(\frac{W}{2}\right),$$

or, changing the origin of W by π,

$$t = \tan\left(\frac{W}{2}\right). \tag{8.65}$$

Thus, substituting this value for t in eqn. (8.64), the solution of the field in the z-plane is

$$z = c\cos W,$$

or, writing W in terms of w and shifting the origin of the w-plane,

$$z = c\sin\left(\frac{-j2\pi w}{q}\right) = c\sin w'. \tag{8.66}$$

These are of the same form as eqns. (7.47) and (7.48) developed as the solution for the field outside an equipotential elliptical boundary. The plate is, of course, the limiting ellipse of the family described by [see eqn. (7.49)]:

$$\frac{x^2}{c^2 \cosh^2 \left(\dfrac{2\pi\psi}{q}\right)} + \frac{y^2}{c^2 \sinh^2 \left(\dfrac{2\pi\psi}{q}\right)} = 1,$$

and the map of its field is shown in Fig. 8.15(b). Equation (8.66) differs from equation (7.48) only in that it includes the magnitude of the charge upon the plate (or elliptical boundaries).

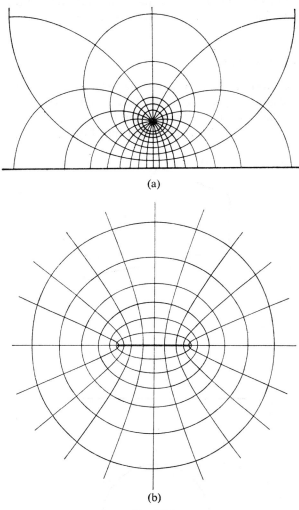

(a)

(b)

Fig. 8.15

The form of t-plane field developed here is used whenever the field exterior to a finite, charged, conducting boundary is obtained by transformation from the real axis. It is used again in section 9.3, where the field of a charged bus-bar is examined. (The form of the t-plane field which is necessary for the determination of the fields of line currents or charges external to finite boundaries is discussed in section 10.2.2.)

AC 13

8.4 Transformations from a circular to a polygonal boundary

In certain problems it is desirable to transform the boundary of a polygon, not from an infinite straight line, but from a circle. This may be because it is then easier to determine and express the solution for the field (see sections 8.4.2 and 10.2), or because the Poisson integral (see section 10.6.1) is to be used. There are four transformation equations which connect circular with polygonal boundaries—the interior or exterior of one boundary may be transformed into the interior or exterior of the other—and they may be derived by combining the bilinear transformation (see section 7.1) with eqn. (8.1) (Schwarz–Christoffel) or eqn. (8.63).

8.4.1. *The transformation equations*

Consider first the derivation of the equation which transforms the perimeter of the unit circle into a polygonal boundary, in such a way that the *interior* regions of the two boundaries correspond with one another. Let the polygon be in the z-plane, Fig. 8.16(a), and the circle in the t-plane, Fig. 8.16(c). The circle is first transformed into the real axis of a third

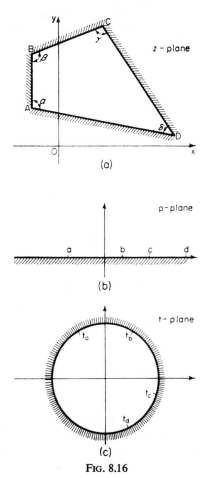

FIG. 8.16

complex plane, the p-plane, Fig. 8.16(b), and then the real axis of the p-plane is transformed into the polygonal boundary. The equation transforming the interior of the circle into the upper half of the p-plane is [see eqn. (7.17)]

$$p = j\frac{1-t}{1+t}, \tag{8.67}$$

which may be differentiated to give

$$\frac{dp}{dt} = -\frac{2j}{(1+t)^2}. \tag{8.68}$$

Also, the equation transforming the upper half of the p-plane into the interior of the polygon is [see eqn. (8.1)]

$$\frac{dz}{dp} = S(p-a)^{(\alpha/\pi)-1}(p-b)^{(\beta/\pi)-1}\ldots,$$

which may be written for brevity as

$$\frac{dz}{dp} = S\prod(p-a)^{(\alpha/\pi)-1}, \tag{8.69}$$

where the symbol $\prod$ is taken to mean the product of terms of the form $(p-a)^{(\alpha/\pi)-1}$. Now the equation which transforms the circle into the polygon is given by the evaluation of

$$\frac{dz}{dt} = \frac{dz}{dp}\frac{dp}{dt},$$

in terms of the above expressions for dz/dp and dp/dt: substituting for these derivatives from eqn. (8.69) and (8.68) respectively, and for p from eqn. (8.67) gives

$$\frac{dz}{dt} = S\prod\left(j\frac{1-t}{1+t}-a\right)^{(\alpha/\pi)-1}\frac{2j}{(1+t)^2}. \tag{8.70}$$

By making $S' = 2jS$, and noting that

$$(1+t)^{-\Sigma((\alpha/\pi)-1)} = (1+t)^2,$$

eqn. (8.70) may be written

$$\frac{dz}{dt} = S'\prod[j(1-t)-a(1+t)]^{(\alpha/\pi)-1},$$

or

$$\frac{dz}{dt} = S'\prod\left[\left(\frac{j-a}{j+a}-t\right)(a+j)\right]^{(\alpha/\pi)-1}.$$

But, as the term $(a+j)$ is a constant, and as $(j-a)/(j+a)$ is the point in the t-plane, t_a, corresponding with $p = a$ (which may be seen by writing eqn. (8.67) for t, in terms of p), the final form of the transformation equation is

$$\frac{dz}{dt} = S\prod(t_a-t)^{(\alpha/\pi)-1}$$

or

$$\frac{dz}{dt} = S(t_a-t)^{(\alpha/\pi)-1}(t_b-t)^{(\beta/\pi)-1}(t_c-t)^{(\gamma/\pi)-1}\ldots, \tag{8.71}$$

13*

where the constant S replaces $S'(a+j)^2$. This equation is identical in form with the Schwarz–Christoffel equation, except that the order of subtraction in each term is reversed. It must be noted, however, that the points t_a, t_b, t_c, ..., corresponding with the vertices of the polygon, are, in general, complex with unit modulus.

The equation transforming the region *exterior* to the circle into that *exterior* to a polygon may be derived in a similar way to the above by using eqns. (8.63) and (8.67), and by noting that eqn. (8.67) transforms the exterior of the circle into the *lower* half of the p-plane. The equation is

$$\frac{\mathrm{d}z}{\mathrm{d}t} = St^{-2}(t-t_a)^{(\alpha/\pi)-1}(t-t_b)^{(\beta/\pi)-1}(t-t_c)^{(\gamma/\pi)-1}\ldots, \tag{8.72}$$

where the angles α, β, γ, ..., are the exterior angles of the polygon and, again, the points t_a, t_b, ..., are complex with unit modulus.

The remaining two equations connecting the unit circle with a polygon—the one from the interior of a polygon into the exterior of the circle, and the one from the exterior of a polygon into the interior of the circle—are very rarely used and are not given here. If required, they may be derived simply in a similar manner to that demonstrated above.

The use of the transformations from the unit circle does not make possible the solution of any problem which cannot be solved by transformation from the real axis; and, for both methods, similar difficulties are experienced in the integration of the differential equations—for a given polygon, the same class of function is involved in each. For the interior region of a polygon it is usually preferable to transform from the real axis because of the much simpler form of the field. For the exterior regions of a polygon, however, it is often preferable to use the transformation from the circle, as the necessary field solutions are easier to visualize. This point is demonstrated in the following example, and is discussed further in section 10.2. It is also demonstrated in the solutions for the field of two finite, intersecting, charged plates.[13]

The general solution for the field inside the unit circle, due to any distribution of potential on its perimeter, is given by the Poisson integral which is discussed in section 10.6.1. With this integral it is possible to obtain the solution for the field within a polygon, due to *any* distribution of potential on the boundary, provided that the polygon can be transformed into the unit circle.

8.4.2. The field of a line current and a permeable plate of finite cross-section

Consider the solution for the field of a line current near a permeable plate of finite cross-section, Fig. 8.17, by transformation of the region exterior to the plate into that exterior to the unit circle. (The field exterior to the circle is of the form shown in Fig. 7.11.) Let the boundary of the plate be in the z-plane, with its ends at $z = l+jm$ and $z = -l-jm$, Fig. 8.18(a), and let the unit circle be in the t-plane, Fig. 8.18(b). The exterior angles of the polygon (at the ends of the plate) are both 2π, and the points, t_a and t_b, lying on the circle and corresponding with the vertices of the polygon, have coordinates which are, by symmetry, equal but of opposite sign. Thus, from eqn. (8.72),

$$\frac{\mathrm{d}z}{\mathrm{d}t} = St^{-2}(t-t_a)(t-t_b),$$

and this, when integrated, gives for the equation connecting the planes

$$z = S\left(t+\frac{t_a^2}{t}\right)+k.$$

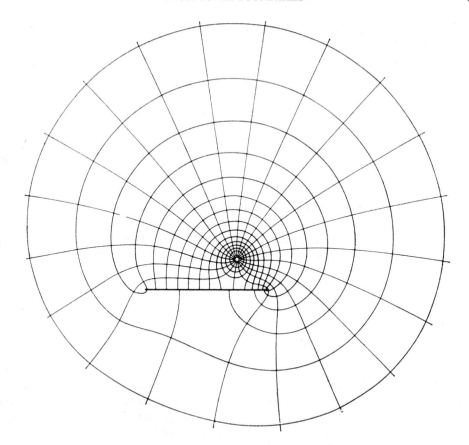

Fig. 8.17

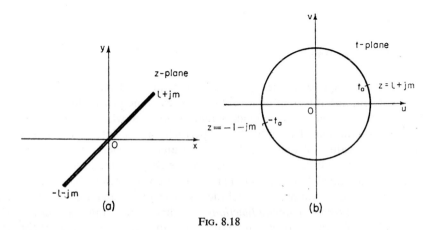

Fig. 8.18

Substituting, in this equation, the corresponding values of z and t for the vertices of the polygon, yields $k = 0$ and

$$S = \frac{l+jm}{2t_a},$$

where t_a may have any convenient value. Hence the transformation from the circle to the plate is

$$z = \frac{l+jm}{2t_a} \left(t + \frac{t_a^2}{t} \right). \tag{8.73}$$

[Compare this equation with eqn. (7.36).]

To obtain the field generated by a line current in the z-plane it is clear that the required field in the t-plane is also generated by a line current at the point corresponding, through eqn. (8.73), with the point of the current in the z-plane. The solution takes its simplest form when the current is positioned on the real axis of the z-plane at some point, c, chosen so that, in conjunction with the choice of $l+jm$, the z-plane field has the desired proportions; for then, the current in the t-plane also lies on the real axis at some point c'. From eqn. (7.40) the solution for the field in the t-plane is

$$\varphi + j\psi = \frac{i}{2\pi} \log \left(\frac{t^2+1}{t} - \frac{c^2+1}{c} \right), \tag{8.74}$$

and elimination of t between eqns. (8.74) and (8.73) gives the solution in the z-plane.

The same solution could be obtained by using eqn. (8.63) to transform the boundary of the plate from the real axis of the t-plane.[14] Then, though the transformation equation is simple [it is developed in section 8.3.2, eqn. (8.64)], the reader will appreciate that the necessary form of the t-plane field is not evident without careful consideration. It is discussed in section 10.2.2.

The solution to this field is of value in the estimation of the influence which the plate has on the inductance of the current. The inductance can be found by comparing the difference of flux function between two points, one lying on the conductor surface and the other at a great distance from the conductor and the plate, for the two cases, with and without, the plate.[14]

8.5. Classification of integrals

A wide variety of simple examples of the transformation of polygonal boundaries has now been considered, and it is helpful at this stage to examine briefly, in general terms, the relationship between a given type of boundary and the integral which is involved in deriving its transformation equation. The difficulty in performing the integration is significantly dependent upon the values of angle at the vertices and, in general, it increases with the number of vertices. Three broad classes of integral are encountered and they are discussed below, where it is assumed that the boundary under consideration has, when possible, been reduced, by the use of symmetry, to include only the minimum number of vertices.

Integrals expressible in terms of simple functions. The first type of integral yields a solution that is expressible wholly in terms of simple functions and all the examples of this chapter are typical of it. The general form of this class of integral may be written

$$\int \frac{P(t)}{Q(t)} \sqrt{\frac{t-a}{t-b}}\, \mathrm{d}t = \int \frac{P'(t)}{Q(t)} \frac{\mathrm{d}t}{\sqrt{(t-a)(t-b)}} \tag{8.75}$$

where P, P', and Q are rational, algebraic functions of t. Inspection shows that a boundary having any number of angles of value $-\pi$, 0, and 2π, and up to two angles which are odd multiples of $\pi/2$, may be expressed in this, or simpler, form. The constants of the transformation equation for such a boundary can usually be determined simply though it has been seen (section 8.2.7) that, in certain cases, graphical methods may be required.

Integrals expressible in terms of non-simple functions. In practice, polygons with three or four vertex angles which are odd multiples of $\pi/2$ (and any number of value $-\pi$, 0, or 2π) frequently occur. They lead to integrals which are termed *elliptic*. Such integrals may be evaluated using elliptic functions and they are discussed in detail in the next chapter.

Also, a number of polygons having two vertex angles which are not multiples of $\pi/2$ give rise to integrals which may be expressed in the form of Euler or, in particular cases, elliptic integrals. They are discussed in section 10.4.

Integrals requiring numerical evaluation. When the boundary is such as to give an integral not falling into any of the above classes, analytical methods fail, and numerical methods are used. See section 10.5.

References

1. H. A. SCHWARZ, Über einige Abbildungsaufgaben, *J. reine Angew. Math.*, **70**, 105 (1869).
2. E. B. CHRISTOFFEL, Sul Problema delle Temperature Stazionarie e la Rappresentazione di una Data Superficie, *Ann. Mat. Pura Appl.* **1**, 95 (1867).
3. W. ROGOWSKI, Die elektrische Festigkeit am Ronde des Plattenkondensators, ein Beitrag zur Theorie der Funkenstrecken und Durchführungen, *Arch. Elektrotech.* **12**, 1 (1923).
4. F. W. CARTER, Air-gap and interpolar induction, *J. Instn. Elect. Engrs.* **29**, 925 (1900).
5. G. M. STEIN, Influence of the core form upon the iron losses of transformers, *Trans. Am. Instn. Elect. Engrs.* **67** I, 95 (1948).
6. I. A. TERRY and E. G. KELLER, Field pole leakage flux in salient pole dynamo electric machines, *J. Instn. Elect. Engrs.* **83**, 845 (1938).
7. F. W. CARTER, Air-gap induction, *Elect. World, NY* **38**, 884 (1901).
8. L. DREYFUS, Die Anwendung des Mehrphasenfrequenzumformers zur Kompensierung von Drehstrom-asynchronmotoren, *Arch. Elektrotech.* **13**, 507 (1924).
9. F. W. CARTER, The magnetic field of the dynamo-electric machine, *J. Instn. Elect. Engrs.* **64**, 115 (1926).
10. J. KUCERA, Magnetische Zahnstreuungen bei elektrischen Maschinen, *Elektrotech. Maschinenb.* **58**, 329 (1940).
11. R. HERZOG, Berechnung des Streufeldes eines Kondensators, dessen Feld durch eine Blende begrenzt ist, *Arch. Elektrotech.* **29**, 790 (1935).
12. W. G. BICKLEY, Two-dimensional problems concerning a single closed boundary, *Phil. Trans.* **228** A, 235 (1929).
13. W. B. MORTON, Electrification of two intersecting planes, *Phil. Mag.* **1**, 337 (1926), and Irrotational flow past two intersecting planes, *Phil. Mag.* **2**, 900 (1926).
14. P. J. LAWRENSON, A note on the analysis of the fields of line currents and charges, *Proc. Instn. Elect. Engrs.* **109** C, 86 (1962).

Additional References

COHN, S. B., Shielded coupled strip transmission line, *Inst. Radio Engrs. Trans.* MTT-3, 29 (1955).

COHN, S. B., Thickness correction for capacitive obstacles and strip conductors, *Inst. Radio Engrs. Trans.* MTT-8, 638 (1960).

DAHLMAN, B. A., A double ground plane strip line for microwaves, *Inst. Radio Engrs.*, MTT-3, 52-7 (Oct. 1955).

DOUGLAS, J. H. F., Reluctance of some irregular magnetic fields, *Trans. Am. Instn. Elect. Engrs.* 34 I, 1067 (1915).

GETSINGER, W. J., A coupled strip line configuration using printed-circuit construction that allows very close cupling, *Inst. Radio Engrs. Trans.* MTT-9, (1961).

KREYSZIG E., Schlitzblendenkondensatoren. Ein Beitrag zur Praxis der Brechung von Potentialfeldern, *Z. angew. Phys.* **7,** 13 17 (1955).

CHAPTER 9

THE USE OF ELLIPTIC FUNCTIONS

9.1. Introduction

The solutions by conformal transformation of many problems of practical importance involve the use of elliptic functions. These are general functions of which trigonometric and hyperbolic functions are particular cases, and they can be used to evaluate integrals which are of an elliptic form. In this chapter elliptic functions are discussed only as they are required for the application of conformal transformation techniques. However, to enable the reader unfamiliar with these functions to use them with confidence, a careful explanation of the fundamental principles is given.

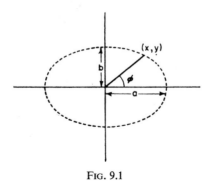

FIG. 9.1

Elliptic functions are so called because they were first encountered in the calculation of the length of the arc of an ellipse. Consider this calculation for an ellipse, Fig. 9.1, described by the parametric equations

$$x = a \sin \varphi \quad \text{and} \quad y = b \cos \varphi.$$

Now the length L of any arc is given by $\int \sqrt{dx^2 + dy^2}$, where the limits of the integral define the ends of the arc. Thus for an elliptical arc defined by the angles φ_1 and φ_2,

$$L = \int_{\varphi_1}^{\varphi_2} \sqrt{a^2 \cos^2\varphi + b^2 \sin^2\varphi} \; d\varphi$$

$$= \int_{\varphi_1}^{\varphi_2} [a^2 - \sin^2\varphi(a^2 - b^2)]^{1/2} \; d\varphi.$$

189

This can be written as

$$L = a \int_{\varphi_1}^{\varphi_2} (1 - k^2 \sin^2\varphi)^{1/2} \, d\varphi, \tag{9.1}$$

where

$$k = \frac{\sqrt{a^2 - b^2}}{a},$$

is the eccentricity of the ellipse and is less than unity. This integral cannot be evaluated in terms of elementary functions, except by a series, and it is known as an elliptic integral of the second kind. Since $k < 1$ and $\sin\varphi \leqslant 1$, the integrand can be expanded by the binomial theorem: thus

$$(1 - k^2 \sin^2\varphi)^{1/2} = 1 - \frac{k^2}{2} \sin^2\varphi - \frac{k^4}{8} \sin^4\varphi - \dots .$$

This series is uniformly convergent, and consequently it can be integrated term by term, using a recursion formula for $\int \sin^n x \, dx$.

In similar ways other types of elliptic integral, and the elliptic functions to which they are related, can be evaluated. Tables of the values of elliptic functions, prepared from such series, are readily available and are frequently used in calculation. However, when a digital computer is employed, it is usually preferable to use the series directly rather than to feed the information from the tables into the machine store. A list of available tables and the various series required for numerical work is given in the appendices.

9.2. Elliptic integrals and functions

There are three basic *elliptic integrals* with which all others can be connected and they were defined by Legendre as being of the first, second, and third kinds. The *elliptic functions* are derived from the first of these integrals as explained below.

9.2.1. *The elliptic integral of the first kind*

The simplest and most useful of the three basic integrals is that of the first kind, defined by

$$z = \int_0^{\varphi} \frac{d\varphi}{\sqrt{(1 - \sin^2\theta \, \sin^2\varphi)}}, \tag{9.2}$$

where $|\sin\theta| < 1$, θ is called the *modular angle*, and φ the *amplitude*. Putting $t = \sin\varphi$ and $k = \sin\theta$ in eqn. (9.2) gives

$$z = \int_0^{t} \frac{dt}{\sqrt{(1 - t^2)(1 - k^2 t^2)}}. \tag{9.3}$$

This form of the elliptic integral of the first kind is due to Jacobi, and k is called the *modulus*. For certain values of k (or $\sin\theta$) z reduces to an elementary function of t: if $k = 0$, $z = \sin^{-1} t$ and if $k = 1$, $z = \tanh^{-1} t$.

For a general upper limit, the integral is said to be *incomplete*, and it is given the symbol $F(t, k)$ or $F(\varphi, \theta)$. When the upper limit is given by $t = 1 (\varphi = \pi/2)$ the integral is *complete* and it is given the symbol $K(k)$ or $K(\theta)$.

Derived from the modulus k is the *complementary modulus* k', defined by

$$k' = \sqrt{1-k^2}. \tag{9.4}$$

The complete integral to modulus k' is denoted by the symbol K', and it can be shown, using eqn. (9.3) and substituting for k' in terms of k, that

$$K' = \int_1^{1/k} \frac{dt}{j\sqrt{(1-t^2)(1-k^2t^2)}}, \tag{9.5}$$

and thus that

$$K+jK' = \int_0^{1/k} \frac{dt}{\sqrt{(1-t^2)(1-k^2t^2)}}. \tag{9.6}$$

This equation can be useful when determining the constants of a transformation, and is employed later in section 9.2.4.

9.2.2. *The principal Jacobian elliptic functions*

It was shown by Jacobi that certain elliptic functions, very useful in analytical work, could be derived simply from the first elliptic integral. Consider again eqn. (9.3): z is a function of t and k; but equally t is a function of z and k and, in the notation of Jacobi,

$$t = \operatorname{sn}(z, k). \tag{9.7}$$

This equation defines a new function sn[†] which is one of the principal Jacobian elliptic functions. When the value of the modulus is obvious, this equation is usually abbreviated to

$$t = \operatorname{sn} z.$$

There are two other principal functions, cn z and dn z,[‡] defined such that

$$\operatorname{sn}^2 z + \operatorname{cn}^2 z = 1 \tag{9.8}$$

and

$$\operatorname{dn}^2 z = 1 - k^2 \operatorname{sn}^2 z. \tag{9.9}$$

sn z, cn z, and dn z are the three principal Jacobian elliptic functions.

9.2.3. *The elliptic integral of the second kind*

Legendre defined the elliptic integral of the second kind, $E(\varphi, \theta)$, by

$$E(\varphi, \theta) = \int_0^{\varphi} \sqrt{1-\sin^2\theta \sin^2\varphi}\, d\varphi. \tag{9.10}$$

Putting

$$t = \sin\varphi \quad \text{and} \quad k = \sin\theta$$

[†] sn is an abbreviation for sine amplitude or sin am, and is pronounced "ess-en".
[‡] cn and dn are pronounced "see-en" and "dee-en".

gives the Jacobian form

$$E(t, k) = \int_0^t \frac{\sqrt{1-k^2t^2}}{\sqrt{1-t^2}} \, dt. \tag{9.11}$$

When the upper limit of the integral is $\varphi = \pi/2$ (or $t = 1$), the integral is said to be complete, and it is given the symbol $E(\theta)$ [or $E(k)$].

The elliptic integral of the second kind can be expressed in terms of Jacobian elliptic functions by the substitution

$$\sin\varphi = \operatorname{sn}\alpha. \tag{9.12}$$

It can be shown (see Whittaker and Watson, p. 492) that

$$\frac{d}{d\alpha}(\operatorname{sn}\alpha) = \operatorname{cn}\alpha \, \operatorname{dn}\alpha,$$

and, hence, by differentiating eqn. (9.12), that

$$\cos\varphi \, d\varphi = \operatorname{cn}\alpha \, \operatorname{dn}\alpha \, d\alpha.$$

This gives

$$d\varphi = \frac{\operatorname{cn}\alpha \, \operatorname{dn}\alpha}{\cos\varphi} \, d\alpha, \tag{9.13}$$

and substituting in eqn. (9.10) from eqn. (9.13) yields

$$E(\varphi, \theta) = \int_0^\varphi \frac{\operatorname{cn}\alpha \, \operatorname{dn}\alpha}{\cos\varphi} \sqrt{1-\sin^2\theta \sin^2\varphi} \, d\alpha. \tag{9.14}$$

Also, from the definitions of $\operatorname{dn}\alpha$ and $\operatorname{cn}\alpha$ and using eqn. (9.12) it is evident that

$$\operatorname{dn}\alpha = \sqrt{1-\sin^2\theta \sin^2\varphi} \quad \text{and} \quad \operatorname{cn}\alpha = \cos\varphi.$$

Finally, therefore, substituting in eqn. (9.14) gives

$$E(\varphi, \theta) = \int_0^\varphi \operatorname{dn}^2\alpha \, d\alpha, \tag{9.15}$$

which expresses the second elliptic integral in terms of Jacobian elliptic functions.

9.2.4. *Two finite charged plates*

The application of elliptic functions to the analysis of a physical problem is first illustrated by considering the field of two charged conducting plates of equal length and lying on a straight line (Fig. 9.2). This field is often transformed into more complicated ones: see, for example, references 1 and 2.

Figure 9.3(a) shows the two conducting plates placed along the real axis of the t-plane. Figure 9.3(b) shows, in the w-plane, a rectangular boundary two opposite sides of which correspond to the conducting plates, the other two sides being flux lines. The uniform field,

$$w = \psi + j\varphi,$$

inside the rectangle is to be transformed into the field in the upper half of the t-plane.

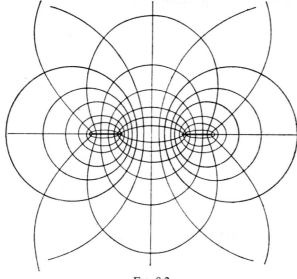

FIG. 9.2

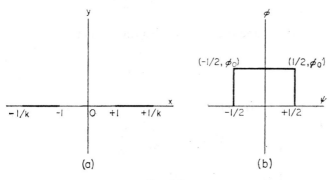

FIG. 9.3

Let the plates in the t-plane lie between $t = +1$ and $+1/k$ and between $t = -1$ and $-1/k$, so that the proportions of the boundaries in that plane are fixed by the choice of the value of k. (This choice of the t-plane constants leads, as demonstrated below, to definite integrals which can be directly expressed in terms of complete elliptic integrals, k being the modulus of the integrals.) Further, let the potential between the plates be unity, and let the corresponding points in the w- and t-planes be as follows:

$$t = +1, \quad \varphi = 0 \quad \text{and} \quad \psi = \tfrac{1}{2};$$

$$t = +\frac{1}{k}, \quad \varphi = \varphi_0 \quad \text{and} \quad \psi = \frac{1}{2};$$

$$t = -1, \quad \varphi = 0 \quad \text{and} \quad \psi = -\tfrac{1}{2};$$

and

$$t = -\frac{1}{k}, \quad \varphi = \varphi_0 \quad \text{and} \quad \psi = -\frac{1}{2}.$$

The Schwarz–Christoffel equation connecting the planes is

$$\frac{dw}{dt} = \frac{S}{\sqrt{1-t^2}\,\sqrt{1-k^2t^2}},$$ (9.16)

which can be written

$$w = \int_0^t \frac{S\,dt}{\sqrt{1-t^2}\,\sqrt{1-k^2t^2}}.$$ (9.17)

This is an elliptic integral of the first kind and it can thus be expressed in terms of Jacobian functions as

$$t = \operatorname{sn}\left(\frac{w}{S}\right),$$ (9.18)

a form which is neat and simple to use for determining a relationship between the w- and t-plane constants. Writing eqn. (9.6) in terms of Jacobian functions

$$\operatorname{sn}(K+jK') = \frac{1}{k}.$$ (9.19)

When $t = 1/k$, eqns. (9.18) and (9.19) are equivalent, so that

$$\frac{w}{S} = K+jK'.$$ (9.20)

But, when $t = 1/k$, $w = \frac{1}{2}+j\varphi_0$, and so

$$K + jK' = \frac{1}{2S}+\frac{j\varphi_0}{S}.$$ (9.21)

Hence, equating the real parts of this equation,

$$S = \frac{1}{2K},$$ (9.22)

and the transformation equation and field solution can be written

$$t = \operatorname{sn}(2Kw, k).$$ (9.23)

This solution can be used, for example, to find the capacitance between plates.
   Equating the imaginary parts of eqn. (9.21),

$$\frac{\varphi_0}{S} = K',$$

and substituting for S from eqn. (9.22), gives

$$\varphi_0 = \frac{K'}{2K}.$$ (9.24)

Since the potential difference between the plates is unity, the flux φ_0 passing between these plates is equal to the capacitance between them. The variation of capacitance with the ratio of plate separation to plate length is shown in Fig. 9.4.

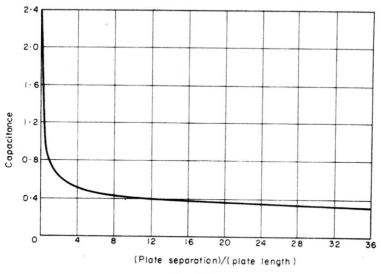

FIG. 9.4

A map of the flux and equipotential lines can be determined by calculating the real and imaginary parts of sn $2Kw$ and substituting values of φ and ψ. This is facilitated by writing sn $2Kw$ in the form sn $(\alpha+j\beta)$ and using the expansion (see Whittaker and Watson)

$$\text{sn}\,(\alpha+j\beta) = \frac{\text{sn}\,\alpha \cdot \text{dn}\,\beta}{1-\sin^2\beta\,\text{dn}^2\alpha} + j\,\frac{\text{cn}\,\alpha\,\text{dn}\,\alpha\,\text{sn}\,\beta\,\text{cn}\,\beta}{1-\sin^2\beta\,\text{dn}^2\alpha}. \tag{9.25}$$

The above and similar transformations, involving boundaries consisting of finite plates, have been used in the calculation of the impedance of waveguides and transmission lines.[3-6]

9.2.5. *Elliptic integrals of the third kind*

In practice, integrals are sometimes encountered which involve three variables (as opposed to two, t and k, in the integrals of the first and second kinds). One of these integrals was defined by Legendre as the elliptic integral of the third kind and given the symbol

$$\Pi(t, k_1, k),$$

where

$$\Pi(t, k_1, k) = \int_0^t \frac{dt}{(1-k_1^2 t^2)\,\sqrt{(1-t^2)(1-k^2 t^2)}}. \tag{9.26}$$

Putting $t = \text{sn}\,u$ and $k_1 = k\,\text{sn}\,\alpha$ gives the same integral in terms of Jacobian functions as

$$\Pi(u, \alpha) = \int_0^u \frac{du}{1-k^2\,\text{sn}^2\alpha\,\text{sn}^2 u}, \tag{9.27}$$

to modulus k. Jacobi, however, defined the elliptic integral of the third kind in a different way as

$$= k^2 \operatorname{sn} \alpha \operatorname{cn} \alpha \operatorname{dn} \alpha \int_0^u \frac{\operatorname{sn}^2 u \, du}{1 - k^2 \operatorname{sn}^2 \alpha \operatorname{sn}^2 u}, \tag{9.28}$$

which is here given the symbol $\Pi_J(u, \alpha)$. This form is the more convenient for evaluation, but integrals are more easily recognized in Legendrian form. The two forms are related by

$$\Pi = u + \frac{\operatorname{sn} \alpha}{\operatorname{cn} \alpha \operatorname{dn} \alpha} \Pi_J. \tag{9.29}$$

In general the evaluation of integrals of the third kind is very difficult, and the only feasible way of doing this involves the use of various auxiliary functions. These are the *Jacobian zeta function*, $Z(u)$, where

$$Z(u) = E(u) - u \frac{E}{K}, \tag{9.30}$$

and the *principal Jacobian theta function*, $\Theta(u)$, defined by

$$\Theta(u) = 1 + 2 \sum_{n=1}^{\infty} (-1)^n q^{n^2} \cos \frac{\pi n u}{k}, \tag{9.31}$$

in which q, known as *Jacobi's nome*, is given by

$$q = e^{-\pi k'/k}. \tag{9.32}$$

The Jacobian form of the third elliptic integral can be expressed in terms of the zeta function, or in terms of a combination of the zeta and theta functions, see section 9.4. These expressions, which are not derived here (see Whittaker and Watson), but are merely listed for use later, are

$$\Pi_J(u, \alpha) = \tfrac{1}{2} \int_0^u [Z(u-\alpha) - Z(u+\alpha) + 2Z(\alpha)] \, du, \tag{9.33}$$

and

$$\int_0^u Z(u) \, du = \log \Theta(u); \tag{9.34}$$

and, by combining these,

$$\Pi_J(u, \alpha) = \frac{1}{2} \log \frac{\Theta(u-\alpha)}{\Theta(u+\alpha)} + uZ(\alpha). \tag{9.35}$$

9.3. The field outside a charged rectangular conductor

An interesting example of the application of elliptic functions, which serves to demonstrate the variety of methods by which they can be manipulated to produce a numerical answer, occurs in the analysis of the field outside a charged rectangular conductor.[†] The

† The difficult analysis was first accomplished by Bickley.[7]

transformations of the boundary of the conductor, both from an infinite straight line and from a unit circle, are developed. Only one field, that for a charged rectangular conductor (compare section 5.3), is examined here; but others involving more complicated basic field solutions are treated in the next chapter.

9.3.1. *The transformation from an infinite, straight line*

The analysis of this problem by transformation from the infinite straight line can be divided into several stages. First, the rectangular boundary is related to the infinite straight line by the Schwarz–Christoffel equation. Next, the resulting transformation equation is manipulated to give expressions for the constants. It is then expressed in alternative forms to demonstrate the various ways of calculating corresponding points in the two planes. Finally, the solution for the field in the upper half plane is derived.

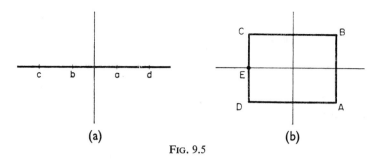

(a)   (b)

FIG. 9.5

The transformation. Let the rectangular boundary in the z-plane, Fig. 9.5(b) be transformed into the real axis of the t-plane, Fig. 9.5(a) in such a way that the upper half of the t-plane corresponds with the region exterior to the rectangle. Let the points in the t-plane corresponding to the corners of the rectangle be $t = a, b, c$, and d; and let the point, E, on the real axis of the z-plane, correspond to the meeting point at infinity of the extremes of the real axis in the t-plane. Applying eqn. (8.63),

$$\frac{dz}{dt} = \frac{S\sqrt{(t-a)(t-b)(t-c)(t-d)}}{(t^2+1)^2}.$$
(9.36)

But since the t-plane is symmetrical,

$$a = -b \quad \text{and} \quad c = -d,$$

and, therefore, eliminating b and d from eqn. (9.36) gives the transformation equation

$$\frac{dz}{dt} = \frac{S\sqrt{(t^2-a^2)(t^2-c^2)}}{(t^2+1)^2}.$$
(9.37)

Now there are two parameters, the lengths of the sides of the rectangle, defining the z-plane configuration, and, therefore, as always, of the three parameters in eqn. (9.37), one can be chosen arbitrarily, and, in fact, it is conveniently expressed in terms of the others. Putting $c = -1/a$, the transformation equation becomes

$$\frac{dz}{dt} = \frac{S\sqrt{(t^2-a^2)(t^2-1/a^2)}}{(t^2+1)^2},$$
(9.38)

and this is a desirable form as will be apparent later. Variation of a in the range from 0 to 1 gives all the possible proportions to the rectangle: when $a = 0$, $c = \infty$, and AB and CD are zero; and when $a = 1$, $c = 1$, and AD and BC are zero.

The evaluation of z from eqn. (9.38) can be performed in several ways, three of which are discussed: the first is to express the integral in terms of Jacobian functions, the second is to expand the integrand into a series, and the third is to express the integral in terms of the three Legendrian elliptic integrals.

Expressions in terms of Jacobian functions. By putting

$$t = -\tan \frac{u}{2} \tag{9.39}$$

and

$$a = \tan \frac{\alpha}{2} \tag{9.40}$$

in eqn. (9.38), dz/dt becomes equal to

$$\frac{S[\tan^2(u/2) - \tan^2(\alpha/2)]^{1/2} [\tan^2(u/2) - \cot^2(\alpha/2)]^{1/2}}{\sec^4(u/2)}. \tag{9.41}$$

By differentiating eqn. (9.39), rearranging eqn. (9.41), and combining these, it can be shown that

$$\frac{dz}{du} = -\frac{S}{2} \operatorname{cosec} \alpha [2(\cos 2u - \cos 2\alpha)]^{1/2}. \tag{9.42}$$

Now, by letting

$$k = \sin \alpha \tag{9.43}$$

and

$$k \operatorname{sn} \beta = -\sin u, \tag{9.44}$$

where $\operatorname{sn} \beta$ is to modulus k, the right-hand side of eqn. (9.42) can be put in Jacobian form. Using the definition of $\operatorname{dn} \beta$, eqn. (9.9), and substituting from equation (9.44) leads to

$$\cos u = \operatorname{dn} \beta. \tag{9.45}$$

Substituting from eqns. (9.43)–(9.45) in eqn. (9.42), having expanded $(\cos 2u - \cos 2\alpha)$, gives

$$\frac{dz}{du} = -\frac{S}{2k} \sqrt{2} [\operatorname{dn}^2 \beta - k^2 \operatorname{sn}^2 \beta - (1 - 2k^2)]^{1/2}, \tag{9.46}$$

which from the definition of $\operatorname{dn} \beta$

$$= -\frac{S}{2k} \sqrt{2} (1 - 2k^2 \operatorname{sn}^2 \beta - 1 + 2k^2)^{1/2},$$

and from the definition of $\operatorname{cn} \beta$, eqn (9.8),

$$= -\frac{S}{2k} 2k \operatorname{cn} \beta$$

$$= -S \operatorname{cn} \beta. \tag{9.47}$$

The next step is to differentiate eqn. (9.44) to obtain

$$\frac{\mathrm{d}\beta}{\mathrm{d}u} = \frac{\cos u}{-k \operatorname{cn}\beta \operatorname{dn}\beta},$$

and to substitute from eqn. (9.45) in the above giving

$$\frac{\mathrm{d}\beta}{\mathrm{d}u} = -\frac{1}{k \operatorname{cn}\beta}. \tag{9.48}$$

Thus, from eqns. (9.47) and (9.48),

$$\mathrm{d}z = S \operatorname{cn}\beta k \operatorname{cn}\beta \, \mathrm{d}\beta$$
$$= Sk(1 - \operatorname{sn}^2\beta) \, \mathrm{d}\beta,$$

which, from eqn. (9.9), equals

$$\frac{S}{k}(k^2 + \operatorname{dn}^2\beta - 1) \, \mathrm{d}\beta,$$

or (from the definition of k')

$$= \frac{S}{k}(\operatorname{dn}^2\beta - k'^2) \, \mathrm{d}\beta.$$

This expression can be integrated simply remembering that

$$\int_0^\beta \operatorname{dn}^2\beta \, \mathrm{d}\beta = E(\beta),$$

where $E(\beta)$ is the elliptic integral of the second kind. **Therefore,**

$$z = \frac{S}{k}(E(\beta) - k'^2\beta) + c_1,$$

which from eqn. (9.30)

$$= \frac{S}{k}\left(Z(\beta) + \beta\frac{E}{K} - k'^2\beta\right) + c_1,$$

where $Z(\beta)$ is the Jacobian zeta function. Finally, **rearranging the right-hand side gives**

$$z = \frac{S}{k}\left[Z(\beta) + \left(\frac{E}{K} - k'^2\right)\beta\right] + c_1. \tag{9.49}$$

This is the transformation equation in Jacobian form, and from it is now derived a relation between the constants in the two planes.

Let the sides of the rectangle be $2x_1$ and $2y_1$. The **correspondence** of points in the two planes is to be noted:

$$\left.\begin{array}{ll} t = 0, & z = x_1; \\ t = a, & z = x_1 - jy_1; \\ t = \pm\infty, & z = -x_1. \end{array}\right\} \tag{9.50}$$

and

14*

Substituting in equations (9.39), (9.40), (9.43), and (9.44) this gives:

$$\left.\begin{array}{llll} t = 0, & u = 0 & \text{and} & \beta = 0; \\ t = a, & u = -\alpha & \text{and} & \beta = K; \\ t = \pm\infty, & u = \pm\pi & \text{and} & b = 2jK'. \end{array}\right\} \tag{9.51}$$

and

Hence, corresponding values of β and z can be found from eqns. (9.50) and (9.51) and substituted in eqn. (9.49) to obtain expressions for the constants. Substituting in eqn. (9.49) for each of the corresponding values of β and z results in the following relationship:

$$c_1 = x_1,$$

since $Z(0) = 0$;

$$\frac{S}{k}(E - k'^2 K) = -jy_1,$$

since $Z(K) = 0$; and

$$\frac{S}{k}\left[-\frac{j\pi}{K} + 2jK'\left(\frac{E}{K} + k'^2\right)\right] = -2x_1,$$

since $Z(2jK') = -j\pi/K$.
Simplifying these gives

$$c_1 = x, \tag{9.52}$$

$$S = -\frac{jy_1 k}{E - k'^2 K}, \tag{9.53}$$

and

$$x_1 = \frac{jS}{kK}\left(\frac{\pi}{2} - K'E + k'^2 KK'\right). \tag{9.54}$$

But it can be shown (see **Legendre**) that

$$KE' + K'E - KK' = \frac{\pi}{2}, \tag{9.55}$$

and using this, eqn. (9.54) becomes

$$x_1 = \frac{jS}{k}(E' - K' + k'^2 K')$$

$$= \frac{jS}{k}(E' - k^2 K'),$$

giving

$$S = -\frac{jx_1 k}{E' - k^2 K'}. \tag{9.56}$$

Finally, eliminating S from eqns. (9.53) and (9.56) yields

$$\frac{y_1}{x_1} = \frac{E - k'^2 K}{E' - k^2 K'}, \tag{9.57}$$

and from this equation the ratio of the sides of the rectangle can be obtained for any value of the modulus k. The range of interest is for values of this ratio between 0 and 1; the case $y_1/x_1 = n$ is the same as $y_1/x_2 = 1/n$, with the sides of the rectangle interchanged, and substituting $k = k'$ in eqn. (9.57) gives a solution for the reciprocal of the ratio of the sides. (The limiting case, $k = k'$, corresponds with $y_1/x_1 = 1$, and eqn. (9.57) then gives $k = 1/\sqrt{2}$). Figure 9.6 shows the relationship between the ratio of the sides and the modulus k.

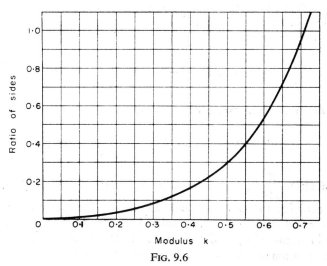

FIG. 9.6

Finally, eqn. (9.49) can be written as

$$z = \frac{-jx_1}{E' - k^2 K'} \left[Z(\beta) + \left(\frac{E}{K} - k'^2 \right) \beta \right] + x_1,$$ (9.58)

and if

$$C_2 = E - k'^2 K$$ (9.59)

and

$$C_3 = E' - k^2 K',$$ (9.60)

then eqn. (9.57) becomes

$$\frac{y_1}{x_1} = \frac{C_2}{C_3}$$ (9.61)

and eqn. (9.58) becomes

$$z = -\frac{jx_1}{C_3} \left[Z(\beta) + \frac{C_2 \beta}{K} \right] + x_1$$

$$= -\frac{jx_1}{C_3} \left[Z(\beta) + \frac{C_2 \beta}{K} + jC_3 \right]$$

$$= \frac{y_1}{C_2} \left[C_3 - j \left(Z(\beta) + \frac{C_3}{K} \beta \right) \right].$$ (9.62)

To summarize, the procedure for transforming a particular rectangular boundary is as follows. The value of k for the particular ratio of the sides (taking the smaller of the two) is read off from Fig. 9.6 and C_2 and C_3 are calculated from eqns. (9.59) and (9.60). Next, eqns. (9.43) and (9.40) are used to calculate k and a. Then the value of u is found from eqn. (9.39) for a given value of t, and is used in eqn. (9.44) to find β. Finally, β is employed in eqn. (9.62) to find z.

Series evaluation of the integral. The above procedure for finding z is tedious, and a much simpler method is now demonstrated. In this the above method is used to find α, and eqn. (9.56) is used to find S, but, for the general evaluation of corresponding points (in the t and z-planes), the right-hand side of eqn. (9.42) is expanded in a series, the coefficients of which can be derived in terms of Legendre polynomials.

Equation (9.42) can be rewritten

$$\frac{dz}{du} = -\frac{S}{2}\operatorname{cosec}\alpha\, e^{ju}(1-2e^{-2ju}\cos 2\alpha + e^{-4ju})^{1/2}, \tag{9.63}$$

and the radical term in this expression can be expanded in a series:

$$(1-2e^{-2ju}\cos 2\alpha + e^{-4ju})^{1/2} = \sum_{n=0}^{\infty} a_n e^{-2nju}. \tag{9.64}$$

A similar form of series $\sum_{n=0}^{\infty} P_n h^n$ is obtained by expanding $(1-2\mu h+h^2)^{1/2}$, and Legendre showed that the coefficients P_n can be calculated readily. These coefficients are known as *Legendre coefficients* and they have been tabulated. Putting $\cos 2\alpha = \mu$ in eqn. (9.64), a_n can be expressed in terms of P_n, and so evaluated. However, Bickley[7] shows that the coefficients can be determined more readily from the recurrence formula

$$(n+1)a_{n+1}-2(n-1)\mu a_n+(n-2)a_{n-1} = 0. \tag{9.65}$$

Thus, noting that $a_0 = 1$ and $a_1 = -\mu$, dz/du can be expressed by the equation

$$\frac{dz}{du} = -\frac{S}{2}\operatorname{cosec}\alpha\left[\sum_{n=0}^{\infty} a_n e^{-(2n-1)ju}\right]$$

$$= -\frac{S}{2}\operatorname{cosec}\alpha\left[e^{ju}+\sum_{n=1}^{\infty} a_n e^{-(2n-1)ju}\right]. \tag{9.66}$$

Integrating this equation gives

$$z = \frac{1}{2}jS\operatorname{cosec}\alpha\left[e^{ju}-\sum_{n=1}^{\infty}\frac{a_n}{(2n-1)}e^{-(2n-1)ju}\right]+c_4, \tag{9.67}$$

and, since S and α are known, z can be found for a given u. Therefore, for a particular value of t, u can be found from eqn. (9.39), and then z can be found from eqn. (9.67).

Expression of integrand in Legendrian form. It is instructive to express

$$\int \frac{\sqrt{(t^2-a^2)(t^2-1/a^2)}}{(t^2+1)^2}\,dt, \tag{9.68}$$

in terms of the principal elliptic integrals. Although in general this technique is difficult, in this case it is found that the integral does not involve the elliptic integral of the third kind and can be evaluated reasonably.

First, putting the radical part in the denominator gives

$$\frac{(t^2-a^2)(t^2-1/a^2)}{(t^2+1)^2} \frac{1}{\sqrt{(t^2-a^2)(t^2-1/a^2)}},$$

and the rational part of this can be put in partial fraction form as

$$1+\left(a^2+\frac{1}{a^2}+2\right)\left[\frac{1}{(t^2+1)^2}-\frac{1}{(t^2+1)}\right].$$

Thus, the integral can be separated into three parts, I_1, I_2, and I_3, where

$$I_1 = \int \frac{dt}{\sqrt{(t^2-a^2)(t^2-1/a^2)}}, \tag{9.69}$$

$$I_2 = -\left(a^2+\frac{1}{a^2}+2\right)\int \frac{dt}{(t^2+1)\sqrt{(t^2-a^2)(t^2-1/a^2)}}, \tag{9.70}$$

and

$$I_3 = \left(a^2+\frac{1}{a^2}+2\right)\int \frac{dt}{(t^2+1)^2\sqrt{(t^2-a^2)(t^2-1/a^2)}}. \tag{9.71}$$

The substitution $t = ax$ is now made in I_1, I_2, and I_3. Thus

$$I_1 = \int \frac{a\,dx}{\sqrt{(1-x^2)(1-a^4x^2)}}$$

$$= aF(x, a^2), \tag{9.72}$$

an elliptic integral of the first kind, and I_2 reduces to an elliptic integral of the third kind as

$$I_2 = -\left(a^2+\frac{1}{a^2}+2\right)\int \frac{1}{(1+a^2x^2)} \cdot \frac{a\,dx}{\sqrt{(1-x^2)(1-a^4x^2)}}$$

$$= -\left(a^2+\frac{1}{a^2}+2\right)\Pi(x, a, a^2). \tag{9.73}$$

I_3 becomes

$$\left(a^2+\frac{1}{a^2}+2\right)\int \frac{a\,dx}{(1+a^2x^2)^2\sqrt{(1-x^2)(1-a^4x^2)}}, \tag{9.74}$$

which can be expressed in the form

$$D(AI_4-I_5+P)/C, \tag{9.75}$$

(see Edwards), where

$$\left.\begin{array}{l} D = a^2+\dfrac{1}{a^2}+2, \\[2mm] A = 2a^2+\dfrac{2}{a^2}+4, \\[2mm] C = 2a^2+\dfrac{2}{a^2}+4, \\[2mm] P = \dfrac{x\sqrt{(1-x^2)(1-a^4x^2)}}{(1+a^2x^2)}; \end{array}\right\} \tag{9.76}$$

$$I_4 = \int \frac{1}{\sqrt{(1-x^2)(1-a^4x^2)}} \frac{dx}{(1+a^2x^2)}$$
$$= \Pi(x, a, a^2). \tag{9.77}$$

$$I_5 = \left(1+\frac{1}{a^2}\right) \int \frac{dx}{\sqrt{(1-x^2)(1-a^4x^2)}} - \frac{1}{a^2} \int \frac{\sqrt{(1-a^4x^2)}}{\sqrt{(1-x^2)}} dx$$
$$= \left(1+\frac{1}{a^2}\right) F(x, a^2) - \frac{1}{a^2} E(x, a^2). \tag{9.78}$$

Thus, combining the expressions for I_1, I_2, and I_3 gives

$$\int \frac{\sqrt{(t^2-a^2)(t^2-1/a^2)}}{(t^2+1)^2} dt = aF(x, a^2) - \frac{1}{2}\left(1+\frac{1}{a^2}\right) F(x, a^2)$$
$$+ \frac{1}{2a^2} E(x, a^2) + \frac{1}{2} \frac{x\sqrt{(1-x^2)(1-a^4x^2)}}{(1+a^2x^2)}$$
$$= \left(a-\frac{1}{2a^2}-\frac{1}{2}\right) F(x, a^2) + \frac{1}{2a^2} E(x, a^2)$$
$$+ \frac{1}{2} \frac{x\sqrt{(1-x^2)(1-a^4x^2)}}{(1+a^2x^2)}, \tag{9.79}$$

in which $x = t/a$.

The evaluation of $F(x, a^2)$ and $E(x, a^2)$ for complex x is tedious, and the above is not a good method for finding the relation between t and z.

The field solution. The boundary of the conductor is transformed into the real axis of the t-plane as described above, and the required field in the t-plane can be seen from the following two considerations. Firstly, since the conductor boundary is equipotential, the real axis of the t-plane must also be equipotential. Secondly, since all the flux emanating from the boundary in the z-plane terminates at infinity, Fig. 9.7(b), all the flux leaving the real axis in the t-plane must terminate at the point $t = j$ corresponding to $z = \infty$. Hence the necessary t-plane field is as shown in Fig. 9.7(a), and this is recognized as that of equal and opposite charges at the points $t = \pm j$ (see section 3.2.1). This field has the solution

$$w = \frac{q}{2\pi} \log\left(\frac{t-j}{t+j}\right), \tag{9.80}$$

where q is the charge per unit length of the conductor.

A quantity of some interest is the field strength, at the conductor surface, given by $dw/dt \times dt/dz$. Substituting the expression for dw/dt obtained by differentiating eqn. (9.80) and that for dt/dz from eqn. (9.38), gives

$$\left|\frac{dw}{dz}\right| = \frac{q(t^2+1)}{\pi S \sqrt{(t^2-a^2)(t^2-1/a^2)}}. \tag{9.81}$$

At the points corresponding to $t = 0$ and ∞ (the mid-points of the horizontal sides),

$$\left|\frac{dw}{dz}\right| = \frac{q}{\pi S},$$

and, of course, at the conductor corners dw/dz is infinite.

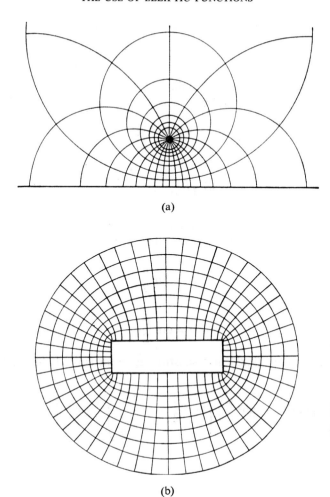

(a)

(b)

FIG. 9.7

In the next chapter other t-plane fields are considered to give, in the z-plane, the field of a current near a permeable rectangle and the flux distribution due to a rectangular conductor in a uniform applied electric field. However, it is interesting at this stage to consider an alternative transformation, that from the unit circle, for obtaining the field outside a rectangular boundary.

9.3.2. The transformation from a circular boundary

The transformation. Let the circular boundary of unit radius be put in the t-plane, and let the rectangular one be in the z-plane, both boundaries being symmetrical about the real and imaginary axes (Fig. 9.8). Further, let the corners of the rectangle be lettered a, b, c, and d, and the corresponding points in the t-plane a', b', c', and d'. These points are symmetrical about the axes. and therefore if at the point a', $t = \alpha + j\beta$, then, at b', $t = -\alpha + j\beta$, at c', $t = -\alpha - j\beta$, and at d', $t = \alpha - j\beta$.

Using eq. (8.72) the transformation between the z and t-planes is

$$\frac{dz}{dt} = \frac{S}{t^2}[t^2-(\alpha+j\beta)^2]^{1/2}[t^2-(\alpha-j\beta^2)]^{1/2}. \tag{9.82}$$

Since the circle in the t-plane is of unit radius $\alpha^2+\beta^2 = 1$, from which it can be seen that $a+j\beta = 1/(\alpha-j\beta)$. Thus substituting for $(\alpha-j\beta)$ in eqn. (9.82) gives

$$\frac{dz}{dt} = \frac{S}{t^2}[t^2-(\alpha+j\beta)^2]^{1/2}\left[t^2-\frac{1}{(\alpha+j\beta)^2}\right]^{1/2}, \tag{9.83}$$

and putting $\gamma = \alpha+j\beta$ in this equation,

$$\frac{dz}{dt} = \frac{S}{t^2}\sqrt{(t^2-\lambda^2)\left(t^2-\frac{1}{\gamma^2}\right)}. \tag{9.84}$$

In order to perform the integration, the right-hand side of this equation is put in Jacobian form, but to facilitate this, it is first expressed in a trigonometrical form by making the substitutions

$$t = \tan\frac{\theta}{2} \tag{9.85}$$

and

$$\gamma = \tan\frac{\varphi}{2}. \tag{9.86}$$

Hence, differentiating eqn. (9.85) and substituting from eqns. (9.85) and (9.86) in eqn. (9.84) gives

$$\frac{dz}{d\theta} = \frac{S}{\tan^2\frac{\theta}{2}}\sqrt{\left(\tan^2\frac{\theta}{2}-\tan^2\frac{\varphi}{2}\right)\left(\tan^2\frac{\theta}{2}-\frac{1}{\tan^2\frac{\varphi}{2}}\right)}\frac{1}{2}\sec^2\frac{\theta}{2} \tag{9.87}$$

$$= \frac{S}{2\sin^2\frac{\theta}{2}}$$

$$\times\sqrt{\left[\frac{\sin^2\frac{\theta}{2}\cos^2\frac{\varphi}{2}-\sin^2\frac{\varphi}{2}\cos^2\frac{\theta}{2}}{\cos^2\frac{\theta}{2}\cos^2\frac{\varphi}{2}}\right]\left[\frac{\sin^2\frac{\theta}{2}\sin^2\frac{\varphi}{2}-\cos^2\frac{\theta}{2}\cos^2\frac{\varphi}{2}}{\cos^2\frac{\theta}{2}\sin^2\frac{\varphi}{2}}\right]}$$

$$= \frac{S}{1-\cos\theta}$$

$$\times\sqrt{\left[\frac{(1-\cos\theta)(1+\cos\varphi)-(1-\cos\varphi)(1+\cos\theta)}{(1+\cos\theta)(1+\cos\varphi)}\right]\left[\frac{(1-\cos\theta)(1-\cos\varphi)-(1+\cos\theta)(1+\cos\varphi)}{(1+\cos\theta)(1-\cos\varphi)}\right]}$$

$$= \frac{S}{1-\cos^2\theta}\sqrt{\frac{4(\cos^2\theta-\cos^2\varphi)}{(1+\cos\varphi)(1-\cos\varphi)}}$$

$$= \frac{S}{\sin^2\theta}\sqrt{\frac{4(\cos^2\theta-\cos^2\varphi)}{(1-\cos^2\varphi)}}$$

$$= \frac{S}{\sin^2\theta\sin\varphi}\sqrt{2(\cos 2\theta-\cos 2\varphi)}. \tag{9.88}$$

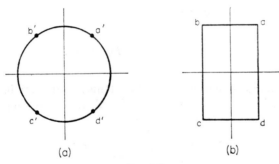

FIG. 9.8

To express this in Jacobian form put

$$k = \sin\varphi \tag{9.89}$$

and

$$\sin\theta = k \operatorname{sn} u. \tag{9.90}$$

Now by definition

$$\operatorname{dn} u = \sqrt{(1 - k^2 \operatorname{sn}^2 u)},$$

which, from eqn. (9.90), makes

$$\operatorname{dn} u = \cos\theta. \tag{9.91}$$

Also, differentiating eqn. (9.90) gives

$$\cos\theta \, d\theta = k \operatorname{cn} u \operatorname{dn} u \, du,$$

and thus

$$d\theta = \frac{k \operatorname{cn} u \operatorname{dn} u \, du}{\cos\theta}$$

or, using eqn. (9.91),

$$d\theta = k \operatorname{cn} u \, du. \tag{9.92}$$

Substituting from eqns. (9.89)–(9.91) in eqn. (9.88) yields

$$\frac{dz}{d\theta} = \frac{2kS \operatorname{cn} u}{k^3 \operatorname{sn}^2 u}, \tag{9.93}$$

and substituting in eqn. (9.93) for $d\theta$ gives

$$\frac{dz}{du} = \frac{2S \operatorname{cn} u}{k^2 \operatorname{sn}^2 u}(-k \operatorname{cn} u)$$

$$= \frac{2S}{k}\left(\frac{\operatorname{sn}^2 u - 1}{\operatorname{sn}^2 u}\right)$$

$$= \frac{2S}{k}\left(1 - \frac{1}{\operatorname{sn}^2 u}\right). \tag{9.94}$$

But it can be shown (see Edwards) that

$$\int \frac{du}{\operatorname{sn}^2 u} = -\frac{\operatorname{cn} u \operatorname{dn} u}{\operatorname{sn} u} + u - E(u) + \text{constant},$$

and so eqn. (9.94) can be integrated to give

$$z = \frac{2S}{k}\left[\frac{\mathrm{cn}\,u\,\mathrm{dn}\,u}{\mathrm{sn}\,u} + E(u)\right] + c_1. \tag{9.95}$$

This is the transformation equation and the constants S and c_1 are found by substitution of known corresponding points. For purposes of evaluation it is simplest to express $E(u)$ in terms of the zeta function, whence

$$z = \frac{2S}{k}\left[\frac{\mathrm{cn}\,u\,\mathrm{dn}\,u}{\mathrm{sn}\,u} + Z(u) + u\,\frac{E}{K}\right] + c_1. \tag{9.96}$$

The field solution. Using this transformation to derive the field of a charged rectangular conductor in the z-plane, the necessary field in the t-plane is that of a charged circular conductor of unit radius centred about the origin. This field has been considered earlier, see section 2.4.4, and its solution is

$$w = \psi + j\varphi = \frac{1}{2\pi}\log t. \tag{9.97}$$

(Note that on the conductor boundary $|t| = 1$, and so the potential of the conductor is set at zero.)

Comparing eqn. (9.97) with the expression for the field in the t-plane when the transformation from an infinite straight line is used, it is seen that eqn. (9.97) is simpler. The evaluation of corresponding points through the transformation equations involves a similar amount of work in the two methods.

9.4. The field in a slot of finite depth

An example involving the use of elliptic integrals of the third kind occurs in the analysis of the influence of slots or grooves on the field in the air gap of an electrical machine. Since for most machines the air-gap width is small compared with the slot dimensions, it is sufficient to consider the field of a single isolated slot. The transformation required for the solution of this problem has the same form as that required for the solution of the field of a succession of infinitely deep slots in a machine air gap, and both cases have been discussed briefly by Coe and Taylor.[8] Since the example involving a succession of slots has been described in detail by Gibbs, p. 190, the field of a shallow slot has been chosen for discussion here.

An accurate method for the evaluation of harmonics in the flux density wave form for the case considered by Gibbs, has been described by Freeman.[9]

The analysis of the field of a shallow slot can be divided into several parts as follows. First, the Schwarz–Christoffel differential equation is used to relate the boundary of the z-plane, which represents the boundary of a slot of finite depth opposite to a plane surface, Fig. 9.9(b), with the real axis of the t-plane, Fig. 9.9(a). Next the resulting integral is expressed in Jacobian form when it is seen to be of the third kind. Then the scale constant is determined by the method of residues, and the remaining t-plane constants are evaluated in terms of the dimensions of the slot. Finally, a solution is found for the t-plane field which gives, in the z-plane, a magnetic potential difference between the plane and the slotted

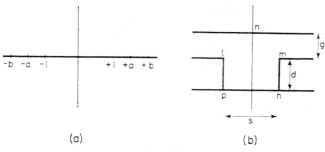

$$(a) \qquad\qquad (b)$$

FIG. 9.9

boundary (this difference in potential being the resultant effect of the currents flowing in the other slots of the machine).

The transformation. Let the origin in the z-plane be chosen as the centre of the slot bottom and let the origins in the two planes correspond so that both fields are symmetrical about their imaginary axes. Also let the points in the z-plane lettered l, m, p, and h correspond to the points $t = -a$, $t = +a$, $t = -1$, and $t = +1$. The points $t = -b$ and $t = +b$ correspond in the z-plane to points infinitely far along the gap on each side of the slot, and $t = \pm \infty$ becomes the point n in the z-plane. Let the gap width be g, the slot width be s, and the slot depth be d.

Employing the Schwarz–Christoffel eqn. (8.1) to relate the two planes gives

$$\frac{dz}{dt} = \frac{S\sqrt{(t^2-a^2)}}{\sqrt{(t^2-1)(t^2-b^2)}}$$

$$= -\frac{aS}{b^2}\sqrt{\frac{1-t^2/a^2}{1-t^2}}\,\frac{1}{(1-t^2/b^2)}, \tag{9.98}$$

and writing $k = 1/a$ and $k_1 = 1/b$, this equation takes the standard elliptic form

$$\frac{dz}{dt} = \frac{S'}{(1-k_1^2 t^2)}\sqrt{\frac{1-k^2 t^2}{1-t_2}}. \tag{9.99}$$

It can be put in Jacobian form by making

$$t = \operatorname{sn} u \tag{9.100}$$

to modulus k, when

$$dt = \operatorname{cn} u \operatorname{dn} u \, du. \tag{9.101}$$

So substituting from eqns. (9.100) and (9.101) in eqn. (9.99) and replacing $\operatorname{dn} u$ by $\sqrt{(1-k^2 \operatorname{sn}^2 u)}$ gives

$$z = S'\int_0^u \frac{1-k^2 \operatorname{sn}^2 u}{1-k_1^2 \operatorname{sn}^2 u}\,du$$

$$= S'\int_0^u 1 + \frac{(k_1^2-k^2)\operatorname{sn}^2 u}{1-k_1^2 \operatorname{sn}^2 u}\,du$$

$$= S'\left[u + (k_1^2-k^2)\int_0^u \frac{\operatorname{sn}^2 u}{1-k_1^2 \operatorname{sn}^2 u}\,du\right]. \tag{9.102}$$

Now $b > a$ and so $k > k_1$; it is thus valid to put $\operatorname{sn}\alpha = k_1/k$, which makes

$$k_1^2 - k^2 = -k^2 \operatorname{cn}^2\alpha.$$

Substituting for k_1 from this equation in eqn. (9.102) gives

$$z = S'\left(u - k^2 \operatorname{cn}^2\alpha \int_0^u \frac{\operatorname{sn}^2 u\,du}{1 - k^2 \operatorname{cn}^2\alpha \operatorname{sn}^2 u}\right)$$

$$= S'\left(u - \frac{\operatorname{cn}\alpha}{\operatorname{sn}\alpha \operatorname{dn}\alpha} k^2 \operatorname{cn}\alpha \operatorname{sn}\alpha \operatorname{dn}\alpha \int_0^u \frac{\operatorname{sn}^2 u\,du}{1 - k^2 \operatorname{sn}^2\alpha \operatorname{sn}^2 u}\right),$$

which, using eqn. (9.28), can be written

$$z = S'\left[u - \frac{\operatorname{cn}\alpha}{\operatorname{sn}\alpha \operatorname{dn}\alpha} \Pi_J(u, \alpha)\right], \tag{9.103}$$

to modulus k, where Π_J is Jacobi's elliptic integral of the third kind.

The value of the scale constant S' is found in the usual way by employing the method of residues. Thus, considering the vertices of the z-plane polygon at the points corresponding to $t = \pm 1/k_1$ and employing eqn. (8.26) in which $D = jg$ yields

$$S' = -\frac{2k_1 g}{\pi} \sqrt{\frac{1 - 1/k_1^2}{1 - k^2/k_1^2}}$$

$$= -\frac{2g}{\pi} \sqrt{\frac{k_1^2 - 1}{1 - k^2/k_1^2}}.$$

Substituting for k_1 and using the definitions of $\operatorname{cn}\alpha$ and $\operatorname{dn}\alpha$ gives

$$S' = -\frac{2g}{\pi} \frac{\operatorname{sn}\alpha \operatorname{dn}\alpha}{\operatorname{cn}\alpha}, \tag{9.104}$$

and putting this value of S' in eqn. (9.103) gives

$$z = -\frac{2g}{\pi} \frac{\operatorname{sn}\alpha \operatorname{dn}\alpha}{\operatorname{cn}\alpha} u + \frac{2g}{\pi} \Pi_J(u, \alpha) \tag{9.105}$$

to modulus k.

It now remains to express α and k in terms of the slot width, slot depth, and gap width. Since $t = \operatorname{sn} u$, at $t = 1/k$, then, $u = K + jK'$ from eqn. (9.6). Thus as t changes from 0 to $1/k$, u changes from 0 to $K + jK'$, and z from 0 to $s/2 + jd$. Putting the corresponding values of u and z in eqn. (9.105) it is found that

$$\frac{s}{2} + jd = -\frac{2g}{\pi} \frac{\operatorname{sn}\alpha \operatorname{dn}\alpha}{\operatorname{cn}\alpha} (K + jK') + \frac{2g}{\pi} \Pi_J(K + jK', \alpha). \tag{9.106}$$

But, substituting $u = K + jK'$ in eqn. (9.35) gives

$$\Pi_J(K + jK', \alpha) = \frac{1}{2} \log \frac{\Theta(K + jK' - \alpha)}{\Theta(K + jK' + \alpha)} + (K + jK') Z(\alpha), \tag{9.107}$$

and to determine α it is necessary to evaluate the theta functions in this equation. To do this, three relationships are used,[†] but not proved here:

$$Z(K+jK'-\alpha) = Z(K-\alpha) + \frac{\text{cn}(K-\alpha)}{\text{sn}(K-\alpha)}\,\text{dn}(K-\alpha) - \frac{j\pi}{2K}, \tag{9.108}$$

$$\int_0^u Z(u)\,du = \log\Theta(u), \tag{9.109}$$

$$\int \frac{\text{cn}\,u\,\text{dn}\,u}{\text{sn}\,u}\,du = \log\,\text{sn}\,u. \tag{9.110}$$

Now, integrating eqn. (9.108), substituting $K+jK'-\alpha = u$ in eqn. (9.109), and $u = K-\alpha$ in eqn. (9.110) the resulting relationships can be used to give

$$\log\Theta(K+jK'-\alpha) = \log\Theta(K-\alpha) + \log\,\text{sn}(K-\alpha) - \frac{j\pi(K-\alpha)}{2K} + \text{constant.} \tag{9.111}$$

A similar expression for $\Theta(K+jK'+\alpha)$ is

$$\log\Theta(K+jK'+\alpha) = \log\Theta(K+\alpha) + \log\,\text{sn}(K+\alpha) - \frac{j\pi(K+\alpha)}{2K} + \text{constant.} \tag{9.112}$$

Subtracting these

$$\log\frac{\Theta(K+jK'-\alpha)}{\Theta(K+jK'+\alpha)} = \log\frac{\Theta(K-\alpha)}{\Theta(K+\alpha)} + \log\frac{\text{sn}(K-\alpha)}{\text{sn}(K+\alpha)} + \frac{j\pi(2a)}{2K}. \tag{9.113}$$

However, it can be shown (see Byrd and Friedman, p. 314) that the first two terms of the right-hand side of this equation are identically zero, and so substituting in (9.107) from (9.113) gives

$$\Pi_J(K+jK',\alpha) = \frac{j\pi\alpha}{2K} + (K+jK')\,Z(\alpha).$$

Combining this with eqn. (9.106) yields

$$\frac{s}{2} + jd = -\frac{2g}{\pi}\frac{\text{sn}\,\alpha\,\text{dn}\,\alpha}{\text{cn}\,\alpha}(K+jK') + \frac{2g}{\pi}\left[\frac{j\pi\alpha}{2K} + (K+jK')\,Z(\alpha)\right],$$

and separating the real and imaginary parts,

$$\frac{s}{2} = -\frac{2g}{\pi}\frac{\text{sn}\,\alpha\,\text{dn}\,\alpha}{\text{cn}\,\alpha}K + \frac{2g}{\pi}KZ(\alpha),$$

and

$$d = -\frac{2g}{\pi}\frac{\text{sn}\,\alpha\,\text{dn}\,\alpha}{\text{cn}\,\alpha}K' + \frac{2g}{\pi}\frac{\pi\alpha}{2K} + \frac{2g}{\pi}K'Z(\alpha).$$

Thus

$$\frac{s}{g} = \frac{4K}{\pi}\left[Z(\alpha) - \frac{\text{sn}\,\alpha\,\text{dn}\,\alpha}{\text{cn}\,\alpha}\right] \tag{9.114}$$

[†] These and other standard relationships are in general difficult to derive and are best left to the specialist. However, a comprehensive list is to be found in the book by Byrd and Friedman.

and

$$\frac{d}{g} = \frac{2}{\pi} K' \left[Z(\alpha) - \frac{\text{sn } \alpha \text{ dn } \alpha}{\text{cn } \alpha} \right] + \frac{\alpha}{K} .$$

(9.115)

These equations give the relationship between the ratios s/g and d/g and the parameters α and k. If values of the latter are chosen, then, using tables, the values of the elliptic functions can be found and substituting in eqns. (9.114) and (9.115) gives s/g and d/g. However, in practice, it is required to find α and k for given ratios of the slot dimensions. To do this, it is necessary to obtain values of the slot dimensions for a range of values of the parameters, and then to plot graphs and to read off the values of the parameters for the required slot dimensions.

Field solution. To represent a potential difference between the plane surface and the slotted surface in the z-plane, the t-plane must have the section of the real axis between $t = 1/k_1$ and $t = -1/k_1$ at one potential, and the remainder of the real axis at another potential. The field is equivalent to that of two equal currents of different sign, one at $t = 1/k_1$ the other at $t = -1/k_1$, and, if the potential difference between the boundaries is ψ_d, the solution is

$$w = \frac{\psi_d}{\pi} \log \left(t - \frac{1}{k_1} \right) - \frac{\psi_d}{\pi} \log \left(t + \frac{1}{k_1} \right)$$

$$= \frac{\psi_d}{\pi} \log \left(\frac{tk_1 - 1}{tk_1 + 1} \right).$$

(9.116)

This gives $\psi = \psi_d$ along the real axis between $t = -1/k_1$ and $t = +1/k_1$, and $\psi = 0$ along the remainder of the real axis.

An interesting point which can be resolved by using the above solution is whether it is permissible or not to analyse the air-gap field in a machine on the simplifying assumption that the slots are infinitely deep. A measure of the effect of the limited depth of the slot is provided by the quantity of flux entering the slot bottom and this quantity, given by the change in w between the points corresponding to $t = -1$ and $t = +1$, equals

$$\frac{2\psi_d}{\pi} \log \left| \frac{k_1 - 1}{k_1 + 1} \right|.$$

This is found to be negligible (less than 1 per cent) when the ratio of slot depth to width is greater than unity. The flux density distribution on the plane surface is also of considerable importance, being required in the evaluation of pole-face loss. It is found in the usual way by evaluating dw/dz from the product $dw/dt \times dt/dz$. Thus, differentiating equation (9.116) leads to

$$\frac{dw}{dt} = \frac{2\psi_d}{\pi k_1 (t^2 - 1/k_1^2)} ,$$

and combining this with eqns. (9.99) and (9.104) makes

$$\left| \frac{dw}{dz} \right| = \left| \frac{\psi_d k_1 \text{ cn } \alpha}{g \text{ sn } \alpha \text{ dn } \alpha} \sqrt{\frac{1 - t^2}{1 - k^2 t^2}} \right| ,$$

(9.117)

where $\text{sn } \alpha = k_1/k$. Substituting $t = \pm \infty$ yields the density at the centre of the plane surface as

$$\frac{\psi_d}{\pi} \frac{\operatorname{cn} \alpha}{\operatorname{dn} \alpha}.$$

This is the minimum value of density on the plane surface, and therefore the amplitude of variation of flux density along the plane surface is

$$\frac{\psi_d}{g} \left(1 - \frac{\operatorname{cn} \alpha}{\operatorname{dn} \alpha} \right). \tag{9.118}$$

9.5. Conclusions

Any integral of the form

$$\int \frac{R_1(t)\, dt}{R_2(t)\, \sqrt{(t-a)\,(t-b)\,(t-c)}}$$

or

$$\int \frac{R_2(t)\, dt}{R_1(t)\, \sqrt{(t-a)\,(t-b)\,(t-c)\,(t-d)}},$$

in which $R_1(t)$ and $R_2(t)$ are rational functions of t, is elliptic. It can be expressed in terms of the basic elliptic integrals, though the required manipulation and subsequent numerical evaluation may be extremely difficult and laborious. The first of the above two forms of integral arises in the transformation of a boundary having three right angles, and the second four right angles. An integral containing more than four rooted factors in the denominator (occurring when the boundary has more than four right angles) is said to be hyperelliptic, and it cannot be expressed in terms of the basic elliptic integrals. Elliptic integrals also occur in other transformations, e.g. in the treatment of some boundaries having rounded corners[10] and of others having angles not multiples of $\pi/2$.[11] Both of these types of transformation are discussed in the next chapter. It is not possible to discuss here the general problem of manipulation, and instead the reader is referred to the excellent treatises by Edwards and by Tannery and Molk.

When attempting to assess the degree of difficulty associated with the analysis of a particular problem, one cannot in general apply any simple test. However, it is possible to distinguish broadly the values of the three methods of handling an elliptic integral which have been described, its expression in terms of Jacobian elliptic functions, its reduction to Legendre's three integrals, and the expansion of the integrand as a series. Of these, the first is the best for analytical manipulation. The second is of very limited value both for manipulation and evaluation, and the third is unsuitable for manipulation but is sometimes good for numerical evaluation. The example of the transformation of a rectangular boundary considered in section 9.3 is suitable for the demonstration of all three methods, but such examples are rare.

The considerations involved in the numerical evaluation of an analytical solution are rather more straightforward. When the evaluation is carried out by hand, the only generally practical method involves the use of the tables; the chief problem is that there are then large intervals over certain ranges and as a result interpolation is very difficult. The alternative to the above method involves the use of a digital computer to evaluate the function from a series expansion, see Appendix II. This is preferable to feeding tables into the computer store. Freeman[9] has applied the series developed by King, which are most suitable for evaluation of elliptic integrals and functions and which converge more rapidly than the other better-known series.

In Chapter 10 a powerful numerical method is described and this can be used for all classes of problem (including of course elliptic and hyperelliptic). Where analysis is impossible or seems likely to be difficult, this method can be used to great advantage.

There is an alternative notation for elliptic functions to that of Jacobi due to Weierstrass (see Whittaker and Watson), and it has been used by several authors including Frank and Mises, Karman and Burgers, and Bateman.

References

1. F. W. Carter, The magnetic field of the dynamo-electric machine, *J. Instn. Elect. Engrs.* **359,** 1115–39 (1926).
2. J. B. Izatt, Characteristic impedance of two special forms of transmission line, *Proc. Inst. Elect. Engs.* **111,** 1551 (1964).
3. Symposium on microwave strip circuits, *Inst. Radio Engrs. Trans.* MTT-3 (1955).
4. S. B. Cohn, Shielded coupled strip transmission line, *Inst. Radio Engrs. Trans.* MTT-3, 29 (1955).
5. D. Park, Planar transmission lines, *Inst. Radio Engrs. Trans.* MTT-3, I, 8–12 (1955) and II, 7–11 (1955).
6. W. H. Hayt, Jr., Potential solution of a homogeneous strip line of finite width, *Inst. Radio Engrs. Trans.* MTT-3, 16–18 (1955).
7. W. G. Bickley, Two-dimensional potential problems for the space outside a rectangle, *Proc. Lond. Math. Soc.,* ser. 2, 37 (2), 82 (1932).
8. R. T. Coe and H. W. Taylor, Some problems in electrical machine design involving elliptic functions, *Phil. Mag.* **6,** 100 (1928).
9. E. M. Freeman, The calculation of harmonics due to slotting in the flux-density waveform of a dynamo-electric machine, *Proc. Inst. Elect. Engrs.* 109 C, 581 (1962).
10. J. D. Cockroft, The effect of curved boundaries on the distribution of electrical stress round conductors, *J. Inst. Elect. Engrs.* **66,** 385 (1928).
11. Y. Ikeda, Die konformen Abbildungen der Polygone mit zwei Ecken, *J. Fac. Sci. Hokkaido. Univ.,* ser. 2, **2** (2), 1 (1938).

Additional References

Anderson, G. M., The calculation of the capacitance of coaxial cylinders of rectangular cross-sections, *Trans. Am. Inst. Elect. Engrs.* **69** II, 728 (1950).
Bates, R. H. T., The characteristic impedance of the shielded slab line, *Trans. Inst. Rad. Engrs.* MTT-4, 28 (1956).
Bergmann, S., *Math. Z.* **19,** 8 (1923).
Chen, T. S., Determination of the capacitance, inductance, and characteristic impedance of rectangular lines, *Trans. Inst. Rad. Engrs.* MTT-8, 510 (1960).
Chisholm, R. M., Characteristic impedance of through and slab lines, *Trans. Inst. Rad. Engrs.* MTT-4, 166 (1956).
Cohn, S. B., Analysis of the metal-strip delay structure for microwave lenses, *J. App. Phys.* **20,** 251 (1949).
Cohn, S. B., Characteristic impedance of broadside coupled strip transmission line, *Trans. Inst. Rad. Engrs.* MTT-8, 633 (1960).
Cohn, S. B., Thickness corrections for capacitance obstacles and strip conductors, *Trans. Inst. Rad. Engrs.* MTT-8, 638 (1960).
Davy, N., The field between equal semi-infinite rectangular electrodes or magnetic pole-pieces, *Phil. Mag.,* ser. 7, **35,** 819 (1944).
Foster, K., The characteristic impedance and phase velocity of high *Q* triplate line, *Br. Inst. Rad. Engrs.* **18,** 718 (1958).

GETZINGER, W. J., A coupled strip line configuration using printed-circuit construction that allows very close coupling, *Trans. Inst. Rad. Engrs.* MTT-9, 535 (1961).

GETZINGER, W. J., Coupled rectangular bars between parallel plates, *Trans. Inst. Rad. Engrs.* MTT-10, 65 1 (1962).

GREENHILL, A. G., Solution by means of elliptic functions of some problems in the conduction of electricity and heat in plane figures, *Q. J. Pure Appl. Math.* **17,** 289 (1881).

HERBERT, C. M., *Phys. Rev.* II, 17, 157 (1921).

IKEDA, Y., Die konformen Abbildungen der Polygone mit zwei Ecken, *J. Fac. Sci. Hokkaido Univ.*, ser. 2, **2** (2), 1 (1938).

LANGTON, N. H., and DARY, N., Two-dimensional field above and below an infinite corrugated sheet, *Br. J. Appl. Phys.* **3,** 156 (1952).

LEVY, R., New coaxial to strip transformers using rectangular lines, *Trans. Inst. Rad. Engrs.* MTT-9, 273 (1961).

MORSZTYN, K., The application of conformal transformations and elliptic functions to the analysis of a synchronous non-salient-pole machine, Monash Univ. Eng. Report, MEE 66-1 (1966).

MORTON, N. B., Two-dimensional fields specified by elliptic functions, *Phil. Mag.* **2,** 827 (1926).

PALMER, H. B., Capacitance of a parallel-plate capacitor, *Elect. Engrs. Lond.* **56,** 363 (1937).

PETERSOHN, H., Electrostatic problems, *Z. Physik.* **38,** 727 (1926).

ROSE, M. E., Magnetic field corrections in the cyclotron, *Phys. Rev.* **53,** 715 (1938).

TERRY, A., and KELLER, E. G., Field pole leakage flux in salient-pole dynamo-electric machines, *J. Instn. Engrs.* **83,** 845 (1938).

CHAPTER 10

GENERAL CONSIDERATIONS

10.1. Introduction

Each of the earlier chapters in Part III is devoted, by the detailed presentation of simple examples, to introducing the reader to the basic techniques of transformation methods. In contrast, this chapter is concerned with more advanced topics, and a minimum of detail is given in the worked examples. The possible extensions and generalizations of the basic methods are examined, and these can be divided into two groups—those concerned with boundary shape and those concerned with the type of field source.

The boundary shapes discussed so far are either wholly curved or wholly polygonal with angles which are integral multiples of $\pi/2$. It is shown here how boundaries, which are partly curved and partly polygonal, or which are polygonal with angles not multiples of $\pi/2$, can be treated. A significant extension in the range of boundary shapes can also be achieved by the use of numerical integration of functions of a complex variable. The use of this technique for the determination, by iteration, of the constants in the Schwarz–Christoffel equation, and also for the subsequent calculation of the field, is described.

The discussion of field type gives particular emphasis to the fields of line currents near boundaries, to fields exterior to closed boundaries, and to the general solution of the Dirichlet problem (i.e. the problem in which potential is specified at all points on a closed boundary). The latter involves the use of the Schwarz complex potential function or the Poisson integral.

10.2. Field sources

There is no previous general discussion of line sources (i.e. currents, charges, and poles) or their combination (e.g., in doublets) in the presence of boundaries, and this type of field is now considered in terms of the magnetic field of line currents. There are two reasons for discussing such fields. First, they have immediate applications in the solution of the fields of currents near complicated boundaries. (This method of solution does not seem to be in common use, though attention has been recently drawn to it.[1]) Secondly, a knowledge of them is necessary for the solution of certain fields exterior to closed boundaries.

First, consider a line current i remote from all boundaries, the field of which is expressed by

$$w = \frac{i}{2\pi} \log(t - \alpha). \tag{10.1}$$

By superposition it is possible to express the field of a combination of m currents, with magnitude i_n as

$$w = \frac{1}{2\pi} \sum_{n=0}^{m} i_n \log(t-\alpha_n). \qquad (10.2)$$

Consideration of these equations raises the matter of the existence of currents at infinity, a point best explained in terms of the hydrodynamical analogy of the above field. The equation of the laminar flow of liquid between a combination of sources and sinks is the sum of logarithmic terms, as in eqn. (10.2) and, it is easily seen that the algebraic sum of the sources at finite points is equal, in magnitude, to the sink at infinity. The corresponding relationship is also true for currents, which must have a return path at infinity except, of course, when the algebraic sum of the currents at finite points is zero. This return path does not affect the field equation therefore since it is at infinity; but, when transformed, it may occur at a finite point and then it must be taken into account. This feature has already appeared in the discussion of two currents inside a permeable tube, section 7.1.3. It is discussed further in section 8.3.2, where the field of a charged plate is obtained; the charge at infinity, in the z-plane, is transformed to the point $t = +j$, in the t-plane.

In general, solutions can be simply obtained for the fields of currents (or charges) near boundaries which are impermeable, infinitely permeable or partly impermeable and partly infinitely permeable. This is conveniently done by transformation from the infinite straight line, and the only limitation is that it must be possible to find the appropriate transformation equation. Consideration is now given to the field solutions for the above types of boundary, and it will be appreciated from the last paragraph that a distinction must be made between infinite and finite boundaries.

10.2.1. *Infinite boundaries*

For convenience, the infinite straight-line boundary is taken to lie along the real axis and the line current at a point on the imaginary axis.

Line current near an infinitely permeable plane. As shown in section 3.2.1, the field of a line current, at a distance a from an infinitely permeable plane extending to infinity, may be expressed as

$$w = \frac{1}{2\pi} \log(t^2 + a^2). \qquad (10.3)$$

Line current near an impermeable plane. If the plane is impermeable to flux, the field equation becomes, see section 3.2.1,

$$w = \frac{1}{2\pi} \log \frac{t-ja}{t+ja}. \qquad (10.4)$$

The field of a current near an impermeable plate, for example, is found by combining eqn. (10.4) and eqn. (8.64) and a map of this field is shown in Fig. 10.1.

Line current near partly permeable and partly impermeable boundary. The field of a current near a boundary consisting of two parts, one permeable and the other impermeable (Fig. 10.2), may be obtained from the field of four line currents—two positive and two negative. Figure 10.3 shows, in the p-plane, the position of these currents the field of which is given by

$$w = \frac{1}{2\pi} \log \frac{(p-a)^2 + b^2}{(p+a)^2 + b^2}. \qquad (10.5)$$

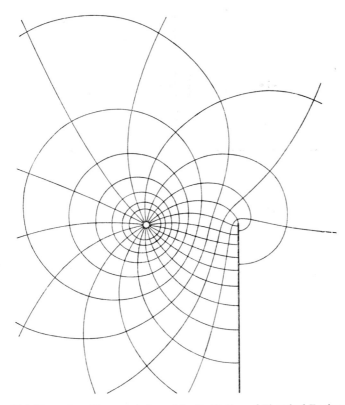

FIG. 10.1 (Reproduced by permission of the Institution of Electrical Engineers.)

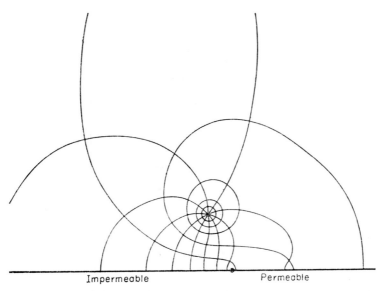

Impermeable   Permeable

FIG. 10.2

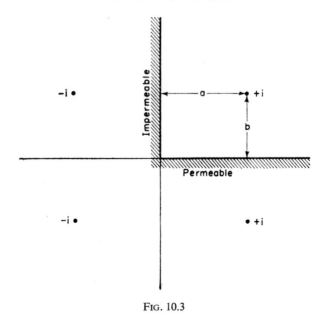

FIG. 10.3

FIG. 10.4 (Reproduced by permission of the Institution of Electrical Engineers.)

It is seen that there are two lines of symmetry—one a flux line and the other an equipotential line, and any quadrant in the p-plane can be transformed to give the upper half of the t-plane by the equation

$$p = 2 \sqrt{t}. \tag{10.6}$$

Thus the field expression in the t-plane is

$$w = \frac{1}{2\pi} \log \frac{(2 \sqrt{t} - a)^2 + b^2}{(2 \sqrt{t} + a)^2 + b^2}. \tag{10.7}$$

The field of a current near a permeable corner. As an example of the use of these field equations, the field of a line current close to an infinitely permeable corner (Fig. 10.4), is developed. Let the t- and z-planes be as shown in Fig. 10.5, the current being at the point

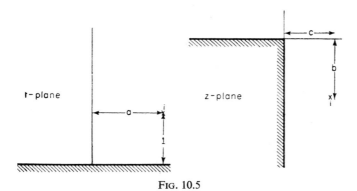

FIG. 10.5

$(a+j)$ in the t-plane, and $(c-jb)$ in the z-plane. Further, let the vertex of the corner be at the origin in the z-plane, corresponding to the origin in the t-plane. By a change of origin of $+a$ in eqn. (10.3), it is seen that the required field in the t-plane is given by

$$w = \frac{i}{2\pi} \log [(t-a)^2 + 1]. \tag{10.8}$$

The Schwarz–Christoffel equation relating the two planes is

$$\frac{dz}{dt} = S \sqrt{t}, \tag{10.9}$$

which, when integrated, gives

$$z = \tfrac{2}{3} S t^{3/2}. \tag{10.10}$$

The constant of integration is zero since the origins in the two planes correspond, and the remaining constants, S and a, are found in terms c and b by substitution in eqn. (10.10) for corresponding points. Then, eliminating t from eqns. (10.8) and (10.10) gives the required field solution.

10.2.2. *Finite boundaries*

Of the three types of boundary mentioned in the introduction to this section the one which is partly impermeable and partly infinitely permeable is of little interest for finite boundaries. It is therefore omitted from the discussion though the field solution can be easily found if required.[1] However, the influence of closed boundaries on a uniform applied field is often of interest and the use of doublets for the treatment of such influences is considered. As pointed out earlier, in the transformation from the infinite half-plane to the outside of a finite polygon, the point at infinity in one plane becomes a finite point in the other, and so a field source at infinity occurs at a finite point when transformed.

Line current near a permeable boundary. To obtain the field of a line current outside a closed polygonal boundary of infinite permeability in the z-plane, the field required, in the t-plane, is that due to the corresponding line current and its return which occurs at the point $t = +j$, and the images of both of these in the boundary, which is an equipotential line. If the point in the t-plane corresponding to the position of the current is $t = t_i$, there is an image of like sign at $t = \bar{t}_i$, where $\bar{t}_i$ is the conjugate of t_i; also, the return current at $t = +j$ has an image of like sign at $t = -j$. Hence the field equation is

$$w = \frac{1}{2\pi} \log \frac{(t-t_i)(t-\bar{t}_i)}{(t+j)(t-j)}$$

$$= \frac{1}{2\pi} \log \frac{(t-t_i)(t-\bar{t}_i)}{t^2+1}. \tag{10.11}$$

Line current near an impermeable boundary. If the polygon is impermeable to flux, the boundaries in both planes are flux lines and the two images in the infinite, straight boundary change their sign as compared with the previous case. The equation expressing the field in the t-plane is, then,

$$w = \frac{1}{2\pi} \log \frac{(t-t_i)(t+j)}{(t-\bar{t}_i)(t-j)}. \tag{10.12}$$

Charged boundaries. The field of a charged plate is obtained earlier, see section 8.3.2. As explained there, the expression for the t-plane field used to obtain the field of finite, charged boundaries is

$$w = \frac{1}{2\pi} \log \frac{t-j}{t+j}. \tag{10.13}$$

Uniform applied field: use of doublet. To obtain the field of a polygonal boundary in an applied uniform field, from the infinite, straight-line boundary, requires the use of doublets, see section 3.3.2: the applied uniform field in the z-plane is achieved by placing a doublet at infinity. This doublet is transformed to the point $t = +j$ in the t-plane where it also has an image at the point $t = -j$. Hence, if the boundary is equipotential the field in the t-plane is given by

$$w = \frac{1}{t+j} + \frac{1}{t-j}$$

$$= \frac{2t}{t^2+1}. \tag{10.14}$$

Figure 10.6 shows a map of the equipotential lines for a rectangular conductor in an applied uniform electric field. If the boundary is a flux line the field takes the form

$$w = \frac{1}{t+j} - \frac{1}{t-j}$$

$$= \frac{-2j}{t^2+1}. \qquad (10.15)$$

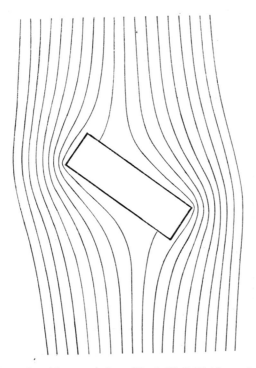

FIG. 10.6 (Reproduced by permission of Prof. W. G. Bickley and the London Mathematical Society.)

10.2.3. *Distributed sources*

It is possible to represent a source, which is distributed over an area or along a line, by a finite number of line sources and, in this way, to obtain a good approximation for certain practical problems. However, because of the difficulty of locating points away from the boundaries, see section 10.5, the method is of most value for problems in which the currents are distributed along the boundary. For example, boundaries which are not equipotential may be represented as consisting of a number of small equipotential segments, the potential difference between each being obtained by a line current at the point of division. An application of this method would be to the field of a salient pole with a distributed winding down the side.

10.3. Curved boundaries

Several simple transformation equations which can be used to give curved boundaries are described in Chapter 7. The treatment of more difficult transformations is given here, though it should be first emphasized that there is no general method of obtaining analytical solutions for curved boundaries. The two principal methods which can be used are both based upon modifications of the Schwarz–Christoffel equation: one is used to round the vertices of a polygon and the other to curve some or all of its sides.

Before discussing these two methods, however, it is useful to recall the device used in connection with the field of a Rogowski electrode, see section 8.2.3: an equipotential line (or flux line) in the field of a simple boundary shape is used to represent an actual boundary of a complicated shape. In this way, Carter[2] examined the field near the salient pole-tip by using the simple boundary shown in Fig. 10.7, representing the pole shape by an equipotential line shown dotted. In a similar way he also represented, in the same paper, the field of a semi-closed slot. There is no routine method of approach to such problems; instead, a combination of intuition and trial and error is required, choosing first a simplified boundary and then an appropriate field line.

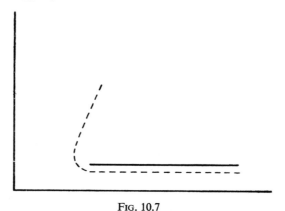

FIG. 10.7

Curved boundaries are, in general, of limited practical application in electric and magnetic field problems, but exceptions to this occur in the study of breakdown of insulation under high voltages and of special forms of high frequency transmission lines.

10.3.1. *Rounded corners*

It is possible to obtain a transformation equation which makes the vertices of a polygon rounded instead of sharp, by modifying the appropriate product terms in the Schwarz–Christoffel equation. The shape of the curve(s) cannot be chosen precisely to correspond with any given curve(s) but, by trial and error, a good approximation to the required curve can usually be obtained. The method consists in rounding one or more corners by replacing the terms of the form $(t-a)^{\alpha/\pi-1}$, (which give rise to sharp corners) by terms of the form

$$[(t-a')^{\alpha/\pi-1}+\lambda(t-a'')^{\alpha/\pi-1}].$$

The points a' and a'' in the t-plane correspond to the limits of the curve, and λ is a factor which determines the shape of the curve—for $t > a'$ or $t < a''$, dz/dt is constant, but for $a' > t > a''$, dz/dt varies continuously with t. The total change in direction over the curved section is the same as that produced by the equivalent, normal vertex factor. By consideration of the new term, it is clear that, providing $\alpha \geqslant \pi$, it does not introduce poles, zeros or sudden changes in dz/dt and so it does not give rise to any additional vertices or discontinuities in the boundary. For problems having $\alpha < \pi$, it is necessary, in order to avoid the introduction of poles, to transform the negative imaginary half plane, for which the corresponding vertex angle is $(2\pi - \alpha)$.

As a demonstration of the method, consider the field between an infinite charged plane and a conducting block with a rounded corner, used by Drinker[3] in the analysis of the performance of magnetic reading heads. Let this configuration be in the z-plane with the dimensions shown in Fig. 10.8; it is seen that the corner is rounded between the points

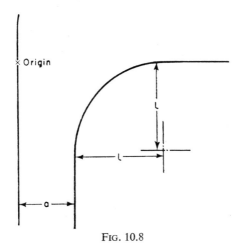

Fig. 10.8

$z = a + l$ and $z = a - jl$. The real axis of the t-plane is to be transformed into this configuration, the precise shape of the rounded corner being fixed by the choice of the constant λ. In this case, the value is chosen to give a minimum value of the maximum field strength on the curved boundary. (The shape of corner so chosen makes breakdown between the two surfaces occur at a maximum voltage.)

Let the corresponding points in the two planes be:

$$z = 0, \qquad t = -1;$$
$$z = -j\infty, \qquad t = 0;$$
$$z = a - jl, \qquad t = b;$$
and
$$z = a + l, \qquad t = c.$$

If the corner were sharp instead of rounded, the Schwarz–Christoffel equation would be of the form

$$\frac{dz}{dt} = \frac{S}{t}(t - \alpha)^{1/2}. \tag{10.16}$$

The rounding of the corner is thus achieved by modifying the term $(t - \alpha)^{1/2}$ to give the equation

$$\frac{dz}{dt} = \frac{S}{t}(\sqrt{t-b}+\lambda\sqrt{t-c}), \tag{10.17}$$

integration of which leads to

$$z = 2S\left[\sqrt{t-b}-\sqrt{b}\tan^{-1}\left(\frac{t-b}{b}\right)^{1/2}\right.$$
$$\left.+\lambda\left(\sqrt{t-c}-\sqrt{c}\tan^{-1}\left(\frac{t-c}{c}\right)^{1/2}\right)\right]+c_2. \tag{10.18}$$

Substituting for corresponding points in the two planes and using the method of residues leaves one relationship connecting the constants to be found from a consideration of the field gradient.

To obtain a potential difference between the two sections of the boundary in the z-plane requires a field in the t-plane of the form

$$w = \frac{1}{\pi}\log t. \tag{10.19}$$

The flux density distribution in the z-plane is given by

$$\left|\frac{dw}{dt}\frac{dt}{dz}\right|$$

and λ is chosen to give the least value of the maximum density on the curved boundary. This maximum is found by differentiating dw/dz with respect to λ and equating the result to zero.

A transformation involving rounded corners has been employed by Cockroft[4] in connection with dielectric breakdown.

10.3.2. Curvilinear polygons

It is possible to transform the infinite straight line into a polygon with curved sides by including in the Schwarz–Christoffel equation a "curve factor", $C(t)$, giving an equation of form

$$\frac{dz}{dt} = SC(t)\,\Pi(t-t_n)^{(\alpha/\pi)-1}. \tag{10.20}$$

The angles at the vertices are unaffected by the factor $C(t)$ which has to be chosen, by trial and error, to give an approximation to the desired shapes of the curved sections. The form of the curve factor must be such that it introduces no singularities or zeros into the equation, for these would result in additional vertices or in discontinuities. It is not possible to give any standard procedure for finding the appropriate curve factor, but several types of factor have been considered by Page[5] and Leathem.[6]

A simple boundary of this type is considered in section 7.2.1, where a semicircular boss on a straight line is discussed. It is shown that the equation

$$z = t+\sqrt{t^2-1} \tag{10.21}$$

transforms the real axis of the t-plane into the required boundary shape, the part corresponding to $|t| \leqslant 1$ being circular and the part corresponding to $|t| \geqslant 1$ being straight. The curve factor is of special type, since it makes only one side of the polygon curved. Its form can be seen by differentiating eqn. (10.21) and writing the resulting expression as

$$\frac{dz}{dt} = \left(t + \sqrt{t^2 - 1}\right)(t+1)^{-1/2}(t-1)^{-1/2}; \tag{10.22}$$

the first term is the curve factor $C(t)$, the remaining two terms giving the defining vertex angles of $\pi/2$ at $t = +1$ and $t = -1$. A more general form of the curve factor,

$$C(t) = t + \lambda \sqrt{t^2 - 1}, \tag{10.23}$$

gives the semi-elliptical boss [compare eqn. (7.34)].

In conclusion, it should be noted that the search for a suitable factor to give a satisfactory approximation to any particular curvilinear polygon is often lengthy and sometimes unrewarding.

10.4. Angles not multiples of $\pi/2$

The treatment of polygonal boundaries given in the earlier chapters is restricted to cases in which all the angles are multiples of $\pi/2$ (including the zero multiple). Whilst these include the great majority of practically useful boundary shapes, there are a number of useful and interesting boundaries which have vertex angles with other values. They have not received attention earlier because, apart from the trivial case of polygons with only one vertex, the transformation equation cannot be expressed in terms of simple functions.

All polygons defined by two vertices involve integrals of the Euler type (see Copson) and some of these can be reduced to elliptic form. When the polygon is defined by more than two vertices, analytical methods fail, but it can be handled by using numerical integration (see section 10.5).

10.4.1. Two-vertex problems

Consider any polygon defined by two vertex angles α and β, the length of the side joining them being l. For a reason which appears later, let this side be on the real axis in the z-plane. Then the polygon, which may be closed at infinity or at a finite point, can be transformed from the real axis of the t-plane by the Schwarz–Christoffel equation

$$\frac{dz}{dt} = St^{(\alpha/\pi)-1}(t-1)^{(\beta/\pi)-1}, \tag{10.24}$$

where the points $t = 0$ and $t = 1$ correspond to the vertices in the z-plane. When integrated this is of the form

$$z = S \int t^{(\alpha/\pi)-1}(t-1)^{(\beta/\pi)-1}\, dt, \tag{10.25}$$

the constant of integration being zero when the origins in the two planes are made to correspond. This integral is Eulerian and of the first kind, and the definite form of it, between real limits of t, can be expressed in terms of the gamma function, which is defined by

$$\Gamma(u) = \int_0^\infty e^{-t} t^{(u-1)} \, dt. \tag{10.26}$$

(For the evaluation of the gamma function the reader should consult Copson, p. 209.) To determine the value of the scale constant S, the definite integral is most simply evaluated between the limits 0 and 1, and it is for this reason that the points in the t-plane corresponding to the vertices are so chosen. Hence,

$$l = S \int_0^1 t^{(\alpha/\pi)-1} (t-1)^{(\beta/\pi)-1} \, dt. \tag{10.27}$$

The solution of this standard definite integral (which is the beta function), see Copson, p. 213, gives

$$S = \frac{l}{(-1)^{(\beta/\pi)-1}} \frac{\Gamma[(\alpha+\beta)/\pi]}{\Gamma(\alpha/\pi)\,\Gamma(\beta/\pi)}, \tag{10.28}$$

so that the scale constant of the transformation can be found simply in terms of the finite dimension of the polygon. However, to find the field in the z-plane, other than on the boundary, it is necessary to employ the method of numerical integration of a complex variable.

It can be shown that the Eulerian integral, eqn. (10.25), becomes elliptic for particular values of the vertex angles. All of the known cases occur when both the angles are multiples of either $\pi/3$, $\pi/4$, or $\pi/6$ and many of them are discussed in reference 7. When all the vertices occur in the finite region, the boundary is, of course, triangular, and the four cases which are elliptic have angles of:

$$\frac{\pi}{2}, \quad \frac{\pi}{4}, \quad \text{and} \quad \frac{\pi}{4}; \qquad \frac{\pi}{6}, \quad \frac{\pi}{3}, \quad \text{and} \quad \frac{\pi}{2};$$

$$\frac{\pi}{3}, \quad \frac{\pi}{3}, \quad \text{and} \quad \frac{\pi}{3}; \qquad \frac{\pi}{6}, \quad \frac{\pi}{6}, \quad \text{and} \quad \frac{2\pi}{3}.$$

An important application of the analysis of two-vertex problems occurs in the study of the breakdown of liquids in an electric field between two sharp corners or between one sharp corner and a plane (see Ollendorf; Buchholz, p. 122; and reference 8).

10.5. Numerical methods

It has already been pointed out that analytical methods are often inadequate for polygonal boundaries with many vertices, particularly when these are not integral multiples of $\pi/2$. For these boundaries a numerical method can be used to make possible a significant extension in the range of tractable problems.

Numerical methods of determining the constants in a conformal transformation and the use of numerical procedures to integrate the Schwarz–Christoffel equation were first developed in the study of electrical machine problems.[9–11] The transformation used for one of these problems[9] and the equations expressing the values of the constants are discussed in section 10.5.3. Later, the application of combined analytical and numerical techniques to the calculation of the characteristic impedance of waveguides was proposed,[12] and the treatment of arbitrary singly and doubly connected regions reported.[13, 14] Finally, direct-search techniques of minimization[15] have been applied to the determination of constants and an integration routine with automatically varying step-length employed.[16]

There are two aspects to the solution of a problem—the numerical integration of the Schwarz–Christoffel equation for any complex value of t, and the determination of the t-plane constants for any set of values of the desired dimensions.

10.5.1. Numerical integration of the function $f(t)$

The Schwarz–Christoffel equation may be expressed in the simple form

$$\frac{dz}{dt} = Sf(t) \tag{10.29}$$

or in the integral form

$$z_{nm} = S \int_{t_n}^{t_m} f(t)\, dt, \tag{10.30}$$

where t_n and t_m are points in the t-plane and z_{nm} is the complex distance between the points in the z-plane corresponding to t_n and t_m. This equation can therefore be used to relate any change in z to the corresponding change in t, and the methods of integration which have been found both simple and reliable make use of Simpson's quadratures. Gaussian quadrature (see, for example, Jennings) appears to cope less well with the singularities in $f(t)$.

For the determination of the constants of the transformation, only the absolute lengths of the boundaries need to be considered, and therefore the integration can be simplified and quickened by restricting it to deal with real variables only. For integration away from the boundaries it is possible to use a complex transformation to change the complex limits to real ones,[16] thereby reducing computing time.

In integrating $f(t)$ along a line corresponding to a straight-line segment of the boundary, the function is terminated at singularities corresponding to the vertices. Special attention must be paid to the integration close to singularities and to the determination of the displacement of limits from singularities at which the routine use of Simpson's quadrature can commence. Figure 10.9 illustrates typical variations in $f(t)$ along boundary segments terminated in one case by two zeros, in another by two poles, and in the third by a pole at one end and a zero at the other. Case A is typical of the variation when the limiting points are both zeros; the function is zero at the limits and has infinite slope there, in one case the maximum value being fairly close to the limit. In case B, the limiting points are poles; the function rises rapidly close to the poles but the integral remains finite. In case C there is a zero at one limit, as in case A, but a pole at the other, the nature of which depends on the vertex angle. If the angle is zero, the area under the curve is infinite; for other angles the area is finite.

In the case of poles it is essential to terminate the routine integration procedure at a suitable distance from the pole. Constant displacements of 0·001 to 0·005 have been used[9] and also displacement proportional to the value of t at the pole,[16] the constant of proportionality varying from 0·001 to 0·01. The relative merits of the two methods of choice[17] depends to a significant degree on the choice of the t-plane constants. The integral of

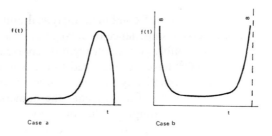

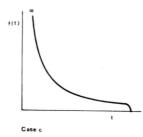

FIG. 10.9

$f(t)$ between the limit and the singularity may be negligible or infinite depending mainly on the vertex angle.

If a function $F(t)$ is defined such that

$$F(t) = \frac{f(t)}{(t-a)^{(\alpha/\pi)-1}},$$
(10.31)

the $\int f(t)\, dt$ between the limit and the singularity becomes

$$\int_{a}^{a+\delta} F(t)\,(t-a)^{(\alpha/\pi)-1}\, dt.$$

Now if δ is very small (such as 0·001), $F(t)$ is virtually constant over the range of integration since it contains no singularities around $t = a$. Hence the integral is equal to

$$F(a)\left|\frac{\pi}{\alpha}\,(t-a)^{\alpha/\pi}\right|_{a}^{a+\delta}$$

or

$$F(a)\,\frac{\pi}{\alpha}\,(\delta)^{\alpha/\pi}.$$
(10.32)

When the vertex angle α is zero, the integral is, of course, infinite, and this arises in the example discussed in section 10.5.3. The integral between the limit and the singularity may be negligible, but this is not the case if α is small (say 0·02). In a routine computation the above expression can be simply evaluated and, if it is appreciable, added to the integral over the remainder of the range corresponding to a side of the polygon.

Outside the regions close to singularities, the numerical integration of the complex function $f(t)$ can be readily performed using Simpson's formulae with a suitable step length.

Clearly the step size has to be sufficiently small to give a reasonable fit to $f(t)$; thus, for example, if the three-point formula is used with a step-length h, the function is fitted by a quadratic over an interval $2h$. One method[16] which has been used is to compare the estimate of the integral obtained using the three-point formula with that obtained with the five-point formula over the same overall range. Initially, the full range of integration is divided into two steps for the three-point formula and four steps for the five-point one. If the solutions differ by more than a prescribed aount, the step lengths are halved and the process repeated until the two estimates lie within the required limit. A tolerance of 0·1 per cent between the two values should be suitable for most problems.

10.5.2. *Solution of the implicit equations*

The relationship between the defining dimensions of a problem and the associated t-plane constants can be expressed as

$$
\left.
\begin{aligned}
z_{12} &= S \int_{t_1}^{t_2} f(t)\, dt, \\[1em]
z_{23} &= S \int_{t_2}^{t_3} f(t)\, dt, \\[1em]
&\;\;\vdots \qquad \vdots \\[1em]
z_{m-1,\,m} &= S \int_{t_{m-1}}^{t_m} f(t)\, dt.
\end{aligned}
\right\}
\tag{10.33}
$$

These equations must be solved for S, t_1, t_2, etc., and where possible the method of residues is used (see section 8.2.5); for the remaining equations, numerical integration is employed. Several iterative techniques for solving these equations can be used, but the discussion will be limited to direct-search methods, since these are particularly suited to general application.

The values of the constants are obtained by minimizing the sum of the squares of the errors in each boundary dimension for successive estimates of the t-plane constants. Using the sum of the squares of the errors leads to continuity in the absolute value and the second derivative of the error function and a minimal distortion of the error. Having given two t-plane constants convenient values and employed any explicit relationships obtainable from the method of residues, it remains to minimize the total error function.

In the direct-search method proposed by Powell[15] in 1964, the properties of quadratic convergence are exploited and the directions of search are continuously changed to maximize convergence rate. Equal emphasis is given to all coordinates and they are maintained mutually orthogonal. After modification, the directions of search are at least as good as, and mostly are much better than, the preceding ones. The minimum with respect to variation of any one coordinate is predicted as the apex of a parabola defined by three points chosen to bracket the minimum. However, when a particular direction is used repeatedly, the properties of the second derivative of a function are used to predict the minimum, and this results in considerable saving in time. At any stage, the step length used in searching for a minimum in one direction is calculated as a function of the total error quantity existing at that stage.

As with all direct-search methods, a local minimum is found, but this is not troublesome since the solution to eqn. (10.33) is unique. An unconstrained minimum is determined which could be troublesome, since variables can become very small and also negative possibly resulting in the total error becoming infinite, or zero at an incorrect point. This difficulty is overcome by transforming all the coordinate variables to log space.

The procedure described above has been found to work efficiently[16] and should be capable of handling virtually any polygon boundary of practical interest without excessive computation time. The starting values of the constants is not of great importance provided these are in the right order. Table 10.1 shows the variation in computation time (on an Atlas computer) for reasonable starting values and for different numbers of defining parameters.

TABLE 10.1. VARIATION IN COMPUTING TIME WITH
NUMBER OF CONSTANTS

Number of constants determined numerically		1	2	3	5
Computation time (mins)	errors < 0·0005%	0·4	3·0	4·0	6·0
	errors < 0·1%	0·34	2·4	3·2	5·2
	errors < 1·0%	0·23	—	2·8	4·8

10.5.3. *The centring force due to displaced ventilating ducts in rotating machines*

As a demonstration of the use of numerical integration, a calculation of the magnetic field between equal stator and rotor ventilating ducts in an electrical machine is described. Since ducts are widely spaced and are unaffected by the ends of the machine, a pair of them can be considered in isolation. With relative displacement of the ducts, the fluxes entering the two sides of a duct become unequal and produce an axial magnetic force of engineering importance.

The displaced ducts are represented in the z-plane as in Fig. 10.10, and the boundary shape is to be obtained by transformation from the real axis of the t-plane. The gap width

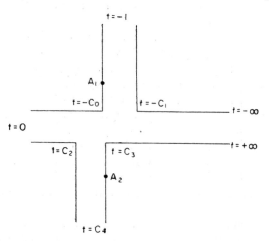

FIG. 10.10 (Reproduced by permission of the Institution of Electrical Engineers.)

is made equal to unity and the ducts are of equal width s. It is evident that the z-plane boundary has skew-symmetry—that is, at points A_1 and A_2 (Fig. 10.10), equidistant from their respective corners the field is the same—and this property is used later. The points $t = -c_0, -c_1, c_2,$ and c_3 correspond to the corners of the ducts, the points $t = -1$ and c_4 correspond to the meeting points at infinity of the sides of the ducts and the point $t = 0$ is infinitely far down the gap.

For unit magnetic potential difference between the two surfaces, each containing a duct, the required expression for the t-plane field is

$$w = \frac{1}{\pi} \log t, \tag{10.34}$$

and the equation connecting the two planes is

$$\frac{dz}{dt} = K \frac{\sqrt{(t+c_0)(t+c_1)(t-c_2)(t-c_3)}}{t(t+1)(t-c_4)}. \tag{10.35}$$

Since the boundary has skew-symmetry, the flux entering the right-hand side of the top duct is equal to the flux entering the left-hand side of the bottom one, and so by considering the flux entering corresponding lengths of the real axis of the t-plane, it can be seen that

$$-\frac{1}{\pi} \log c_0 = \frac{1}{\pi} \log \frac{c_3}{c_4},$$

and so

$$c_0 = \frac{c_4}{c_3}. \tag{10.36}$$

Similarly, by considering the flux entering the other two sides of the ducts it is seen that

$$c_1 = \frac{c_4}{c_2}. \tag{10.37}$$

By the method of residues it can be shown that the duct width is given by

$$s(c_4-1) = \sqrt{(c_1-1)(1-c_0)(1+c_2)(1+c_3)}, \tag{10.38}$$

and that

$$K = \frac{1}{\pi}. \tag{10.39}$$

The four relationships, eqns. (10.36)–(10.39), are the only explicit ones which can be established for use in the determination of the six unknown constants. To determine the constants two further relationships have to be established.

The condition that the two surfaces on either side of a duct are collinear is expressed by

$$\int_{-c_0}^{-1+\delta} |f(t)\,dt| = \int_{-c_1}^{-1+\delta} |f(t)\,dt|, \tag{10.40}$$

a suitable value for δ being $0\cdot001$. The duct displacement d is given by

$$\int_{-\delta}^{-c_0} |f(t)\,dt| - \int_{\delta}^{c_z} |f(t)\,dt| = d. \tag{10.41}$$

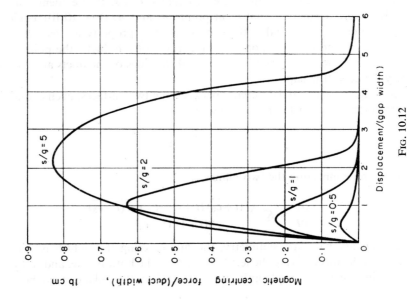

Fig. 10.12

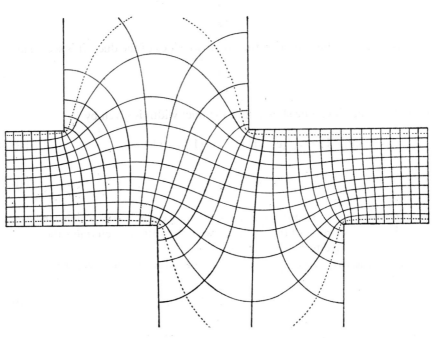

Fig. 10.11 (Reproduced by permission of the Institution of Electrical Engineers.)

Equations (10.36)–(10.41) can now be used to give the constants in terms of any desired values of the defining dimensions.

By using numerical integration any desired points of intersection of flux and equipotential lines can be found and a field map constructed. The points of intersection of the desired lines are first found (analytically) in the *t*-plane and the corresponding points in the *z*-plane are then found, most economically by integrating successively between the points in the *t*-plane. Figure 10.11 shows a field map which was obtained accurately in this way without the use of excessive computing time.

The axial centring force (the force in the direction of the gap) is found by numerically integrating the square of the flux density down each side of each duct and subtracting the two results. Figure 10.12 shows the variation of force/duct width, with displacement.

10.6. Non-equipotential boundaries

Consideration previously in the book has been mainly restricted to equipotential or flux line boundaries, though a variety of boundaries which are combinations of both has been considered. This section discusses the three basic classes of problem as defined by the field conditions on the (closed) boundary: the first class has a potential distribution specified at all points on the boundary, the second has the normal component of field gradient specified, and the third is a combination of the two previous classes. The first class of problem is treated in section 10.6.1 and the two previous classes. The first class of problem is treated in section 10.6.1 and the second and third classes in section 10.6.2.

10.6.1. *Boundary value problems of the first kind*

A problem in which values of potential are specified at all points along the boundary is said to be of the *first kind* (or of the *Dirichlet* type). It is possible to obtain a solution to any such problem for the interior of the unit circle or the infinite half-plane, and then to use this solution, by means of transformation methods, for a wide range of other boundary shapes. *Field inside the unit circle.* The solution inside the unit circle is obtained by means of the *complex potential function of Schwarz.* If $\psi'(\theta')$ describes the specified potential distribution along the periphery of the circle, θ' being the angle subtended at the centre, and if p is the complex variable of position, then the solution for the field inside the unit circle is expressed by

$$\psi + j\varphi = \frac{1}{2\pi} \int\limits_0^{2\pi} \frac{e^{j\theta'} + p}{e^{j\theta'} - p} \psi'(\theta') \, d\theta'. \tag{10.42}$$

The real part of this equation can also be put in a standard form by putting $p = re^{j\theta}$ to give

$$\psi = \frac{1}{2\pi} \int\limits_0^{2\pi} \frac{1 - r^2}{1 - 2r \cos(\theta - \theta') + r^2} \psi'(\theta') \, d\theta'. \tag{10.43}$$

This is the *Poisson integral* and it can be used for evaluating potential distributions for the case of a single, circular boundary.

In using either of these equations it is required only to express the potential distribution along the circular boundary in terms of θ' and to solve the resulting definite integral for every point p at which the field is to be determined. This integral often has no analytical solution, but numerical methods can be used (see section 10.5.1). If the complex potential function has to be evaluated at many points, the numerical method involves much computation, though it is unlimited in its scope.

A useful problem, capable of analytical solution, arises when the boundary consists of equipotential sections. If there are m of these, each with potential ψ_n, and if the limits of the nth one subtend angles θ'_n and θ'_{n-1} at the centre, the complex potential function is, from eqn. (10.42),

$$\psi + j\varphi = \frac{1}{2\pi} \sum_{n=1}^{m} \psi_n \int_{\theta'_{n-1}}^{\theta'_n} \frac{e^{j\theta'} + p}{e^{j\theta'} - p}\, d\theta'. \tag{10.44}$$

This gives, when integrated,

$$\psi + j\varphi = \frac{-1}{2\pi} \sum_{n=1}^{m} (\theta'_n - \theta'_{n-1}) \psi_n - \frac{j}{\pi} \sum_{n=1}^{m} \psi_n \log \frac{e^{j\theta'_n} - p}{e^{j\theta'_{n-1}} - p}, \tag{10.45}$$

which may be simply used to give immediately a solution to any problem of this type. For example, the solution for the unit circle with two equal equipotential sections, of potential $+\psi_A$ and $-\psi_A$, is given, by substituting $\theta'_0 = 0$, $\theta'_1 = \pi$ and $\theta'_2 = 2\pi$ as

$$\psi + j\varphi = -\frac{2j}{\pi} \psi_A \log \frac{p+1}{p-1} + 2\psi_A. \tag{10.46}$$

Field in the upper half-plane. The above solution obtained by Schwarz can be modified to give the solution of the field in the upper half-plane due to a potential distribution along the real axis. Thus, by applying the bilinear transformation

$$t = j\left(\frac{1-p}{1+p}\right), \tag{10.47}$$

the complex potential function becomes

$$\psi + j\varphi = \frac{j}{\pi} \int_{-\infty}^{+\infty} \frac{1 + v't}{(1 + v'^2)(t - v')} \psi(v')\, dv', \tag{10.48}$$

in which $\psi(v')$ is the potential distribution along the real axis of t. The real part of eqn. (10.48) leads to the equivalent of the Poisson integral for the infinite half-plane, namely,

$$\psi = \frac{1}{\pi} \int_{-\infty}^{\infty} \frac{v}{(u - v')^2 + v^2} \psi(v')\, dv'. \tag{10.49}$$

Again, if there are m equipotential sections along the real axis, the integrals can be performed analytically; and if the nth section is at potential ψ_n and lies between the points v'_n and v'_{n+1}, the potential function becomes

$$\psi + j\varphi = \frac{j}{\pi} \sum_{n=1}^{m} (\psi_n - \psi_{n+1}) \log \frac{\sqrt{1 + v'^2_n}}{t - v'_n} + \psi_m. \tag{10.50}$$

This equation could be obtained by considering the field of a combination of line sources, at the points of potential division, along the real axis. For two sections, at potentials $+\psi_A$ and $-\psi_A$, divided at the origin, equation (10.50) reduces to the familiar form

$$\psi + j\varphi = -\frac{2j}{\pi}\psi_A \log t. \tag{10.51}$$

Equation (10.50) has been used (see Weber) to give the electric field of a transformer winding, represented as three sheets (Fig. 10.13). This boundary can be transformed simply

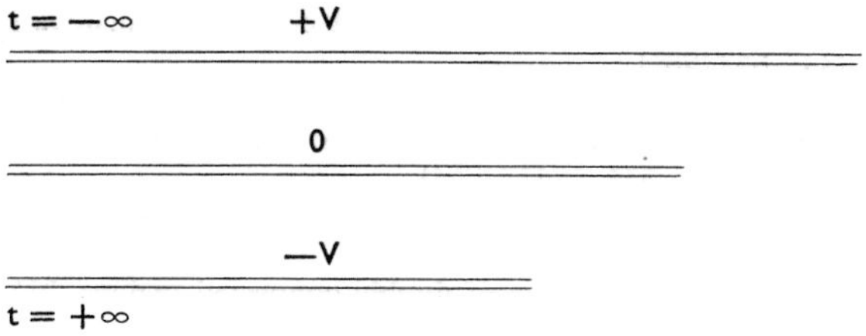

FIG. 10.13

from the real axis of the t-plane. If the relative electric potentials on the windings are $+V$, 0, and $-V$, the field in the t-plane is given by eqn. (10.50) as

$$w = \psi + j\varphi = \frac{j}{\pi}\left(V\log\frac{\sqrt{2}}{t+1} + V\log\frac{\sqrt{1+q^2}}{1-q}\right) - V, \tag{10.52}$$

the sections being separated by the points $t = -1$ and $t = +q$.

10.6.2. *Boundary value problems of the second and mixed kinds*

If the boundary conditions specify the normal components of gradient, the problem is said to be of the *second* kind (or of the *Neumann* type). If the boundary conditions describe potentials on some parts of the boundary and potential gradient over the rest, the problem is said to be of *mixed* kind. In general, transformation methods are unsuited to both these classes of problem, which for simple boundary shapes are best treated by the methods of Chapter 4. However, in the special case where the boundary is formed partly by flux lines (that is, lines with a constant normal gradient) and partly by equipotential lines, solutions are often available and appear frequently elsewhere in the book. For example, in section 10.2, the solution for the fields of currents in the infinite half-plane is given for second and mixed kinds of boundary condition on the real axis; and in section 9.2.4, the solution is given for two equal finite plates (mixed conditions on the real axis). Similar solutions have also been obtained for the unit circle, see Weber and reference 18, but these are not of great practical interest.

References

1. P. J. LAWRENSON, A note on the analysis of the fields of line currents and charges, *Proc. Instn. Elect. Engrs.* **109** C, 86 (1962).
2. F. W. CARTER, The magnetic field of the dynamo-electric machine, *J. Inst. Elect. Engrs.* **359**, 1115 (1926).
3. S. DRINKER, Short wavelength response of magnetic reproducing heads with rounded gap edges, *Phillips Res. Rep.* **16**, 307 (1961).

4. J. D. Cockroft, The effect of curved boundaries on the distribution of electrical stress round conductors, *J. Instn. Elect. Engrs.* **66**, 385 (1928).
5. W. M. Page, Some two-dimensional problems in electrostatics and hydrodynamics, *Proc. Lond. Math. Soc.*, ser. 2, II, 313 (1912).
6. J. G. Leathem, Some applications of conformal transformation to problems in hydrodynamics, *Phil. Trans.* A, **215**, 439 (1915).
7. Y. Ikeda, Die konformen Abbildungen der Polygone mit zwei Ecken, *J. Fac. Sci. Hokkaido Univ.*, ser. 2, **2** (2), 1 (1938).
8. L. Dreyfus, Über die Anwendung der konformen Abbildung zur Berechnung der Durchschlags und Überschlagsspannung zwischen kantigen Konstruktionsteilen unter Öl, *Arch. Elektrotech.* **13**, 123 (1924).
9. K. J. Binns, The magnetic field and centring force of displaced ventilating ducts in machine cores, *Proc. Instn. Elect. Engrs.* **108** C, 64 (1961).
10. K. J. Binns, Pole-entry flux pulsations, *Proc. Instn. Elect. Engrs.* **109** C (1962).
11. K. J. Binns, Calculation of some basic flux quantities in induction and other doubly-slotted electrical machines, *Proc. Inst. Elect. Engrs.* **111**, 1847 (1964).
12. P. A. Laura, Characteristic impedance of a rectangular transmission line, *Proc. Instn. Elect. Engrs.* **113**, 1595 (1956).
13. P. A. Laura, Conformal mapping of a class of doubly connected regions, NASA Tech. Re., Reep., Grant No. G125–61 (1965).
14. M. K. Richardson, A numerical method for the conformal mapping of finite doubly connected regions with application to the torsion problem of hollow bars, Ph.D. thesis, Univ. of Alabama (1965).
15. M. J. D. Powell, An efficient method for finding the minimum of a function of several variables without calculating derivatives, *Comput. J.* **7**, 115 (1964).
16. P. J. Lawrenson and S. K. Gupta, Conformal transformation employing direct-search techniques of minimisation, *Proc. Instn. Elect. Engrs.* **115**, 427 (1968).
17. K. J. Binns, Numerical methods of conformal transformation, *Proc. Instn. Elect. Engrs.* **118**, 909 (1971).
18. J. Hodgkinson, A note on a two-dimensional problem in electrostatics, *Q. J. Math.* **9**, 5 (1938).

Additional References

Construction and application of conformal maps, *Appl. Math. Ser.*, US Bur. Stand., nos. 18 and 42.
Joyce, W. B., Separation of variables solution from the Schwarz–Christoffel transformation. *Q. J. Appl. Math.*
Langton, N. H. and Davy, Two-dimensional field in a semi-infinite slot terminated by a semi-circular cylinder, *Br. J. Appl. Phys.* **4**, 134 (1953).
Langton, N. H. and Davy, Two-dimensional field above and below an infinite corrugated sheet, *Br. J. Appl. Phys.* **3**, 156 (1952).
Love, A. E. H., Some electrostatic distributions in two dimensions, *Proc. Lond. Math. Soc.*, ser. 2, **22**, 337 (1923).
Powell, M. J. D., A method of minimising the sum of squares of non-linear function without calculating derivatives, *Comput. J.* **7**, 303 (1965).
Richmond, H. W., On the electrostatic field of a plane or circular grating formed of thick rounded bars, *Proc. Lond. Math. Soc.*, ser. 2, **22**, 389 (1924).

PART IV

NUMERICAL METHODS

CHAPTER 11

FINITE-DIFFERENCE METHODS

11.1. Introduction

Finite-difference methods, which have been traced back to Gauss[1] but which have only been generally used since about 1940, can be used to obtain numerical solutions, of any desired accuracy, to any problem within the scope of this book.[†] Further, though some theoretical considerations associated with their development are very difficult, they can be applied to fields regardless of the complexity of the boundary shape, the nature of boundary conditions, or the number of interconnected regions, and they can be extended to handle non-linear fields such as those in the presence of saturating iron. Their application is extremely easy and, indeed, merely involves simple arithmetic. The main disadvantage of finite-difference methods is that, as with all numerical methods, the process of solution must be carried out for each set of parameters defining a problem. (The relative merits of analytical and numerical methods are considered in section 11.7 and also in Chapter 1.)

The solutions obtained by finite-difference methods consist of the values at discrete points spaced in an ordered way over the whole field region of the function describing the field. These values are obtained by replacing the one partial differential equation of the field by many simple finite-difference equations which take the form of linear equations connecting the potential at each point with the potentials at other points close to it. In this way, the solution of the field is reduced to the solution of a set of simple simultaneous algebraic equations for the potential values.

Because of the large number of these equations normally occurring in a problem, it is at present generally unreasonable to solve them using methods involving elimination, determinants, or matrix inversion.

Instead, some form of iterative procedure, exploiting the simple form which the difference equations take, is usually adopted. These procedures can be applied using hand or machine computation, though the latter is now by far the most usual and important. Computation carried out by hand has traditionally been termed *relaxation*,[‡] but now, with the dominance of machine methods and the adoption of the word relaxation in connection with these methods, it is necessary to qualify the earlier approach as *hand* or *manual* relaxation. Hand relaxation is very flexible, but for its full effectiveness depends considerably on the experience and skill of the computer, and, even so, it can be very time consuming. When the computation is carried out by machine, the procedure is often referred to as *iteration*. In contrast with relaxation, iterative schemes are based upon an entirely fixed cycle of operations. They

---

[†] They are also applicable to three-dimensional and time-varying field problems.

[‡] This name is due to Southwell who described the technique by analogy with the relaxation of strains in a stressed, jointed framework.

are designed to fit naturally with, and to take advantage of, digital computing techniques, and in optimum forms and situations give strikingly rapid solutions.

Over the last 10–15 years a large number of different methods has been developed, and the related literature is extremely extensive and often difficult to evaluate comparatively. The latter situation reflects a variety of things: the number of iterations to achieve a solution is not necessarily simply related to the time involved; a method which is successful and convenient with some problem or class of problems may be neither successful nor convenient with others; and, more generally, the subject as a whole is still expanding and still posing a number of difficult problems. Accordingly, the treatment in this chapter is kept as simple as possible, concentrating on those basic features of finite-difference techniques which remain unchanged regardless of the particular method of solution employed, and on the practical application of a method of solution which, though amongst the earliest to be developed, has proved itself to be both quick and reliable over a very wide range of problems. Readers who wish to go more deeply into the underlying theory or to study those techniques which are only referred to briefly here, should study the original papers referenced at the end of the chapter and the books by Forsythe and Wasow, Milne, Todd, Varga, Wachspress, and Walsh.

The basic finite-difference equations are established in the next section, 11.2, and then both hand relaxation and machine iteration are discussed in sections 11.3 and 11.4 respectively. In the remaining sections of the chapter, the finite-difference equations for the more difficult boundary and interface conditions are developed and the accuracy of the solutions is considered. Throughout, the finite-difference equations are developed in terms of Poisson's equation for vector potential, the current density being set to zero when Laplace's equation is considered. (Naturally, the finite-difference equations are identical whether expressed in terms of A, ψ, the scalar potential, or φ, the flux function.)

11.2. Finite-difference representation

11.2.1. *Regular distributions of field points*

In replacing the field equation by a set of finite-difference equations which connect values of the potential function at discrete points, any spatial distributions of the points can be used. However, it is clear that if a completely regular distribution of points is chosen, the same form of finite-difference equation is satisfied at all of the points, and the representation of the problem is much simplified. Such distributions are given by the arrangements of points lying at the "nodes" of any regular *network* or *mesh*.

The first of these meshes consists of squares and is shown in Fig. 11.1(a) in a simple, square field region. It is seen that any node *within* the boundary is positioned with respect to neighbouring nodes (including those on the boundary) as shown in Fig. 11.1(b), and this typical arrangement of nodes is termed a *symmetrical star* or *molecule*.

There are two other perfectly regular meshes. These are the equilateral-triangular one, Fig. 11.2(a), and the equiangular-hexagonal one, Fig. 11.2(b). The authors are not aware of any practical applications of the latter, but the triangular mesh has been exploited in a number of problems particularly where coincidence of the field boundaries and the mesh lines have permitted a more accurate representation than would otherwise have been possible. It is, however, of much less importance than the square net and it is not discussed further in this chapter. For further information on this mesh the reader may consult the books by Allen, Forsythe and Wasow, Richtmeyer, and Southwell.

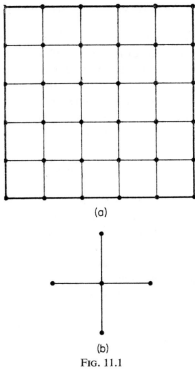

(a)

(b)

Fig. 11.1

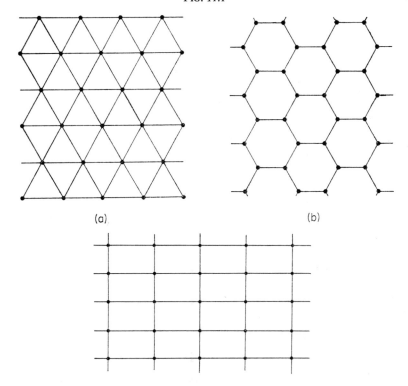

(a)

(b)

(c)

Fig. 11.2

Two *irregular* meshes which must also be mentioned are the *polar* one and the *rectangular* one. The polar mesh, described, for example, in reference 2 and by Smith (p. 137) has obvious advantages for a few problems with circular boundaries, but is not discussed further here. The rectangular mesh, Fig. 11.2(c), is frequently of importance, either in achieving a better fit between field boundaries and mesh lines or in minimizing the necessary number of nodes. It is, of course, closely related to the square mesh and the necessary equations for its use are derived in the next section along with those for the square mesh.

11.2.2. *Basic equations for the square and rectangular meshes*

The symmetrical star of Fig. 11.3(a) is relevant at all nodes in a uniform square mesh. However, adjacent to boundaries where the mesh and boundaries do not coincide, and in regions of a field where meshes of different size meet (arising due to considerations of accuracy and/or economy of nodes), one or more of the "outer" nodes of a star will be separated from the "centre" node 0 by distances different from the regular distance. Similarly, the star which applies even in the uniform region of a rectangular mesh, Fig. 11.3(b), has arms of unequal length reflecting the width to length ratio $q : 1$ of the mesh. In the most general

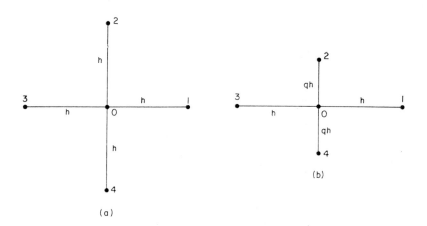

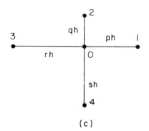

FIG. 11.3

case possible, all the arms of the star are of different lengths, and such an asymmetrical star is shown in Fig. 11.3(c). The difference equation for the potential at the centre of this star is now developed, and from this the equations for simpler stars can be written down as required.

In the asymmetrical star let the points 1, 2, 3, and 4 be distance from 0 respectively equal to ph, qh, rh, and sh, where p, q, r, and s are less than unity and h is the distance between lines in the regular portion of the mesh. The term *mesh length* is usually applied to h; it is equal to the side of a square of the square mesh and the longer of the two arms in the rectangular case. The difference equation is developed by expanding the potential at the point 0 in Taylor's series and deriving expressions for $(\partial^2 A/\partial x^2)_0$ and $(\partial^2 A/\partial y^2)_0$ which are substituted in Poisson's equation. Other methods of developing the difference equations are referred to briefly in section 11.7.

At any point x on the line parallel to the x-axis passing through the point 0 the potential A can be expanded in terms of the potential A_0 at 0 by the use of the Taylor's series

$$
A = A_0 + \left(\frac{\partial A}{\partial x}\right)_0 (x-x_0) + \frac{1}{2!}\left(\frac{\partial^2 A}{\partial x^2}\right)_0 (x-x_0)^2
$$

$$
+ \frac{1}{3!}\left(\frac{\partial^3 A}{\partial x^3}\right)_0 (x-x_0)^3 + \frac{1}{4!}\left(\frac{\partial^4 A}{\partial x^4}\right)_0 (x-x_0)^4 + \ldots \tag{11.1}
$$

Thus, substituting in this equation for the values $x = x_0 + ph$ and $x = x_0 - rh$ yields the values of the potentials at the points 1 and 3 respectively as

$$
A_1 = A_0 + ph\left(\frac{\partial A}{\partial x}\right)_0 + \frac{1}{2!}p^2h^2\left(\frac{\partial^2 A}{\partial x^2}\right)_0 + \frac{1}{3!}p^3h^3\left(\frac{\partial^3 A}{\partial x^3}\right)_0 + \frac{1}{4!}p^4h^4\left(\frac{\partial^4 A}{\partial x^4}\right)_0 + \ldots \tag{11.2}
$$

and

$$
A_3 = A_0 - rh\left(\frac{\partial A}{\partial x}\right)_0 + \frac{1}{2!}r^2h^2\left(\frac{\partial^2 A}{\partial x^2}\right)_0 - \frac{1}{3!}r^3h^3\left(\frac{\partial^3 A}{\partial x^3}\right)_0 + \frac{1}{4!}r^4h^4\left(\frac{\partial^4 A}{\partial x^4}\right)_0 - \ldots \tag{11.3}
$$

Forming the sum of r times eqn. (11.2) and p times eqn. (11.3) gives

$$
rA_1 + pA_3 = (p+r)A_0 + \frac{h^2}{2!}pr(p+r)\left(\frac{\partial^2 A}{\partial x^2}\right)_0 + \frac{h^3}{3!}pr(p^2 - r^2)\left(\frac{\partial^3 A}{\partial x^3}\right)_0
$$

$$
+ \frac{h^4}{4!}pr(p^3 + r^3)\left(\frac{\partial^4 A}{\partial x^4}\right)_0 + \ldots, \tag{11.4}
$$

and, ignoring terms containing h to the power three or more, which is valid when h is small, yields the simple expression for $(\partial^2 A/\partial x^2)_0$,

$$
h^2\left(\frac{\partial^2 A}{\partial x^2}\right)_0 = \frac{2A_1}{p(p+r)} + \frac{2A_3}{r(p+r)} - \frac{2A_0}{pr}. \tag{11.5}
$$

In an exactly similar manner an expression for $(\partial^2 A/\partial y^2)_0$ can be obtained as

$$
h^2\left(\frac{\partial^2 A}{\partial y^2}\right)_0 = \frac{2A_2}{q(q+s)} + \frac{2A_4}{s(q+s)} - \frac{2A_0}{qs}. \tag{11.6}
$$

Hence, substitution of these values for $(\partial^2 A/\partial x^2)_0$ and $(\partial^2 A/\partial y^2)_0$ in Poisson's equation,

AC 17

in which the term $\mu\mu_0 J$ is written for brevity as W, gives

$$2\left[\frac{A_1}{p(p+r)} + \frac{A_2}{q(q+s)} + \frac{A_3}{r(p+r)} + \frac{A_4}{s(q+s)} - \left(\frac{1}{pr} + \frac{1}{qs}\right)A_0\right] + h^2W = 0. \qquad (11.7)$$

This equation (for a small value of h) is, therefore, a valid approximation for, and is to be used in place of, Poisson's equation. But it should be emphasized that, whereas the differential equation is the same at every point in the field, a particular difference equation holds at one point only (though its form is identical at many points).

The completely general form of difference eqn. (11.7) is only rarely required, but from it, simply by substituting the appropriate values of p, q, r, and s, any necessary equation is derivable immediately. Most used of all the equations is that for a node in a uniform square mesh and having the star pattern shown in Fig. 11.3(a). Thus, for the node 0 of this star, putting $p = q = r = s = 1$, the difference equation is

$$A_1 + A_2 + A_3 + A_4 - 4A_0 + h^2W = 0. \qquad (11.8)$$

Also very important is the equation for nodes within a regular region of a rectangular mesh. The star pattern is that shown in Fig. 11.3(b) and the difference equation for node 0 is obtained by putting $p = r = 1$ and $q = s$ in eqn. (11.7). The result is

$$A_1 + \frac{A_2}{q^2} + A_3 + \frac{A_4}{q^2} - 2\left(1 + \frac{1}{q^2}\right)A_0 + h^2W = 0. \qquad (11.9)$$

(Also see section 11.5.3 for an equation involving gradient boundary conditions.) For other asymmetrical stars the values of the multipliers must be determined for each node according to the circumstances.

For points in regions where the field is Laplacian, the appropriate equations for the asymmetrical and symmetrical stars are obtained by setting $W = 0$ in eqns. (11.7)–(11.9) respectively. Also, for this class of field, these equations can be written in terms of the scalar potential ψ or the flux function φ: and so, for example, Laplace's equation in scalar potential for a node within a regular square mesh is

$$\psi_1 + \psi_2 + \psi_3 + \psi_4 - 4\psi_0 = 0. \qquad (11.10)$$

The above difference equations are, of course, only approximations to the field equations since higher order terms of eqn. (11.4) are neglected (this is known as truncation error). It is important to investigate this error carefully, and this is done in section 11.6, but it is sufficient to say at this point that, in practice, h is easily chosen for any problem so that the error is negligibly small.

11.2.3. *Field problem as a set of simultaneous equations*

Consider now the way in which a field problem is treated as a set of simultaneous equations and, to be specific, consider the simple (but typical) Poissonian field region within the square boundary shown in Fig. 11.4. Let the nodes on the boundary be designated by the primed numbers $1'$ to $16'$ and let the interior nodes be numbered 1–9. The value of W and the values of potential on the boundary A_1'–A_{16}' are given, and it is required to find, as the solution, the values of potential A_1–A_9 at the interior points.

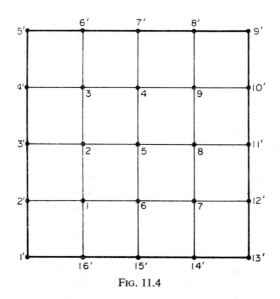

FIG. 11.4

This can be accomplished easily using eqn. (11.8) at each of the interior points. At the interior point 1 the equation gives

$$A_6 + A_2 + A_2' + A_{16}' - 4A_1 + h^2W = 0;$$

at the point 2 it gives

$$A_5 + A_3 + A_3' + A_1 - 4A_2 + h^2W = 0;$$

and it gives similar equations at the remaining points 3–9. The nine potentials A_1–A_9 are therefore connected, amongst themselves and to the boundary conditions, by nine linear algebraic equations, and they can be evaluated by solving these equations simultaneously.

In a similar way any field problem can be solved as a set of simultaneous linear difference equations. For problems with boundaries which are curved (or for some other reason do not coincide with the nodes of the mesh) or with different sizes or types of mesh, the only difference in the treatment as compared with the simple one above is that the coefficients in the equations for the subsequent asymmetrical stars must be derived using the general difference eqn. (11.7) (with $W = 0$ for Laplacian regions).

It is helpful at this stage, and for the later discussion, to examine the set of simultaneous difference equations in matrix form. Clearly, if $\mathbf{M}$ is the matrix of the coefficients of the unknown potentials, $\mathbf{u}$ is the column matrix of unknown potentials A_n, and $\mathbf{v}$ is the column matrix of the sums of the known potentials A_n' and the constant terms h^2W (when present), the equations can be written

$$\mathbf{Mu} = \mathbf{v}. \tag{11.11}$$

The small number of terms in each difference equations results in a correspondingly small number in each row of $\mathbf{M}$ and is the basic feature exploited in an iterative solution. It is also apparent that the magnitudes of the elements of $\mathbf{M}$ depend directly upon the particular difference equations employed (but not at all upon whether the field is Laplacian or Poissonian), as does the basic distribution of elements within $\mathbf{M}$. The reader can readily confirm that $\mathbf{M}$ is always essentially diagonal and is frequently purely so and symmetrical about the leading diagonal.

The properties of the matrix $\mathbf{M}$ completely govern the nature of any iterative scheme applied to the solution of eqn. (11.11) and so are central to any mathematical study of, for example, whether or not convergence is guaranteed, how fast convergence takes place, or what errors can occur. Full consideration of these general questions is impossible and unnecessary in the present treatment, but a number of broad points can be simply stated. Firstly, and in fundamental mathematical terms, iterative solutions will give convergence to the correct solution provided that $\mathbf{M}$ is symmetric and positive definite. If $\mathbf{M}$ does not have these properties, convergence will generally still be possible, but it cannot be *guaranteed* and may take place slowly. For most practical purposes it is sufficient to note that for all Dirichlet problems $\mathbf{M}$ does have the desired properties (no matter what difference equations are employed) and solutions are always "well-behaved"; but, with gradient boundary conditions (and difference equations developed, as is common using Taylor series), $\mathbf{M}$ is normally not symmetrical and convergence is less good than with Dirichlet conditions. The potential values assumed at the start of a solution have no influence upon whether or not convergence takes place,[3] but do have an influence on the time taken to achieve a solution of given accuracy.

In considering the solution of the equations, two important features are evident. Firstly, in any practical problem, the number of simultaneous equations may be very great (1000 is quite common) and, secondly, each equation contains very few terms. The first of these features means that the usual methods of obtaining exact solutions are wholly impracticable simply because of the length of computation which they would involve. For example, the solution of 100 equations by a standard elimination method (which is much quicker than a determinant method) would take at least 10,000 hours by hand computation and some 5–10 minutes by a reasonably fast computer. Fortunately, however, because of the small number of terms in the equations, very much quicker numerical methods are possible. These are iterative in character, being based upon a technique of successive approximation. As an example of the speed of these methods, 100 difference equations can be solved to an accuracy of 0·1 per cent in about 8 hours by hand (relaxation), and in a few seconds by machine. The length of computation is closely related to the accuracy achieved, so that it is possible to terminate the process as soon as adequate accuracy has been achieved.

11.3. Hand computation: relaxation

11.3.1. *Introduction*

The relaxation method of solving the simultaneous, finite-difference equations was first used by Gauss, but not until it was rediscovered and developed by Southwell and his co-workers in the nineteen-thirties and forties was its great flexibility and power appreciated by engineers and physicists.[†] Since the late fifties it has been largely displaced by machine iteration which has permitted the treatment of larger and more varied problems. However, it is still useful for smaller problems or for others with particularly complex boundary shapes, and knowledge of it considerably helps the later discussion of machine methods. Essentially it consists of a continual modification of the potential values until all of the simultaneous equations are satisfied to a sufficient degree of accuracy. It is rather different

[†] For a comprehensive list of references to applications of the method, the majority of which are to mechanical problems, see the book by Allen or the chapter by Higgins of the book edited by Grinter (see bibliography).

from other methods considered in the book for it is not possible to formulate general rules of procedure, and no two analysts would follow precisely the same course of solution in any given problem. However, the method is extremely simple, and a little practice quickly affords the experience necessary for the choice of an appropriate procedure in any problem.

In the following section the concept of residual and the use of the simple basic technique of point relaxation are described. Next, means of accelerating the rate of convergence of the solutions are given and, finally, some general advice on practical details of the relaxation procedure is presented.

11.3.2. *The basic method*

The concept of residual. The whole of the relaxation method turns on the concept of residuals and these are now explained. When the potentials A_0, A_1, ..., A_4 are so chosen as to satisfy a difference equation, say (11.8), the right-hand side is zero. When however, the potentials do not satisfy the equation the right-hand side is not equal to zero but to some quantity, R_0, given by

$$R_0 = A_1 + A_2 + A_3 + A_4 - 4A_0 + h^2W. \qquad (11.12)$$

R_0 is known as the residual of the difference equation at the point 0, and the relaxation method is directed to its systematic reduction to zero for all the equations. When the residuals of all the difference equations are zero, the values of potential correspond to an exact solution but, in practice, it is sufficient, to obtain a good approximation to the exact solution, merely to reduce them to low values.

It should be noted that there is no precise relationship between the magnitude of the residuals and the accuracy of the potentials. However, in practice it is found that solutions are generally satisfactory when all the following conditions are satisfied:

  (a) Individual values of the residuals are reduced to about 0·1 per cent of the mean value of the potentials.
  (b) The algebraic sum of all the residuals is of similar magnitude to the individual residuals.
  (c) The residuals are uniformly mixed, as regards magnitude and sign, over the whole field region.

More detailed considerations of the accuracy of solution are treated in section 11.6, and the basic process by which the residuals are reduced is now discussed.

Point relaxation. The basic operation used in the reduction of residuals is known as point relaxation; it is extremely simple and is as follows. At the node under consideration (assumed initially to be at the centre of a symmetrical star) the value of the residual is calculated using eqn. (11.12) and then this residual is reduced to zero by a change in A_0 of $R_0/4$. This change in A_0 causes four equal changes in residual, one at each of the adjacent nodes of the star, and applying eqn. (11.12) to each of these nodes in turn it is seen that the change they experience is $R_0/4$ (that is, equal to the change in A_0 at the original centre node). Thus a residual of $4R$ at the central node of a star can be reduced to zero by an increase, of magnitude R, in the potential at the node, and by the addition of R to the residuals at each of the "points" of the star. This process is known as "spreading" the residual, and it can be represented and easily memorized by the relaxation pattern shown in Fig. 11.5(a). It is to be realized, of course, that in the spreading of residuals, boundary points are not influenced.

In an exactly similar way relaxation patterns can be constructed for asymmetrical stars from eqn. (11.7). For example, considering the star shown in Fig. 11.3(c) with $r = s = 1$, a unit change in potential at the node 0 gives the pattern of Fig. 11.5(b) (nodes 1 and 2 being assumed to be on a boundary and so of fixed potential), and a unit change at the node 3 gives the pattern of Fig. 11.5(c). (See section 11.3.4 for a consideration of the terms in p and q which introduce nonintegral changes in residual.)

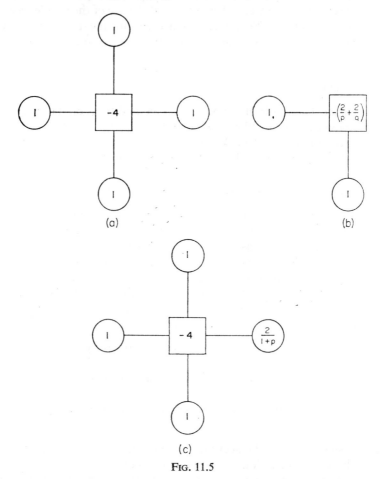

Fig. 11.5

In using point relaxation to solve the equations the most efficient routine is to consider at each stage, the equation (node) which has the largest residual. In practice, residuals (at any stage) are not reduced exactly to zero; it is sufficient to limit the changes in potential to integral amounts, so eliminating the use of decimals, which are unnecessary for the method and which are likely to lead to the introduction of errors (see section 11.3.4).

A simple example. Consider now the application of the above relaxation process to the problem shown in Fig. 11.6. For simplicity it is assumed that the field is Laplacian, and that the boundaries pass through nodes of the mesh. The values of potential at the boundary nodes are as shown, and the initial values of potential at all the interior nodes, lettered a to i, are taken to be zero. For the computation, the values of potential are recorded in the upper left-hand quadrant and the values of residual in the upper right-hand quadrant near each node. It is usual, at any stage, after each change in potential or residual, to record the resultant value and to cross out the previous value.

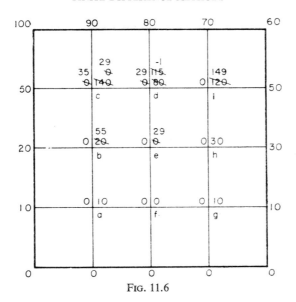

FIG. 11.6

The first step is the evaluation of the residuals at each node using eqn. (11.12) (with $W = 0$) and the assumed values of potential—for example, at node a the value of the residual is $0+0+10+0-4\times0 = 10$. These values are recorded, together with those of potential, as described above, and the initial values are indicated by the bottom two numbers at each node in Fig. 11.6.

Then, using the point relaxation operator, the maximum residual occurring at each stage of the calculation is reduced (temporarily) to zero. Initially the largest residual occurs at node c and is 140: it can be reduced to zero by changing A_c by $140/4 = 35$. The new value of A_c is $0+35$ and is recorded as shown: the new value of R_c is 0, also recorded as shown. The change in A_c influences the nodes d and b (but not those on the boundary where the potential is fixed) and so the residuals there must be altered by the addition of the change made in A_c, i.e. by 35. These changes lead to new residuals, which are also recorded, of 55 and 115 at b and d respectively.

The largest residual now occurs at d and is considered next. To reduce it to zero requires a change in A_d of $115/4 \doteq 29$, and (to save the introduction of decimals) it is convenient to make the change of 29 and to leave a residual of -1. The residuals at nodes c, i, and e are also changed by 29.

In an exactly similar way, reducing the largest residual remaining at each stage (approximately) to zero, and spreading one quarter of it to the adjacent nodes (except those on the boundary), the process is continued until the residuals obey the three conditions above. A very approximate solution (2–6 per cent) is shown in Fig. 11.7.[†]

This example demonstrates the great simplicity of the relaxation method. Using the basic operation in a similar way to that described above, it is possible, allowing where necessary for the use of asymmetrical relaxation patterns, to obtain solutions to any problem. However, it would be found in many problems that the solution would be reached only slowly and consideration is now given to ways in which the method can be made to converge more quickly.

† Greater accuracy and fulfilment of conditions (a) and (b) would involve the use of decimal points or the equivalent, see p. 255, *Decimal points*.

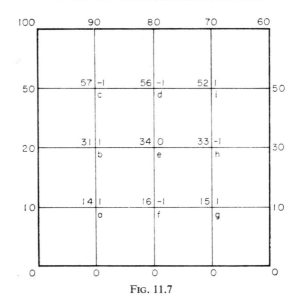

FIG. 11.7

11.3.3. *Accelerating processes*

Introduction. The rate of convergence of the solution can be greatly increased by the use of two important devices: the first involves variations in the degree of relaxation applied at a node (under or over relaxation), and the second consists in relaxing at several points simultaneously (line or block relaxation). Before describing them in detail, however, it must be emphasized that considerable time in every problem can be saved by making an initial estimate of the potentials at the nodes. This estimate does not need to be in any way accurate and even a rough guess can save considerable time.

Over- and under-relaxation. When the potential at a node is changed by an amount sufficient to cause the sign of the residual to reverse, the operation is termed over-relaxation; when the potential is changed by an amount insufficient to reduce the residual to zero, the operation is termed under-relaxation. The use of these operations can greatly accelerate the convergence and, with a little experience, the need for them is easily recognized.

To demonstrate the use of over-relaxation consider the first two steps which were worked in detail for the problem above. The first relaxation reduced the residual at node *c* to zero, but the first one at node *d* caused a residual of 29 to reappear at *c*. In a similar way it is found that the first operation at *b* also causes an increase in residual at *c*. Both of these effects —the so-called wash-back of residuals in which the residual at a point, having been once eliminated, reappears with the same sign due to relaxation at neighbouring nodes—could have been foreseen and the residual at *c* could, in the first operation there, have been over-relaxed. It would then have become negative but, assuming the appropriate degree of over-relaxation were used, it would have been reduced to zero when the operations at nodes *d* and *b* were completed.

It is not possible to give rules for the use of over- or under-relaxation except to indicate the general circumstances in which each is appropriate. Where it can be seen that a residual is surrounded by residuals of like sign, or when wash-back is found to occur, the use of over-relaxation results in improved convergence. In a similar way, in regions where a particular residual is surrounded by residuals of opposite sign, or when relaxation of adjacent residuals causes a reversal in sign of the residual considered, then the use of under-relaxation is help-ful. The degree of relaxation to be applied at any node can only be estimated, the computer having regard to the future influence of relaxation at the other nodes, in particular those immediately adjacent to the one considered.

Line and block relaxation. As a simple example of the advantage of relaxing at several points simultaneously consider again the problem above—in particular, the first operation at each of the nodes, c, d, and i. The changes in residual at each node are roughly equal and also changes at one node cause changes at one or both of the remaining ones. It would, therefore, be economical if the resultant of the three separate changes could be obtained at one step. This, in fact, is accomplished simply by using a line relaxation operator, the pattern for which is obtained by simple addition of three point patterns (Fig. 11.5). The resultant pattern is shown in Fig. 11.8(a) and it is seen that changes in potential of $+1$ at the nodes are associated with changes in residual of -3 at c and i, -2 at d and changes of $+1$ at all adjacent nodes. (There would, of course, be no change at adjacent nodes on the boundary.)

In a similar way line patterns, involving any number of nodes along a line, or block pat-terns, involving any number of nodes in a block, can be constructed (for unit change in potential at each node) by superposition of the point pattern for each node. Several exam-ples are shown in Fig. 11.8(b), (c) and (d), in which the squares relate to points in the line or block. The patterns demonstrate the following general features:

(a) The residual change at a node within a line or block is equal to $(n-4)$, where n is the number of other nodes in the line or block to which the particular node is connected.
(b) The residual change at a node outside the line or block is equal to the number of nodes, within the line or block, to which the particular node is connected directly by the lines of the mesh.
(c) The total change in residual at all nodes inside the line or block is equal to the total change occurring at the nodes immediately adjacent to the line or block.

The use of block (or line) relaxation is clearly indicated when, over groups of nodes, the residuals are of the same sign. In this respect block relaxation and over-relaxation are to some extent alternatives but, because it deals at one operation with a large number of nodes, block relaxation is usually much the more powerful technique.

Its more obvious application, to spread the residuals rapidly at a large number of nodes, is demonstrated in principle by the simple example for the three-node line above. There is, however, a less obvious and more powerful method of application indicated by (c) above. Since the total change of residual in the block is equal to the corresponding change at adjacent points outside, it can be simply used to reduce the total residual inside the block to zero. This, of course, results in positive and negative residuals being well spread through-out the block when judicious use of point relaxation inside the block generally leads quickly to the complete elimination of the residuals.

As an example of the way in which the total residual in a block is reduced to zero, con-sider again the pattern shown in Fig. 11.8(d) and let it be assumed that the block has been arranged to include all nodes having a residual of 4. There are 15 nodes in the block and so the total residual within the block is $15 \times 4 = 60$. The total change in residual within the block for a unit change of potential at each node is given by the sum of the numbers in the squares and is -20. Therefore, by changing the potential at every point by $60/20 = 3$, the sum of residuals within the block is reduced to zero (and 60 units of residual are spread outside the block). The resultant pattern of residuals may indicate that use of a further

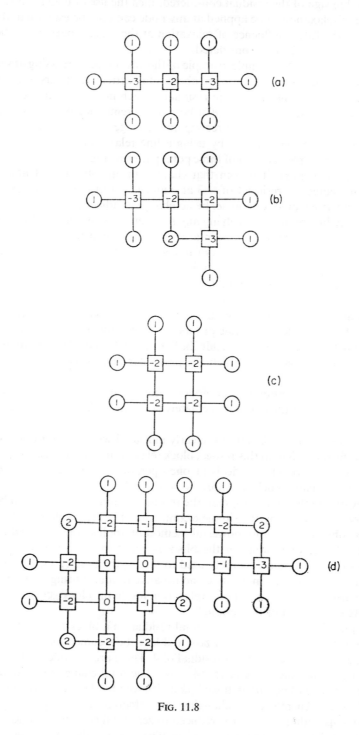

Fig. 11.8

block (of different shape) would be beneficial, perhaps to spread the residuals near to the boundary; or it may indicate that positive and negative residuals are sufficiently well inter-mixed for the use of point relaxation to be required.

11.3.4. *Practical aspects*

As a manual process, the method has the advantages that it affords interest to the comput-er, requiring the continual use of intelligence, and that errors are automatically compen-sated for, so eliminating the need for a continuous check. (However, see *Checking* below.)

From the previous discussion, it is apparent that a rigid set of rules cannot be prescribed for use in any given problem and, indeed, it is the corresponding flexibility that largely accounts for the power of the method in the hands of an experienced computer. (Equally it is the lack of rigid rules which prevents the method from being programmed successfully for machine use.) However, it is possible to indicate a number of points which are helpful in practice, and these are listed below under five self-explanatory headings.

Mesh size. The first requirement is to choose a suitable mesh size and, to minimize the inconvenience of asymmetrical stars, the choice should be made so that the mesh coincides with as much of the boundary as is possible. (This may lead to the choice of a rectangular mesh or of a triangular or polar one or, indeed, of some combination of different meshes.) A fine mesh should not be used in the early stages of the computation as this causes un-necessary work and results in poor convergence; instead, the first computation should be carried out on a coarse mesh and, when satisfactory convergence has been achieved for this mesh, the calculation should be taken up on a finer one. This new mesh length is conveniently made equal to one half of the previous one, with the new nodes lying midway between the old. The values for the potentials at the new nodes are found satisfactorily by linear interpolation (as this gives an accuracy equivalent to that inherent in the differ-ence equations). Successive advances to finer nets should be made until a sufficiently fine mesh is reached (see section 11.6). The use of fine meshes can, of course, be restricted to small regions where accuracy is particularly important. The points made in this paragraph also apply generally in connection with machine methods.

Layout of the computation sheet. When the likely size of the final mesh has been decided, the boundary (and meshes) should be drawn to such a size that there is adequate space for computation on this mesh. It is useful to do the drawing on opaque paper and then to do the calculation on one or more sheets of tracing paper laid over the drawing; in this way the drawing is not damaged when any rubbing out is done.

Decimal points. Decimal points should never be used in the calculations, since they are inconvenient and increase the risk of mistakes. Instead, whenever necessary (perhaps when they appear in the given data or when residuals have been reduced to the order of unity and an advance to a finer net is needed) all numbers should be multiplied by 10 to the appropriate power and the entire calculation carried out in terms of whole numbers.

The only exception to this occurs in the later stages of calculation at nodes involving the use of asymmetrical stars. (The residual changes due to the asymmetry generally involve decimals.) Even for such stars, in the earlier stages of the computation, it is sufficient to work to the integer nearest to the number involving decimals. This is because errors in the residuals (introduced as a result of this) can easily be removed at any stage by recalculating the residuals from the current values of potential. (See *Checking* below).

The computation. As indicated earlier, the aim of the calculation is to make the residuals and their algebraic sum small and to spread them uniformly over the mesh. As much use as possible should be made of block (or line) relaxation; point relaxation should only be begun when it no longer proves worth-while to construct further blocks. First attention should be given to areas with particularly large residuals, and to areas where residuals are surrounded by those of opposite sign—at such positions as the latter both positive and negative residuals can be reduced simultaneously.

Checking. As mentioned above, errors in the computation are largely corrected automatically, but they should be minimized by careful working, since they reduce the rate of convergence. To ensure that no discrepancy occurs between the value of potential and the residual at a point, it is important, at intervals in the calculation, particularly at an advance to a finer net, to recalculate the residuals from the current values of potential.

11.3.5. *The use of point values of potential: capacitance of a three-core rectangular cable*

As a simple example of the type of problem that can easily be handled by relaxation, and to demonstrate the calculation of capacitance from the numerical potential values, consider the electric potential field of a rectangular cable with an earthed sheath and three rectangular cores, as shown in Fig. 11.9. The dimensions are chosen for simplicity so that the boundaries of the conductors and sheath coincide with lines of a square network. The capacitance between the left-hand conductor and all other conducting parts is to be determined. This is simply done in terms of the differences in potential values at adjacent points (see below) and, to ensure satisfactory accuracy (2–3 per cent), experience indicates that an ultimate mesh not coarser than that shown is needed. (The calculation should, of course, be begun with a coarser net.)

The total capacitance of the left-hand conductor is determined by calculating the total flux entering (or leaving) the conductor per unit potential difference between it and all neighbouring conducting parts. For the relaxation solution, the potential of the conductor is conveniently set at 1000, whilst the two remaining conductors and the sheath are set at zero. Then reduction of the residuals to 1 or 2 gives satisfactory accuracy.

The total flux is calculated by integrating the normal component of flux density, determined from the potential gradient, round a suitable contour enclosing the conductor. The contour should be chosen, if possible, to avoid regions where the rate of change of potential gradient is high and the dotted line in Fig. 11.9 is satisfactory. Then, if ψ_i is the potential at a node just inside it and ψ_o is the potential at the adjacent node just outside it, the normal potential gradient on the contour is $(\psi_i - \psi_o)/h$. (See, however, section 11.6.) Thus, if $\varepsilon\varepsilon_0$ is the permittivity of the insulating material and N is the number of meshes cutting the contour, the total flux is

$$\varepsilon\varepsilon_0 \sum_{n=1}^{N} (\psi_{in} - \psi_{on})$$

(assuming the gradient constant over a mesh length h), and the capacitance is

$$0.001\varepsilon\varepsilon_0 \sum_{n=1}^{N} (\psi_{in} - \psi_{on}). \tag{11.13}$$

The capacitance of the conductor or any other particular one of the neighbouring conducting parts is obtained by using eqn. (11.13) for a suitable contour cutting only the flux linked to the particular conductor.

Field maps may be obtained from relaxation solutions by using linear interpolation to obtain the values of potential at points off the net. This has been done for the above prob-

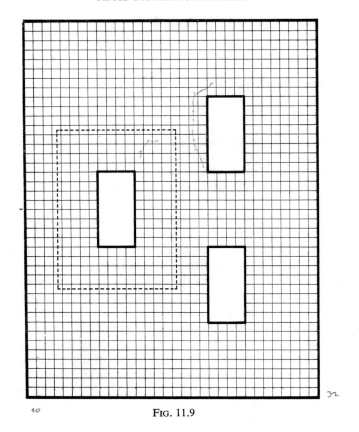

FIG. 11.9

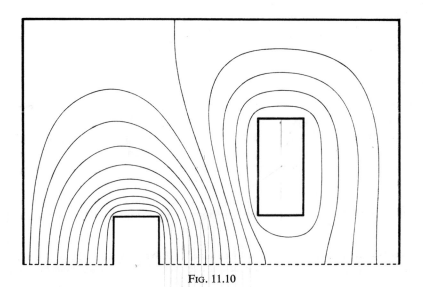

FIG. 11.10

lem with the left-hand conductor at a potential of 1000 and the two lower ones at a potential of -500; the field is then symmetrical and one half of it is shown in Fig. 11.10.[†]

11.4. Machine computation: iteration

11.4.1. *Introduction*

In solving the finite-difference equations using machine computation, the technique, as with hand relaxation, is to modify continually values of potential until all the equations are satisfied to a sufficient degree of accuracy. Considerations of stability and accuracy of solution for both methods are similar, but all machine methods are characterized by two features which distinguish them from hand methods. Firstly, they are designed for use in a completely automatic cycle in which each of the equations (and so each node of the network) is considered in turn throughout the calculation. Secondly, improved values of potential are determined directly from the difference equation, the concept of residual being disregarded in this context. The first of the above features means, of course, that the rate of convergence of a *basic* iterative method is slower than that of the relaxation method in the hands of a skilled computer. However, it also means that iterative methods can be easily programmed for machine use, so eliminating the considerable labour of manual calculation, and yielding solutions with remarkable rapidity.

This suitability for machine use is their great merit, and, indeed, interest in iterative schemes has been closely linked with the development of digital computers. During the last 20 years there has been an intensive study of iterative techniques by mathematicians, and the result has been a steady, and continuing, output of powerful methods. Of these, however, the one known as the *successive over-relaxation method* (SOR) or, sometimes the *extrapolated Liebmann method* is, because of its combination of rapid convergence with simplicity, still supreme at the present time for general purpose application to practical problems.

In the following discussion the basic concepts of iterative schemes—the ordering of the cycle and the examination of convergence—and the two simplest (but theoretically fundamental) methods are treated first. Then a detailed discussion of the successive over-relaxation method is given, and, in the final section, other methods are briefly reviewed and conclusions are drawn.

11.4.2. *Basic considerations and methods*

Designation of nodes. In using any machine method, each node of the mesh is considered in turn in a fixed, repeated cycle. To facilitate this during the calculation and for convenience of discussion, each node is designated by its own number or number pair. In the basic rectangular array of $(p+1) \times (q+1)$ points (on which the boundary is superposed), Fig. 11.11, any node (and its associated quantities) can be distinguished by a number pair (h, k), where $1 \leqslant h \leqslant (p+1)$ and $1 \leqslant k \leqslant (q+1)$, or by a single number $(h-1)(q+1)+(k+1)$, where 1 is taken to be the point at the lower left-hand corner, and columns of points are scanned, in each cycle, from left to right (increasing p) and from bottom to top (increasing q).

[†] This figure has been prepared from a solution obtained by J. Webb, Master's thesis, Queen's University, Belfast, May 1954. For this thesis, and for the loan of reference 19, the authors are indebted to Prof. P. L. Burns.

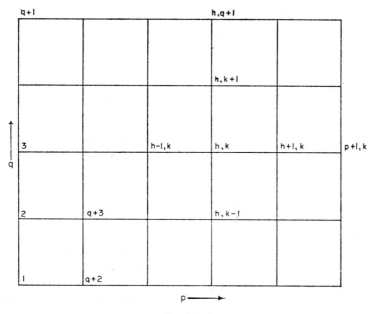

Fig. 11.11

Examination of convergence. Whilst the concept of residual plays no part in the modification of potentials (as shown below), it does provide the most satisfactory check on the convergence of the solution and, in this context, considerations are identical with those for a relaxation solution (see section 11.3.2). However, when the relevant information is stored inside a computer, it is very difficult to observe the *distribution* of the residuals and this particular guide to the convergence of the solution is generally ignored. Because of the ease of computation the maximum residual can be set to a much lower value (say 0·0001 per cent) than in hand relaxation, and so the one condition for the maximum value of the final residuals normally proves satisfactory (though, if thought desirable, the algebraic sum total of all the residuals can also be computed).

It will be realized, of course, that the calculation of residuals, not being basic to the machine scheme, involves a certain amount of additional computation which increases the total solution time. An obvious and satisfactory way of minimizing this increase is by computing and examining the residual only after each ten iterations or so. Alternatively, another widely used procedure is to disregard the residuals entirely and to test for convergence by studying the change in potential (or *displacement*) occurring at each node [eqn. (11.20)]: the iteration procedure is stopped when the changes are sufficiently small (say 10^{-6} per cent of the potential values). This seems to be the most economical test possible and, though open to the objection that small changes can (and sometimes do) mean slow convergence rather than high accuracy, it has been found to give good results when used with care. Other reasons for examining the change in potentials emerge in the next section. For other considerations of accuracy of solution, see section 11.6.

Consideration is now given to the equations used in different iterative procedures for the modification of the potential values. The two methods to be described in this section are introduced primarily to show, in easy stages, the development of the SOR method (discussed in the next section) from the simplest iterative method.

The Jacobi method.[†] In the simplest of the iterative methods, each new value of potential at the centre of a star is determined as that which exactly satisfies the basic difference equation in terms of the previous values of potential at the other points. Thus, if A^n designates the potential after the nth iterative cycle, the Jacobi form of the symmetrical Poisson difference equation, (11.8), is

$$A_0^{n+1} = \tfrac{1}{4}(A_1^n + A_2^n + A_3^n + A_4^n + h^2W), \tag{11.14}$$

or, in more general terms, using the number pair notation (above),

$$A_{h,k}^{n+1} = \tfrac{1}{4}(A_{h+1,k}^n + A_{h,k+1}^n + A_{h-1,k}^n + A_{h,k-1}^n + h^2W). \tag{11.15}$$

(Fig. 11.11.) Equations similar to (11.15) hold, of course, for asymmetrical patterns.

As may be expected, the convergence of the Jacobi method is poor—see Table 11.1 (p. 266) for comparative figures—and the method has the disadvantage that two full sets of potential values (for the nth and $(n+1)$th iterations) need to be stored. Because of these disadvantages, this method is not used in practice.

Gauss–Seidel method. A simple modification of the above routine yields a continuous substitution method (also proposed by Liebmann[5]) in which the most recently computed values of potential are used at each stage. Scanning the nodes column by column, from left to right, and starting at the bottom of each ($q = 1$), the general form of the symmetrical Poisson difference equation is

$$A_{h,k}^{n+1} = \tfrac{1}{4}(A_{h+1,k}^n + A_{h,k+1}^n + A_{h-1,k}^{n+1} + A_{h,k-1}^{n+1} + h^2W). \tag{11.16}$$

This method is economical in that it requires storage of only one complete set of potential values (not two as with Jacobi's method). Also, it converges twice as quickly as Jacobi's method (this may be guessed since each calculation is based upon two old and two new values, but it is discussed more carefully in references 6 and 13 and by Varga, for example) though compared with more sophisticated methods it is still uneconomically slow and it is not discussed further. Consideration is now given to the successive over-relaxation method which, whilst still using only one complete set of potential values, has a very much higher rate of convergence.

11.4.3. *The successive over-relaxation method*

The method. The method, described independently by Frankel[6] and Young,[7] is the most flexible and useful of the rapidly convergent iterative methods, and it is therefore treated here in some detail. As its name implies, it is the equivalent of systematic over-relaxation, point by point, and with (basically) the same degree of over-relaxation at every step. It is derivable from the Gauss–Seidel method by the introduction of a factor α, a new value of potential being determined as the sum of the old value and α times the difference between the value given by eqn. (11.16) and the old value, i.e. by

$$A_{h,k}^{n+1} = A_{h,k}^n + \frac{\alpha}{4}(A_{h+1,k}^n + A_{h,k+1}^n + A_{h-1,k}^{n+1} + A_{h,k-1}^{n+1} + h^2W - 4A_{h,k}^n). \tag{11.17}$$

α is a *convergence* or *relaxation factor* determining the degree of over-relaxation, and it can be shown that it must lie between 1 and 2. When $\alpha = 1$, eqn. (11.17) reduces to eqn. (11.16) for the Gauss–Seidel method, and when $\alpha \geqslant 2$ the process becomes unstable. When α

[†] Also known as the *Richardson*[4] method or the *method of simultaneous displacements.*

lies between these limits, the convergence rate is higher than for $\alpha = 1$ and, for some optimum value α_b, different for every problem, the rate is greatly increased. For example, the Dirichlet problem (specified values of boundary potential) for the square with twenty nodes per side can be solved with a reduction in the maximum error of 10^{-10} times in seventy iterations (as compared with about 840 for the Gauss–Seidel method); on a fast machine this represents about 10 seconds of computing time. For a full comparison of methods, see Table 11.1 (p. 276). Full numerical details for each iteration in the solution of a simple, four node problem are given in Vitkovitch, chapter 3, and Todd.

The convergence of the solution. It has been shown theoretically (see, for example, Varga or references 7 or 13) that, for a stable convergence process, the number of iterations N required for the reduction of the largest error at any node to a fraction ε of some previous value, is given by an expression of the form

$$N \doteqdot -F \log \varepsilon. \tag{11.18}$$

F, known as the *asymptotic rate of convergence*, is in general a function of the boundary shape and conditions, the number of nodes, the particular type of difference equation (see later), and the convergence factor. It is defined by

$$F = -\log \lambda, \tag{11.19}$$

in which $\lambda^\dagger$ may be obtained as the limiting value of the ratio of the absolute values of the maximum changes in potential occurring on successive iterations when the convergence factor is unity. Thus, if

$$U^n = \max |A_m^n - A_m^{n-1}|, \tag{11.20}$$

λ is given by

$$\lambda = \mathop{\mathrm{Lt}}_{n \to \infty} \frac{U^{n+1}}{U^n}. \tag{11.21}$$

Typical curves of N against ε for the same problem but different values of α have the form of those shown in Fig. 11.12. Similar curves occur for the reduction of the largest residual and the potential displacement and, indeed, in practice, it is found that they also apply approximately for the mean of the moduli of the residuals. Of the three curves, the one marked A reduces the maximum error by the greatest amount (over the range of N presented) and so the value of α associated with this curve is the most suitable (of the three)

$\dagger$ λ is the largest eigenvalue, or spectral radius, of the iteration matrix characterizing the SOR scheme. If eqn. (11.11) for the set of difference equations is rewritten

$$(I - L - U)u = v,$$

where I is the unit matrix, and L and U are the lower and upper triangular matrices with null diagonals, the SOR iteration scheme is defined by

$$u^{n+1} = u^n + \alpha(Lu^{n+1} + Uu^n + v - u^n).$$

This can be rewritten as

$$u^{n+1} = \{(I - \alpha L)^{-1} [\alpha U - (\alpha - 1)I]\}u^n + \alpha(I - \alpha L)^{-1} v,$$

and the term in curly brackets is the iteration matrix referred to. It is worth adding that, after $(n+1)$ iterations, the error (matrix) e is

$$e^{n+1} = \{(I - \alpha L)^{-1} [\alpha U - (\alpha - 1)I]\}e^0.$$

Also, when M possess Young's *property A*,[7] it is possible to write

$$\lambda = a_b - 1.$$

AC 18

for computation. If similar curves for the whole range of values of α from 1 to 2 are considered, it is found that, for one particular value, the overall reduction in error after a large number of iterations is greater than for all the others, and this value is defined as the *optimum* convergence factor, α_b, for the problem.

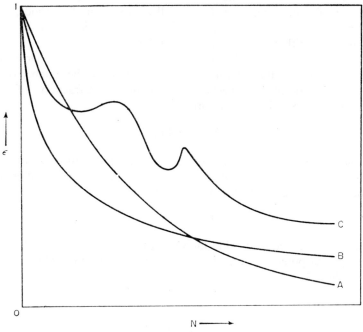

FIG. 11.12

It is to be noted, however, that the rate of convergence achieved with any value of α varies widely from one stage of the computation to another and, further, that in the earlier stages the convergence achieved using the optimum factor is not necessarily the best. These effects are due to the changing distributions of the residuals that occur during the computation and to the suitability of the particular degree of over-relaxation which is being used: for example, when residuals over large regions of the field are of the same sign, then a large value of α (giving a high degree of over-relaxation) produces the most rapid instantaneous convergence; but when positive and negative residuals are uniformly distributed throughout the field, then a low value of α (possibly 1) is best.

The preceding discussion assumes a stable convergence process, and it must be pointed out that unstable processes can also occur. As already stated, when $\alpha \geqslant 2$ the process diverges, but, even when less than 2, high values of α may be found to cause oscillations in the magnitudes of the residuals, and this serves as an indication that α is too large for rapid overall convergence (see curve C, Fig. 11.12).

It should be noted that an equation of the form of (11.18) applies also to the other iterative methods. In section 11.4.5 the rates of convergence of several methods are compared, and values of F for the Dirichlet problem in the square are given.

The determination of the optimum convergence factor. Whilst any value of α between 1 and 2 serves to improve the convergence, to take full advantage of the SOR method it is important to use a value close to the optimum. It is certainly necessary for good results to know α correctly to the second place of decimals and more accurate values are desirable in some cases. Unfortunately, however the determination of the optimum convergence factor causes considerable difficulty. Except for certain simple boundary shapes for the Dirichlet problem, it is not possible to calculate α_b, and possible difficulties associated with an empirical determination are apparent from the previous discussion. (It may also be noted that it is not possible to predict in general the number of iterations required to produce a given improvement in the solution for any value of α, though it is observed in practice that the variation is rather critical.) Consideration is now given to the few theoretical results which are available in respect of α_b, after which these are used as the basis for certain empirical or semi-empirical methods for evaluating α_b in problems with more difficult boundary shapes and conditions.

Frankel and Young have shown that the optimum value of the convergence factor α_b for the Dirichlet problem is given by

$$\alpha_b = \frac{2}{1+\sqrt{1-\lambda}}, \tag{11.22}$$

where λ may be regarded as being defined by eqn. (11.21). Equation (11.22) holds for the Dirichlet problem with any shape of boundary, but its use is not so simple as this might indicate because of the very limited situations for which it is possible to obtain values of λ theoretically. For a rectangular boundary with $(p+1)\times(q+1)$ nodes, λ can be expressed algebraically, and it leads to a value of optimum convergence factor given by

$$\alpha_b = 2-\pi\sqrt{2}\sqrt{\frac{1}{p^2}+\frac{1}{q^2}}, \tag{11.23}$$

where p and q are "large" (generally 15 or greater).[6] This result, of course, covers the case of a square mesh, but an alternative result for the optimum convergence factor for the square with $(p+1)$ nodes per side is

$$\alpha_b = \frac{2}{1+\sin(\pi/p)}. \tag{11.24}$$

(In this connection see Todd, p. 406, and reference 8.)

For all other boundary shapes, λ and α_b can only be determined using methods which are to a greater or lesser extent empirical. There is now quite a number of such methods and these are mostly based upon the combined application of eqns. (11.21) and (11.22). Prominent amongst these are the methods of Varga, described in his book, Kulsrud,[9] and Carré.[10] In the application of these equations, iterations are carried out for the problem in question with $\alpha = 1$ until a reasonably limiting value is obtained for λ. This is then used to predict the value of α_b which is employed thereafter. It has been found by Young[11] that the number n of iterations to determine λ with reasonable accuracy should normally be about 100. 100 iterations may, of course, involve a significant length of computing time and, in practice, this method would not normally be used except possibly for very lengthy or repeated problems.

The above approach can be improved in two ways—by achieving a sufficiently accurate value of λ more quickly and by ensuring that the iterations involved in finding λ are also useful in progressing the solution towards its final state. A suitable procedure is as follows. After an initial iteration with $\alpha = 1$, a larger value of α, but one less than α_b, is used for a small number of iterations (about 10). Use of an α greater than unity results in a more rapid convergence than with $\alpha = 1$ and hence, after the ten iterations, a better estimate of λ.

This estimate is used to evaluate a further value of α which is again used for about ten iterations in obtaining a further improved value of λ. This process is continued until a satisfactory estimate is made for α_b, as determined by a sufficiently small change occurring between successive estimates, or until the whole problem has been solved to a sufficient degree of accuracy (monitored through the values of U^n or some average of all the U^n's). Additionally, Carré[10] in his method introduces an empirical relationship following the calculation of α_b from eqn. (11.22), designed to ensure that the estimated α_b is always less than the true α_b (preventing the occurrence of complex eigenvalues). Carré estimated from experience with several particular problems that the number of iterations likely to be needed with his method, even for relatively small problems, is likely to exceed the minimum possible (using α_b throughout) by only about one-third.

Mention should also be made of a very simple technique due to Young[11] which can be used to give very quickly an estimate of α_b in the absence of more satisfactory procedures. It consists merely in applying eqn. (11.23) *for the rectangle having the same area as the given region*, and is an approach particularly valuable for small scale investigations (see section 11.4.4). It is known (references 7 and 8) that as α becomes greater than α_b the number of iterations required for a given accuracy increases less rapidly than when α becomes smaller than α_b. Therefore, in using the above approach it pays to use a high, rather than a low value of α and, to this end, it may be preferable to base estimates upon the square (rather than rectangle) having the same area as the region considered.

In accordance with the discussion in section 11.2.3, the foregoing applies strictly only to the Dirichlet problem, and when consideration is given to problems involving gradient boundary conditions it is found that not only are there no theoretical results comparable with those of eqns. (11.21)–(11.24), but that (other than when the matrix $\mathbf{M}$ is symmetric and positive definite) no completely sound mathematical basis for the general use of iterative schemes has been established.[†] For a given boundary and mesh type, the effect of changing from potential to gradient boundary conditions is "only" to modify a relatively small number of coefficients in the large number associated with the whole set of difference equations, and this encourages the hope that the more rigorously established results will apply reasonably to the gradient case also. This, indeed, proves to be so and, though convergence rates are slower, it is found that problems involving gradient conditions can be handled in much the same way as the simpler type. Carré's method, for example, works satisfactorily as does the equal area principle, and eqns. (11.21)–(11.24) can be used to give values of α which generally prove to be quite close to the true optimum values as determined by experiment.

Practicals aspects. It is possible, as with hand relaxation, to give some general guidance on the use of the method, and a number of points based on observation and experience are given below:

(a) The number of iterations required for a given reduction in error in any given problem increases roughly linearly with the square root of the number of points.

(b) Convergence is best with "simple" boundary shapes such as the square, and it deteriorates for more complex regions, particularly those involving narrow sections (see, for example, the air gap in Fig. 11.21, p.277).

(c) For a given boundary shape and number of nodes, the convergence is less good for gradient or mixed boundary conditions than it is for Dirichlet conditions.

(d) When an unfavourably high value of α is used the curve of residual against number of iterations oscillates.

[†] This is intended in the context of difference equations derived using Taylor series.

(e) Whilst an initial estimate of potentials is valuable in the longer computations, for shorter ones it may be easier to perform the additional iterations rather than to make the estimate and prepare the data for the computer.

11.4.4 *Current flow in an I-section conductor*

To illustrate the use of the SOR method, the problem of current flowing with uniform density in a highly permeable conductor with an I-shaped cross-section (approximating to that of a railway line) is considered. The conductor boundary, together with a map of flux lines, is shown in Fig. 11.13, and the map can be used for the calculation of the inductance of the conductor due to flux inside the section (see section 2.2.4).

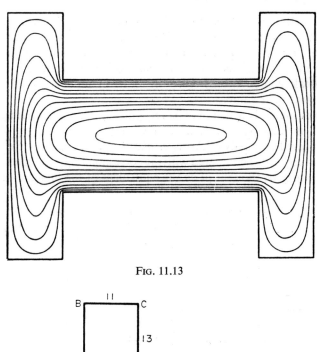

FIG. 11.13

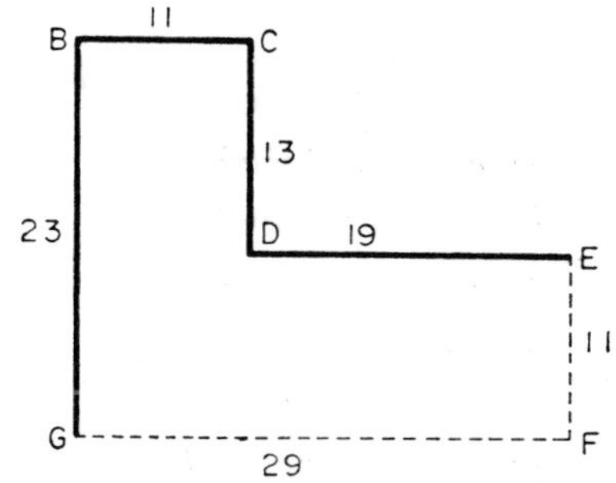

FIG. 11.14

The field is symmetrical about horizontal and vertical lines through its centre: it is sufficient, therefore, to consider one quarter of it (Fig. 11.14), and so doing minimizes the number of iterations for the computation. The number of nodes used for each side of the boundary is indicated in the figure. Along the boundary *GBCDE* the vector potential is constant (see section 5.3) and along the lines of symmetry, *EFG*, the normal gradient of vector potential is zero. The first boundary condition is satisfied by putting A = constant at all relevant nodes and the second is satisfied by using a simple difference equation developed in section 11.5.4, eqn. (11.51). At nodes within the boundary, A satisfies Poisson's equation which is expressed in finite-difference iterative form by eqn. (11.17). Initial values of A at all points were taken to be zero in the calculation.

The value of α can be estimated simply from the method of "the rectangle of equivalent area": inspection shows that an equivalent rectangle has sides with 15 and 29 points ($p = 14$, $q = 28$), and for these values eqn. (11.23) gives the optimum value of α as 1.717. This value is found to give a satisfactory rate of convergence; after only seventy iterations the residual at every point becomes less than 0·0068 (for a mean value of A of 60) indicating that the solution is sufficiently accurate.

11.4.5. *Other rapidly convergent methods*

Whilst the SOR method is still, because of its simplicity, flexibility, and relatively rapid rate of convergence, the most generally useful iterative method so far developed, there are a number of other rapidly convergent methods which have more recently occupied the attention of many mathematicians. These are now briefly indicated, and the reader concerned with large-scale computations, either in terms of number or of size of problem, should carefully assess their value.

In the first group of methods which may be recognized, nodes are no longer treated separately but, instead, in lines or blocks, with or without the use of over-relaxation. Such methods were considered originally by Arms *et al.*,[12] and in the basic form they give improved rates of convergence as compared with the corresponding point methods, as indicated in Table 11.1. They may be thought of as the equivalent of the line-and-block methods of

TABLE 11.1. CONVERGENCE RATES FOR THE DIRICHLET PROBLEM IN THE SQUARE
WITH $(p+1)^2$ NODES

	Time per iterative cycle	Theoretical nos. of iterations[a] $(K = -\log \varepsilon)$	Nos. of iterations $p = 19$ $\varepsilon = 10^{-10}$	Theoretical nos. of iterations with point Chebyshev method
Jacobi	$k \cdot 4p^2$	$K \cdot 2p^2/\pi^2$	1680	$K \cdot p/\pi$
Line Jacobi	$k \cdot 4p^2$	$K \cdot p^2/\pi^2$	840	$K \cdot p/\pi \sqrt{2}$
Gauss–Seidel	$k \cdot 4p^2$	$K \cdot p^2/\pi^2$	840	$K \cdot p/\pi \sqrt{2}$
Line Gauss–Seidel	$k \cdot 4p^2$	$K \cdot p^2/2\pi^2$	420	$K \cdot p/2\pi$
SOR (α_b)	$k \cdot 7p^2$	$K \cdot p/2\pi$	70	—
Line SOR (α_b)	$k \cdot 7p^2$	$K \cdot p/2\sqrt{2}\pi$	50	—
Alternating direction implicit	$k \cdot 7p^2$	$K \cdot p/4\sqrt{2}\pi$	25	—
Alternating direction (optimum)	$k \cdot 9p^2$	$K \cdot \sqrt{p}/4\pi(2)^{1/4}$	7	—

[a] See eq. (11.18).

hand relaxation but, as a little consideration will show, they are not purely iterative in character: corresponding with the need to construct a hand-relaxation operator is the requirement to solve simultaneously and normally directly, for the potentials at the nodes of each block during every iteration. A general discussion of block methods is given in Varga. Of the basic line methods, indicated in the table, only that of line SOR is superior to point SOR, and the gain it brings is not generally such as to justify the loss in flexibility and increase in complexity of programming.

However, when other features are introduced along with the use of blocks, very substantial reductions in numbers of iterations and computing times are possible. Best known at the present time amongst methods of this kind is the alternating-direction implicit method (ADI) of Peaceman and Rachford.[13] In this method, as the name indicates, nodes are treated line by line, but the direction of sweep of lines across the mesh alternates (in each iteration there is a sweep in the x-direction and a sweep in the y-direction). The power of the method is indicated by the figures in the table. Most theoretical results for ADI methods are restricted to rectangular regions, but good performance is obtained in practice with more general regions (see, for example, Young and Erlich[14]). Particularly with very large numbers of nodes and reasonably simple boundaries the ADI method can be specially effective, but difficulties remain in determining the optimum convergence factors (at least one for each direction of sweep and also their variations in non-stationary versions of the technique). General discussions of the above line methods are given by Keller[15] and Heller.[16]

A second group of powerful methods which involves the use of Chebyshev polynomials was first developed by Shortley.[17] Rates of convergence are indicated in the table for point methods. The technique can also be used in conjunction with blocks and the two-line cyclic Chebyshev method (2LCC) is a particularly powerful example which has been found in some cases to give even shorter computing times than the ADI method.[18, 19] See Varga for further details.

Thirdly, purely direct methods of solution are receiving considerable attention at the present time, and it is the opinion of several workers that, when the speed and size of computers have increased sufficiently, such methods will be preferred to iterative ones. Whether this will prove to be the case remains to be seen, but even at the present time there are examples of extremely powerful direct techniques. One which may be noted is due to Hockney:[20] this employs Fourier analysis and is (currently) restricted to very simple, single-region fields and simple boundary conditions but, for such restricted problems, is reported to be ten times faster than the best iterative schemes (2LCC and ADI).

Finally, mention is made of the "boundary contraction" method of Milnes and Potts.[21, 22] Although currently it appears to be receiving little attention, it has some apparent merits.

11.4.6. A special technique

To conclude the discussion of machine methods of solving the difference equations, a very effective technique first reported by Ahamed[23] is briefly described. Ahamed was concerned with the fields of distributed currents and the description here is in the same terms, but other applications will be immediately apparent.

After each iteration the line integral of field strength round a suitable path enclosing the field is evaluated from the potential values. Let its value after the nth iteration be F^n. In general, this value will not be equal (as it should be for an exact solution) to the total current enclosed by the path of integration J, but instead to some multiple $(1/C^n)$ of this. Accordingly, the method consists in "forcing" all the potentials at the end of each iteration by multiplying them by C^n. The process of standard iteration followed by a multiplication by C^n is continued until C^n is sufficiently close to unity. Ahamed suggests $0.999 < C^n < 1.001$, when the procedure is terminated by a small number of standard iterations.

Excellent performance is claimed of the method for both linear and non-linear (saturated) problems by Ahamed, but the results of a careful investigation by Reece[24] suggest that considerable care and/or experience are needed on the part of the analyst. Reece draws attention, for example, to the need for a careful choice of the integration contour and of convergence failure when used with SOR.

A very similar technique (employing information about the physical nature of the true solution) was subsequently used successfully by Stoll.[25]

11.5. Gradient boundary conditions

11.5.1. *Introduction*

The earlier discussion of boundary conditions was restricted to problems in which the potential function was specified at the boundaries, but in this section consideration is given to more general classes of boundary condition. Treatment of these involves considerable detail and has been withheld until this stage, so that the reader could be presented with an overall picture of the use of the methods as soon as possible. The difference equations to be developed are, of course, used in exactly the same ways, for hand or machine iteration, as are the previously developed equations.

The boundary conditions to be treated are those applying to straight or curved boundaries, coincident or non-coincident with the nodes, and separating regions which, in general, have different values of the electric or magnetic constants and different values of current density. There are, evidently, very many different combinations of the above variables and there is neither space nor necessity to detail the difference equations associated with all of them. However, the most important equations are developed in general terms and certain useful special forms are then presented explicitly; the reader can easily derive other forms as they are required.

The boundaries fall into two classes depending upon whether they coincide with nodes or not. Those which do coincide with nodes and which are parallel or diagonal to the network, are treated in section 11.5.2; those which do not coincide with the nodes and which are parallel to, but not coincident with, the network, or which are generally curved, are treated in section 11.5.3. In section 11.5.4 boundary conditions represented by lines of symmetry of the field are discussed. The treatment is concluded with two examples.

11.5.2. *Boundaries coincident with nodes*

For each of the two classes of boundary which coincide with nodes, namely the one parallel to and the one diagonal to the network, two different difference equations are required—one applies for the nodes at corners of the boundary and one for the other nodes on the boundary. For each class consideration is given first to nodes not adjacent to corners and the parallel type of boundaries is treated before the diagonal one.

The most general boundary or interface condition that is likely to be met in practice applies for two regions having different permeabilities with current in one of them, and this combination is assumed for each of the above classes. For the particularly useful class of parallel boundaries, a number of important special equations are also presented explicitly. *Boundaries parallel to the mesh: general case.* Consider the boundary and nodes shown in Fig. 11.15; let the region a to the left carry current distributed with uniform density J, and let the region b to the right be free of current. Also, let A_a, μ_a, and $W_a (= \mu_a \mu_0 J)$ denote the quantities A, μ, and W in region a, and A_b and μ_b denote those in region b.

At all nodes in a, including those on the boundary, Poisson's equation is satisfied, and so eqn. (11.8) gives for node 0

$$A_{a1} + A_{a2} + A_{a3} + A_{a4} - 4A_{a0} + h^2 W_a = 0; \qquad (11.25)$$

and, at all nodes in b, Laplace's equation is obeyed, so that eqn. (11.8) with $W = 0$ gives for the node 0

$$A_{b1} + A_{b2} + A_{b3} + A_{b4} - 4A_{b0} = 0. \qquad (11.26)$$

However, the potentials A_{a1} and A_{b3} have no physical meaning because the nodes 1 and 3 lie outside the regions a and b respectively. Potentials such as these are said to be *fictitious*

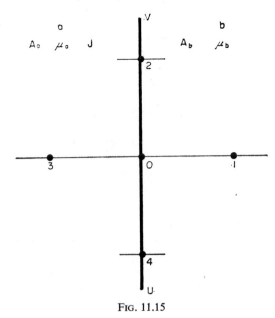

FIG. 11.15

and they occur frequently in the consideration of boundary effects. Though they have no real value it is nevertheless possible to consider them as mathematical quantities for use in the derivation of the equation for the boundary node. This is because the boundary conditions between the two regions, when expressed in finite-difference form, also involve the same fictitious values A_{a1} and A_{b3} which can thus be eliminated using eqns. (11.25) and (11.26).

The boundary conditions give the following two equations: for the continuity of vector potential between the two regions

$$A_a = A_b; \qquad (11.27)$$

and, for the continuity of the tangential component of field strength, across the boundary

$$\frac{1}{\mu_a}\left(\frac{\partial A_a}{\partial n}\right) = \frac{1}{\mu_b}\left(\frac{\partial A_b}{\partial n}\right), \qquad (11.28)$$

where $\partial A/\partial n$ is the gradient of vector potential normal to the boundary. Expressed in terms of the potentials at a boundary node m, the first of these equations gives

$$A_{am} = A_{bm} = A_m, \qquad (11.29)$$

and substituting for $\partial A_a/\partial n$ and $\partial A_b/\partial n$ in the second equation gives

$$\frac{1}{\mu_a}(A_{a1}-A_{a3}) = \frac{1}{\mu_b}(A_{b1}-A_{b3}). \tag{11.30}$$

The fictitious values, A_{a1} and A_{b3}, are eliminated from eqn. (11.30) by direct substitution from eqns. (11.25) and (11.26) respectively, and then using eqn. (11.29) and writing $R = \mu_b/\mu_a$, the difference equation for the typical boundary node, 0, becomes finally,

$$A_{b1}\frac{2}{1+R}+A_2+A_{a3}\frac{2R}{1+R}+A_4-4A_0+\frac{R}{1+R}h^2W_a = 0. \tag{11.31}$$

Boundary between Poissonian and Laplacian regions: constant permeability. A frequently required equation is that for nodes lying on a boundary between two regions of the same permeability, one current-carrying and the other current-free. It is given simply by writing $\mu_a = \mu_b$, that is, $R = 1$ in eqn. (11.31), and is

$$A_{b1}+A_2+A_{a3}+A_4-4A_0+\tfrac{1}{2}h^2W_a = 0. \tag{11.32}$$

Boundary between two Laplacian regions: different permeabilities. The equation for a node lying on the boundary between two current-free regions of different permeabilities is given by writing $W_a = 0$ in eqn. (11.31). For this condition the equation may also be required in terms of scalar potential. Then, because the boundary condition (11.30) is replaced by

$$\mu_a(\psi_{a1}-\psi_{a3}) = \mu_b(\psi_{b1}-\psi_{b3}) \tag{11.33}$$

(for the normal component of flux density), the required value of R is the inverse of that above, and the finite difference equation becomes

$$\psi_{b1}\frac{2}{1+R}+\psi_2+\psi_{a3}\frac{2R}{1+R}+\psi_4-4\psi_0 = 0, \tag{11.34}$$

where $R = \mu_a/\mu_b$ for magnetic fields, and $\varepsilon_a/\varepsilon_b$ for electric ones.

(When $R = 1$ this equation reduces to the simple difference form of Laplace's equation for a symmetrical star.)

Boundary between two Laplacian regions: one infinitely permeable. If the region a is infinitely permeable, the flux crosses the boundary normally. This normal gradient condition is defined by the equation

$$2A_{b1}+A_2+A_4-4A_0 = 0, \tag{11.35}$$

obtained by writing $R = 0$ in equation (11.31).

Corner node with boundaries parallel to the mesh: general case. The above equations apply to all nodes on the boundary except those at corners of the boundary; for these a different equation is required. Consider a right-angled corner of an interface between two regions, a and b, having the same permeabilities and current distributions as those above, see Fig. 11.16. Because of the representation of the field by discrete points the true shape of the boundary cannot be accounted for exactly, and, to derive an equation valid for the corner node, it is necessary to introduce two additional boundaries: these are symmetrically placed with respect to the corner and are represented by the dotted lines including the angles α and β. For the node 0 on the boundary NOP, a single equation applies for any value of α

between $\pi/2$ and $3\pi/2$ and, similarly, one (different) equation holds at the same node on the boundary QOR for any value of β between 0 and $\pi/2$. From these two equations, the equation valid at the node 0 on the actual boundary is taken to be that giving the mean of the two values of A_0.

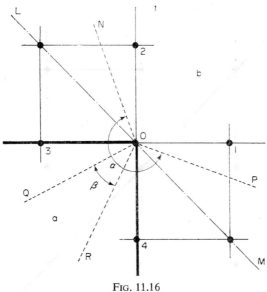

FIG. 11.16

Now the equation for the boundary QOR is simply seen, from a consideration of the value $\beta = 0$ (for which the boundary vanishes), to be

$$A_{b1} + A_{b2} + A_{b3} + A_{b4} - 4A_0 = 0. \tag{11.36}$$

Also, the equation for the boundary NOP is seen, from the particular case of $\alpha = \pi$, to be that for the boundary LM. The difference equation for nodes on such a diagonal boundary is derived below, eqn. (11.44), and using this, the equation for node 0 on the boundary containing the angle α is

$$A_{b1} + A_{b2} + R(A_{a3} + A_{a4}) - 2(1 + R)A_0 + \tfrac{1}{2}Rh^2W_a = 0. \tag{11.37}$$

Thus, combining eqns. (11.36) and (11.37) and remembering the equality of A_{a3} and A_{b3}, and A_{a4} and A_{b4}, the equation for the node 0 at the vertex of a right-angled interface is

$$A_{b1} + A_{b2} + \tfrac{1}{2}(1 + R)(A_3 + A_4) - (3 + R)A_0 + \tfrac{1}{4}Rh^2W_a = 0. \tag{11.38}$$

The corresponding equations for current in the region b are obtained in a similar way.

An important special case derivable from equation (11.38) applies for the node at the corner of an infinitely permeable region occupying three of the quadrants (that is, including an angle of $3\pi/2$) when both regions are current-free. The result, for the region b infinitely permeable, is obtained by putting $R = \infty$ in the above equation, and is

$$A_3 + A_4 - 2A_0 = 0. \tag{11.39}$$

Boundaries diagonal to the mesh: general case. In developing equations for the nodes on boundaries diagonal to the lines of the network, see Fig. 11.17, two approaches, giving equations connecting different sets of nodes, are possible: the equations developed in the previous section can with slight modification be used directly, or a new (and more accurate) set of equations can be derived.

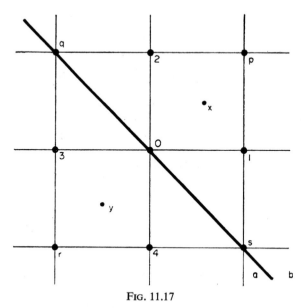

FIG. 11.17

The equations already developed can be applied to the diagonal boundary by rotating the star at the boundary node through 45° and increasing the mesh length to $\sqrt{2}h$. Thus, using the notation of Fig. 11.17, the general equation developed from eqn. (11.31) is

$$A_{bp}\frac{2}{1+R}+A_q+A_{ar}\frac{2R}{1+R}+A_s-4A_0+\frac{2R}{1+R}h^2W_a = 0. \tag{11.40}$$

Because of the effective increase in mesh size, this equation is less accurate than the equations used for nodes off the boundary.

A more accurate equation for the node 0 can be found in terms of the values at nodes 1, 2, 3, and 4, and the method is similar to that used in the derivation of eqn. (11.31), though it is necessary to consider nodes at x and y, where x is the mid-point of the line joining the points 1 and 2, and y is the mid-point of the line joining 3 and 4. At node 0, A_0 satisfies the difference forms of Poisson's and Laplace's equations (involving fictitious values of potential); that is

$$A_{a1}+A_{a2}+A_{a3}+A_{a4}-4A_0+h^2W_a = 0 \tag{11.41}$$

and

$$A_{b1}+A_{b2}+A_{b3}+A_{b4}-4A_0 = 0. \tag{11.42}$$

Also the equality of the tangential components of field strength at 0 requires that

$$\frac{1}{\mu_a}(A_{ax}-A_{ay}) = \frac{1}{\mu_b}(A_{bx}-A_{by}). \tag{11.43}$$

Now A_{a1}, A_{a2}, A_{ax}, A_{b3}, A_{b4}, and A_{by} are fictitious values, but, using linear interpolation, they are simply connected by the equations

$$2A_{ax} = A_{a1} + A_{a2} \quad \text{and} \quad 2A_{by} = A_{b3} + A_{b4}.$$

Thus, noting that for the real values

$$2A_{ay} = A_{a3} + A_{a4} \quad \text{and} \quad 2A_{bx} = A_{b1} + A_{b2},$$

all fictitious values can be eliminated between the above equations and eqns. (11.41)–(11.43) to give for the equation at node 0

$$2(A_{b1} + A_{b2}) + 2R(A_{a3} + A_{a4}) - 4(1 + R)A_0 + Rh^2 W_a = 0. \tag{11.44}$$

From this equation special forms can be derived as required.

Corner node with diagonal boundaries: general case. The equation for the node at a right-angled corner formed by the intersection of two interfaces diagonal to the network is obtained from considerations similar to those given in the preceding subsection for corner nodes. When the quandrant is the current-carrying region, *a*, containing the node 3, the equation is

$$A_{b1} + \tfrac{1}{2}(1 + R)A_{a3} + \tfrac{1}{4}(3 + R)(A_{b2} + A_{b4}) - (3 + R)A_0 + \tfrac{1}{4}Rh^2 W_a = 0. \tag{11.45}$$

11.5.3. *Boundaries non-coincident with nodes*

The general class of boundary which is non-coincident with the nodes of the mesh (including both straight and curved boundary lines) is now to be discussed. However, before commencing the general discussion, the special case of a straight-line boundary, parallel to (but non-coincident with) the lines of the mesh, is treated.

Straight boundary parallel to the mesh. In the treatment of this class of boundary additional nodes are introduced at the points of intersection of the boundary with the mesh (Fig. 11.18).

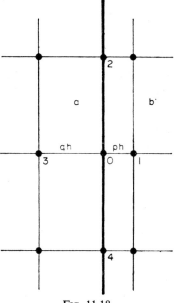

FIG. 11.18

It can easily be shown (see section 11.2.2) that the finite-difference form of Poisson's equation for any one of these additional nodes is

$$A_1 \frac{2}{p(p+q)} + A_2 + A_3 \frac{2}{q(p+q)} + A_4 - 2A_0\left(1+\frac{1}{pq}\right) + h^2W = 0, \qquad (11.46)$$

where p and q are as shown in Fig. 11.18. Using this equation to obtain expressions for the fictitious values A_{a1} and A_{b3} and substituting these in the boundary condition, (11.30), gives, for the boundary nodes, the difference equation

$$A_1 \frac{2}{p(q+pR)} + A_2 + A_3 \frac{2R}{q(q+pR)} + A_4 - 2A_0\left(1+\frac{1}{pq}\right) + h^2 \frac{W_a pR}{(q+pR)} = 0. \qquad (11.47)$$

This assumes, as usual, a distribution of current in region a only.

In using this equation the boundary nodes are considered directly as part of the complete pattern. (This, of course, requires also the use of asymmetrical difference equations for the nodes on the mesh and adjacent to the boundary.)

Note that putting $p = q \left(\neq \frac{1}{2}\right)$ in eqn. (11.46) gives the basic equation for a rectangular mesh [compare eqn. (11.9)]; and that the same substitution in eqn. (11.47) gives the equation for a node on a rectangular mesh between regions of different permeability.

Curved boundaries: general. Several equations have been developed for use with curved boundaries along which the normal gradient is specified (see below) but, so far as the authors are aware, no general equations connecting the fields on both sides of a boundary [corresponding with eqn. (11.31)] have been developed. If they were developed there would inevitably be many difficulties associated with their use: they would be extremely long [see eqns. (11.49) and (11.50) for normal gradient conditions]; they would involve many measurements of length for asymmetrical stars and related geometrical constructions; and, to achieve accurate results, a very fine mesh would be required.

It seems probable that the most satisfactory treatment is to avoid the use of additional difference equations by approximating the boundary shape by many straight-line segments which are each parallel or diagonal to the lines of the mesh. The accuracy of this representation can be estimated from a consideration of the differences between the actual and the assumed boundary shapes.

Because of the above considerations, and because the most general class of boundary condition occurs infrequently in practice, no attempt is made here to develop general equations, attention being concentrated on the particular class of normal gradient conditions which frequently arise.

Curved boundaries: normal gradients. Consider the curved boundary, shown in Fig. 11.19, along which the normal gradient, $\partial A/\partial n$, is specified at all points. The simplest approach is to introduce additional nodes such as F on the far side of the boundary, and to determine the fictitious values of potential, such as A_F, at these nodes, the normal equations being used at the real nodes. A_F may be determined with the aid of one of a number of simple geometrical constructions, and a simple and satisfactory one, suggested by Fox,[26] is as follows.

FH is drawn normal to the boundary which it cuts at the point G. Then the point P is chosen to make $PG = FG$ and the points L and P are joined by a line projected to intersect the line MN in Q. The potentials at F and P are thus related by

$$A_F = A_P + \left(\frac{\partial A}{\partial n}\right)_G FP.$$

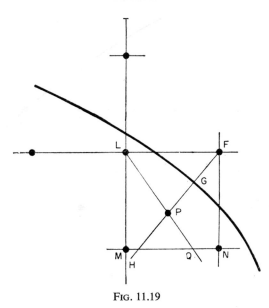

FIG. 11.19

A_P is found by linear interpolation between the values of A_L and A_Q, the value of A_Q being obtained by interpolation between A_M and A_N, and hence

$$A_F = \frac{1}{LQ}\left[PQ\cdot A_L + \frac{LP}{MN}(QN\cdot A_M + MQ\cdot A_N)\right] + \left(\frac{\partial A}{\partial n}\right)_G FP. \qquad (11.48)$$

The same expression (and construction) apply, of course, at all parts of the boundary. Equation (11.48) is mostly used for the important condition of zero tangential field when $\partial A/\partial n = 0$ and $A_F = A_P$.

A more elaborate treatment, which involves the values of the vector potential on the boundary line, has been given by McKibbin.[27] In terms of the quantities shown in Fig. 11.20 he derives the equations

$$[(1+q)\tan\theta - 1]A_0 + \frac{q}{1+q}[1-q\tan\theta]A_4 + \frac{1}{2}q(q+1)(q+2)A_3 - q^2(q+2)A_5$$

$$+ \frac{1}{2}q^2(1+q)A_7 - \left\{\frac{(1+q^2)}{q} - \frac{(1+2q)}{q(1+q)}[1-q\tan\theta]\right\}A_c + \frac{qh}{\cos\theta}\left(\frac{\partial A}{\partial n}\right)_c + \frac{1}{2}qh^2W = 0$$

$$(11.49)$$

and

$$[(1+p)\cot\theta - 1]A_0 + \frac{p}{1+p}[1-p\cot\theta]A_3 + \frac{1}{2}p(p+1)(p+2)A_4 - p^2(p+2)A_5$$

$$+ \frac{1}{2}p^2(1+p)A_6 - \left\{\frac{(1+p^2)}{p} - \frac{(1+2p)}{p(1+p)}[1-p\cot\theta]\right\}A_D + \frac{ph}{\sin\theta}\left(\frac{\partial A}{\partial n}\right)_D + \frac{1}{2}ph^2W = 0.$$

$$(11.50)$$

These equations yield more accurate values for A than the simpler ones above. McKibbin discusses their accuracy and applies them in the calculation of the inductances of conductors in semi-closed and tapering slots.

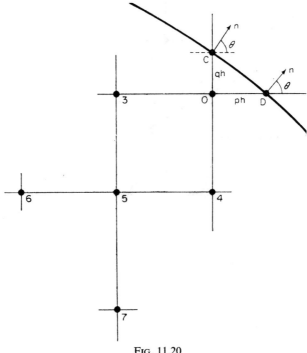

FIG. 11.20

11.5.4. *Lines of symmetry*

When a field has lines of symmetry, it is convenient to consider only a part of the field region, the lines of symmetry being treated as boundaries; in this way the necessary number of nodes is greatly reduced. Special difference equations are required for the nodes lying on the lines of symmetry but they are simply obtained. The equation for lines parallel to the mesh (both coincident and non-coincident with it) and the equation for lines diagonal to the mesh are given below. The solutions apply to the condition in which current is present (giving rise to the term h^2W); and, of course, symmetry requires that the current densities and the permeabilities on the two sides of the line are equal.

Consider first the case of a line coincident with the mesh, for which the basic Poisson difference equation (11.8), applies at all nodes. Let UV in Fig. 11.15 be the line of symmetry of the field. Because of this symmetry $A_{b1} = A_{a3}$, and the equation at 0 for the field in the region b is

$$2A_{b1}+A_{b2}+A_{b4}-4A_0+h^2W = 0. \tag{11.51}$$

The difference equation for a line of symmetry diagonal to the network is derived in a similar way to the above. Figure 11.17 shows the positions of the nodes with respect to the line and for symmetry of the field, it is evident that $A_{b1} = A_{a4}$ and that $A_{b2} = A_{a3}$. Therefore, substituting for these values in the basic Poissonian equation, (11.8), the required equation (applicable in region b) for the node 0 on the line of symmetry is

$$(2A_{b1}+A_{b2})-4A_0+h^2W = 0. \tag{11.52}$$

For the additional nodes introduced on a line of symmetry which is parallel to, but not coincident with, the mesh (Fig. 11.18) the difference equation is given by setting $p = q = \frac{1}{2}$ (since the line bisects the mesh length) and $A_1 = A_3$ in eqn. (11.46). It is, therefore,

$$8A_1 + A_2 + A_4 - 10A_0 + h^2W = 0. \tag{11.53}$$

The above equations are those most frequently used but others are required from time to time. These are derived from the normal form of the appropriate equation by allowing for the equality of values of potential at nodes on opposite sides of the line of symmetry.

11.5.5. *Two examples*

To conclude the discussion of boundary conditions, the application of some of the above boundary-node equations is demonstrated with the aid of two examples.
Conductors in a slot. The first example is in the analysis of the field of two current-carrying conductors in the slot of an electrical machine, of interest in the determination of leakage reactance. The iron surfaces are assumed infinitely permeable, and account is taken of the radial air gap. A field map for the instant when the two currents are equal is shown in Fig. 11.21 (the quantity of flux between continuous lines is 7·5 times that between the dotted

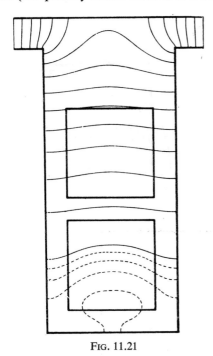

FIG. 11.21

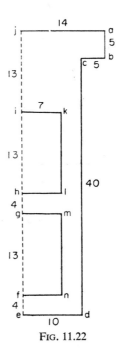

FIG. 11.22

ones). The appropriate boundary conditions are explained with the aid of Fig. 11.22. (This also shows the number of nodes to each dimension for a suitably fine mesh.)

(a) The line *ab* is a flux-line, and so at nodes on it A is a constant chosen to be zero.
(b) On the iron surfaces, since μ is infinite, eqn. (11.35) is appropriate, except at the corner nodes *c*, *d*, *e*, and *j*. At the nodes *e* and *j*, eqn. (11.35) is used in a form allowing for symmetry about the line *ej*, i.e. with $A_3 = A_1$. At node *c*, eqn. (11.38) with $W = 0$ and $R = \infty$ applies, and at *d* eqn. (11.39).

AC 19

(c) Equation (11.51) is required at nodes along the line of symmetry *ej* except at *f, g, h,* and *i* where the symmetrical form (with $A_3 = A_1$) of eqn. (11.31) is used.

(d) For all the nodes on the conductor surfaces the equation is (11.31) except for the nodes *k, l, m,* and *n* where it is (11.38).

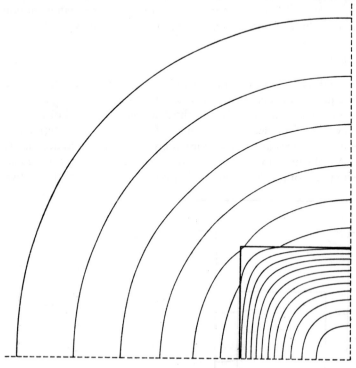

FIG. 11.23

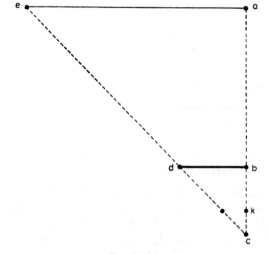

FIG. 11.24

Conductor of finite permeability. The second example is in the analysis of the magnetic field of a current-carrying conductor, of permeability $5\mu_0$, remote from all permeable boundaries. The map of one quarter of the field is shown in Fig. 11.23 and this was calculated using the region shown in Fig. 11.24. For the boundary nodes the equations are as follows:

(a) Values of the potential along *ea* are calculated with acceptable accuracy on the assumption that the whole current is concentrated at *c* (see section 5.2.1). (The alternative to this is to specify zero tangential gradient on a (large) circle, with its centre at *c*, but this involves far more difficulty.

(b) On the line of symmetry *ec*, the required equation for nodes between *c* and *d* is (11.52), and for nodes between *d* and *e*, it is the same with $W = 0$. At node *d*, the symmetrical form of eqn. (11.38), i.e. with $A_3 = A_2$ and $A_4 = A_1$, is used.

(c) The only other equation worthy of note is that for the node *c* and this is easily seen to be

$$A_k - A_c + \tfrac{1}{4} h^2 W = 0, \tag{11.54}$$

where *k* is the node immediately adjacent to *c*.

11.6. Errors

11.6.1. *Introduction*

Solutions obtained by using finite-difference methods may contain errors of two kinds. The first of these is due to the replacement of the differential equation by the finite-difference equations, and it is a function of the mesh size. The second arises because, in general, the difference equations are not solved exactly. The magnitude of either of these cannot be determined exactly at any stage of the solution but there are ways, which are discussed below, of estimating and improving the accuracy of solution.

11.6.2. *Mesh error*

General considerations. As indicated in section 11.2, because of the neglect of higher order terms (truncation error) in the Taylor expansion for the potential, eqn. (11.4), an error, known as the mesh error, is present in the solution of the field equation. It is instructive to observe that this error is exactly the same as that caused by assuming a linear variation of potential between nodes and, indeed, it is easy to develop the familiar difference equations on this assumption [that, for example, $2h(\partial A/\partial x)_0 = A_1 - A_3$].

To examine the magnitude of the error, consider again eqn. (11.4). It is seen that, for the general asymmetrical star, the error in $\partial^2 A/\partial x^2$ is

$$+ \frac{h}{3!} \left[pr(p^2 - r^2) \left(\frac{\partial^3 A}{\partial x^3} \right)_0 \right] + \frac{h^2}{4!} \left[pr(p^3 + r^3) \left(\frac{\partial^4 A}{\partial x^4} \right)_0 \right] + \ldots, \tag{11.55}$$

and an equivalent expression exists for the error in $\partial^2 A/\partial y^2$. The error in the finite-difference form of the field equation therefore contains terms in *h*. For the symmetrical star, *p* and *r* are equal to 1, and the error in $\partial^2 A/\partial x^2$, eqn. (11.55), reduces to

$$\frac{2h^2}{4!} \left(\frac{\partial^4 A}{\partial x^4} \right)_0 + \ldots. \tag{11.56}$$

Therefore the largest term of the error in the finite-difference form of the field equation involves h^2. Consequently (since h is small) the errors introduced by using asymmetrical stars are greater than those introduced by using symmetrical ones. Because of this, particular care is needed to achieve a given accuracy with asymmetrical stars, for example, at nodes adjacent to curved boundaries. Greater care is also necessary with such stars in regions where the rate of change of field strength is high, for example, near to sharp corners of boundaries, for there the assumption of negligible values for the higher derivatives of the potential is least satisfactory.

Whilst it is not possible to determine precisely the magnitude of the maximum error at any node due to the finite-difference approximation, it is possible to obtain a simple expression, giving a value, the upper bound, below which it is known to lie. In terms of the radius ϱ of a circle which just encloses the whole field region, and the maximum M_4 of the absolute values of the fourth-order partial derivatives of the exact solution, this expression for the upper bound of the mesh error is

$$\frac{M_4 h^2 \varrho^2}{24}. \tag{11.57}$$

Since this involves a knowledge of the exact solution, it can only be used by making an estimate of the value of M_4 (see Milne, p. 219) and this is rarely, if ever, worth while. It is mentioned here because the form of variation, with h^2, suggests the use of a method, extrapolation to zero error (see below), for the improvement of a solution.

Consideration is now given to devices for the improvement of solutions involving mesh error, though in passing, it should be noted that general inspection of a given solution can indicate the presence of such error (see section 11.6.3). The first two of the devices to be discussed here, the use of finer meshes and the use of more accurate difference equations, are both helpful for the detection and the reduction of error; unless the necessary mesh size is known from experience, one of them must always be used to ensure that a sufficiently fine mesh is employed ultimately. There are two other techniques, extrapolation to zero error, and the use of difference corrections, which are designed to improve the accuracy of results obtained from a given solution. Also, there are more elaborate expressions for the evaluation of potential gradient, and these are used when high accuracy is required or where the field strength is changing rapidly.

Fine mesh. To ensure that mesh error is negligibly small, the simplest technique is to use a fine mesh; and the only convenient way of ensuring that a sufficiently fine, final net is used, consists in obtaining solutions with nets, the mesh lengths of which are successively reduced (most conveniently by a factor two at each stage) until the condition, that the potentials at nodes common to the final and the penultimate nets differ negligibly, is reached.

More accurate difference equations. A second method of improving the accuracy involves the use of difference equations for which the mesh error is intrinsically smaller than that in the equations so far considered. Many such equations can be devised (see, for example, Milne, p. 130, and Buckingham, p. 502), but the most generally useful of them is that connecting the nine nodes shown in Fig. 11.25, and is easily shown to be

$$4A_1 + A_2 + 4A_3 + A_4 + 4A_5 + A_6 + 4A_7 + A_8 - 20A_0 + 6h^2 W = 0. \tag{11.58}$$

The lowest term in the expression for the truncation error does not involve terms of h of lower order than the sixth, and the upper bound for the mesh error is

$$\frac{M_8 h^6 \varrho^2}{12,096} \tag{11.59}$$

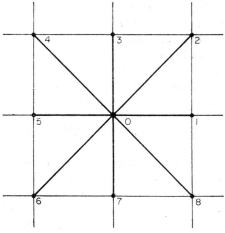

Fig. 11.25

where ϱ is the radius of the circle enclosing the field and M_8 is the maximum of the absolute values of the eighth derivatives of the exact solution.

Equations of this sort can be used both to check (by evaluating residuals) the accuracy of a solution obtained from less accurate difference equations and also to compute more accurate solutions. However, because of the additional labour involved in using them, they should be employed only in the final stages of a solution by hand. With machine iteration, however, because the arithmetical operations are often completed more rapidly than the logical ones ordering the calculations, the accurate equations may be used to advantage throughout. It is found with iterative solutions that the convergence rates with the nine- and five-node equations are almost identical and so, in many cases, it is possible to estimate and compare the times involved in achieving a given accuracy using the nine-node equation or the five-node one with a finer mesh.

Extrapolation to zero error. On the assumption that mesh error is proportional to the square of the mesh size (see *General considerations* above), it is possible to use a process of extrapolation, due to Richardson,[28] to estimate the solution for zero mesh length. Thus, if A is the solution for zero mesh length, and A_h and $A_{h/2}$ are the solutions for mesh lengths of h and $h/2$ respectively,

$$A = A_h - kh^2 = A_{(h/2)} - \frac{kh^2}{4},$$

where k is the constant of proportionality in the error term. Hence, eliminating k from these equations gives the estimated solution for zero mesh length as

$$A = \tfrac{4}{3} A_{(h/2)} - \tfrac{1}{3} A_h. \tag{11.60}$$

This equation often gives good results but, since its validity depends upon an assumption which is only approximately true, it has to be used with care (see reference 8 and Milne, p. 219).

Difference correction. For the five-node symmetrical star, the Poisson equation can be represented in finite-difference form, including the terms normally neglected, as

$$[(A_1 + A_2 + A_3 + A_4 - 4A_0) + h^2 W] - \tfrac{1}{12}(\delta_{ox}^4 + \delta_{oy}^4) + \tfrac{1}{90}(\delta_{ox}^6 + \delta_{oy}^6) - \ldots = 0, \tag{11.61}$$

where, at the point 0, δ_{ox}^n is the nth order difference[†] of values of A in the x-direction, and δ_{oy}^n is the nth order difference in the y-direction. This is, of course, more accurate than the usual form of equation and with it allowance can be made fairly easily for some of the terms normally neglected; though usually, because of the additional computation, only the fourth differences would be allowed for. Having developed the solution for the final net in the usual way, the differences are found at each point (except near the boundary where they cannot be calculated) and these are used in eqn. (11.61), to compute improved values of potential. Further details of this technique are given by Fox,[(29,30)] by Buckingham (p. 559), who suggests the use of second instead of fourth differences, and by Shaw (p. 212) who details an example.

Evaluation of gradient. It has been pointed out that a calculation of gradient based upon a linear variation in potential may result in considerable inaccuracy. Also, inaccuracies may occur since values of gradient are formed as the (usually) small differences between large numbers in which error may be present. To minimize both of these inaccuracies it is, therefore, sometimes necessary to use rather more elaborate formulae for the evaluation of gradient. Such expressions are derived by fitting a polynomial expansion to the values of potential at points lying on a straight line along which the gradient is required.[(31)] For example, in terms of the values A_0, A_1, A_2, and A_3 at equally spaced points on a line parallel to the x-axis

$$
\left.h\left(\frac{\partial A}{\partial x}\right)\right._0 = \frac{1}{3!}(-11A_0+18A_1-9A_2+2A_3),
$$

$$
\left.h\left(\frac{\partial A}{\partial x}\right)\right._1 = \frac{1}{3!}(-2A_0-3A_1+6A_2-A_3),
$$

$$
\left.h\left(\frac{\partial A}{\partial x}\right)\right._2 = \frac{1}{3!}(A_0-6A_1+3A_2+2A_3),
$$

and

$$
\left.h\left(\frac{\partial A}{\partial x}\right)\right._3 = \frac{1}{3!}(-2A_0+9A_1-18A_2+11A_3),
$$

(11.62)

where $(\partial A/\partial x)_n$ is the gradient in the x-direction at the point n. These equations, involving values of potential at four points, yield values of gradient which are generally of sufficient accuracy. However, in reference 31 Bickley presents in addition, similar, but more accurate equations for groups of 3, 5, 6, 7, 9, and 11 points, together with expressions for the error

[†] If A_1 to A_N are the values at a series of nodes, 1 to N, on a straight line, the quantities known as the differences are formed according to the table

A_1				
	$\delta_{1/2}^1$			
A_2		δ_2^2		
	$\delta_{3/2}^1$		$\delta_{3/2}^3$	
A_3		δ_3^2		δ_3^4
	$\delta_{5/2}^1$		$\delta_{5/2}^3$	$\delta_{5/2}^5$
A_4		δ_4^2		δ_4^4
	$\delta_{7/2}^1$		$\delta_{7/2}^3$	$\delta_{7/2}^5$
A_5		δ_5^2		δ_5^4
	$\delta_{9/2}^1$		$\delta_{9/2}^3$	$\delta_{9/2}^5$
A_6		δ_6^2		δ_6^4
	$\delta_{11/2}^1$		$\delta_{11/2}^3$	$\delta_{11/2}^5$

where each value of δ is formed as the difference between the two numbers to its immediate left, for example, $\delta_3^2 = \delta_{3/2}^1 - \delta_{5/2}^1$.

involved in each. Also, Shaw (p. 329) gives tables for the evaluation of gradients when the values of the function are known at any three *unequally* spaced points. These are needed mainly in the determination of gradients at or near curved boundaries.

11.6.3. *Computational errors*

There is no method by which the error present at any stage in the solution of the difference equations can be precisely determined, though it is possible to obtain an expression for the upper bound for them in terms of the maximum residual existing at any stage. For a maximum value, m, of the residuals, and for a radius, ϱ, of the circle which just contains the field region, it can be shown (see, for example, Milne, p. 217) that the error for the five-node difference equation, (11.8), does not exceed

$$\frac{m\varrho^2}{4h^2},$$ (11.63)

and for the nine-node difference equation, (11.58), it does not exceed

$$\frac{m\varrho^2}{24h^2}.$$ (11.64)

Hence, the ultimate size of individual residuals (and their sum) to be aimed at, can be estimated. (As already stated, an even distribution of residuals should also be aimed at in all cases.)

Other and very useful indications of the presence of computational (and in some cases mesh) error can be revealed by comparing the characteristics of a solution at any stage with those which the exact solution must possess. In an accurate solution the following conditions should be fulfilled, and may usefully be examined.

(a) Flux and equipotential lines must satisfy the field conditions at boundaries and lines of symmetry; for example, flux lines must be normal to boundaries of infinite permeability or conductivity, and to lines of symmetry; and at interfaces between media of different permeabilities or permittivities, flux lines should be "refracted" according to eqn. (2.68).

(b) Any symmetry in the field must be apparent.

(c) The general variation of flux distribution must correspond with that estimated from the boundary dimensions and potentials or currents.

(d) Certain obvious features of the flux density distribution should be apparent; for example, values should be high near exterior corners and low near interior corners of permeable or conducting boundaries.

(e) The formation and location of the kernel (or kernels) in fields of distributed currents must be correct. For such fields this gives a particularly good indication of the presence of error (even when other conditions are fulfilled).

(f) The variation of the potential from node to node should be continuous and not irregular (implying the intersection of flux or equipotential lines). Such irregularities are found to occur, particularly, near to interior corners of boundaries.

In the earlier stages of computation, a solution is generally unsatisfactory with regard to all of the above features (which are appropriate to the problem) but, as the residuals are reduced and the accuracy improves, it assumes the correct form. (If it fails to do so, even though the residuals are negligible, then too large a mesh size is indicated.)

11.7. Conclusions

Finite-difference methods can be used to solve *any* steady-state, two-dimensional field problem (or, indeed, any three-dimensional or time-varying one). This should be evident from a consideration of the examples given, and it is to be noted that, though these examples are relatively simple, none could be handled using analytical methods. Also, although consideration has been restricted to linear problems, the methods of solution can be extended to deal with *non-linear* problems, involving saturated iron, for example. This application has recently been studied particularly by Erdélyi and his co-workers and references 32–36 are representative of some of their work. It involves what is essentially a double iterative procedure in which the permeabilities (or reluctivities) as well as the potential (field) values are relaxed. Generally, whilst a small over-relaxation factor is found appropriate for the potentials, an under-relaxation factor is best for the permeabilities. Not surprisingly, since the problem is non-linear, no general mathematical theory has been established, and computing times are naturally longer and convergence is sometimes more troublesome than in the corresponding linear case. Nevertheless, a number of valuable results have been developed.

The solutions obtained by finite-difference methods are approximate, though any desired degree of accuracy can be achieved provided that the computation is continued for a sufficient length of time. In many practical problems, however, the desired accuracy can be achieved very quickly. It is to be remembered that, as with all numerical methods, a separate solution is required for each set of parameters of a problem (though it is sometimes possible to effect a considerable saving in time by making the final solution of one problem the starting point of the computation for a similar one). When a problem can be solved either by analytical or numerical means, it is the length of computation that frequently determines which should be selected. Of course, analytical methods are usually preferable but, for some applications, they may involve such lengthy manipulation and, in some cases, computation, that a numerical method is more economical. This is especially likely when a general computer program—requiring only data defining the boundary shape and conditions—is available for the application of an iterative method.

The choice between a hand and a machine method of solution is dependent upon several factors including the required accuracy of solution, the number of particular cases to be solved, the boundary conditions and degree of complexity of the problem, and upon whether a suitable computer program is already available. Most commonly now, a machine method would be preferred but, for preliminary work, for small-scale investigations where program preparation time would be excessive, or perhaps for particularly complex problems, hand methods can still be valuable.

As an example of the saving in time and effort which may be achieved using machine methods, consider the following data which relate to the solution of the field due to the current in the I-section conductor, Fig. 11.13. Using about 400 nodes and taking the initial values of potential to be zero, reduction of the maximum residual to 0·001 by hand would take perhaps 25–50 hours, whereas (on a fast modern machine) it would take about 10 seconds computing time plus 3–4 hours for the writing and preparation of the program assuming this had to be done from scratch. If a general program were available, however, the total time would be reduced merely to that for the preparation of the machine input data, and the overall time required would be reduced to about 1 hour.

Finally, brief reference is made to two potentially important lines of development which the interested reader might wish to pursue in more depth. The first of these concerns the method used to develop the difference equations. In this chapter, in line with common practice, these equations have been developed using Taylor's series. Other approaches are possible and mention is made particularly of the *integration* method discussed by Varga. In simple situations this leads to the same equations as the Taylor-series approach but, in others, it gives a form of difference equation which permits symmetry to be retained in the **M** matrix [eqn. (11.11)]. Up to the present time, this does not seem to have been widely recognized (at least, outside purely mathematical circles), but it is of the greatest importance. It ensures that, regardless of boundary shape or condition, the convergence is always well behaved and that the optimum convergence factor can be reliably calculated.

The second line of development is that associated with the *finite-element* technique.[37-40] This is claimed to have advantages over the finite-difference method particularly in the ease with which boundary conditions can be handled. In many applications it also appears to be possible to achieve a given accuracy of solution with a relatively very small number of elements and, consequently, computing times. Examples of this are given by Silvester *et al.*, whose work also incorporates the treatment of saturating iron. Somewhat surprisingly, the precise relationship between the element and difference methods has been, and remains, unclear, but no doubt in the near future their relative merits will be established.

References

1. C. F. GAUSS, Brief an Gerling, *Werke*, **9**, 278–81 (Dec. 26, 1823). (Translated by G. E. Forsythe, *Math. Tab.*, *Wash.* **5**, 255 (1951).)
2. V. A. GOVORKOV, Calculation of electric and magnetic fields in polar co-ordinates by the method of potential mesh, *Elektrichestvo* **7**, 51 (1951).
3. G. TEMPLE, Generalised theory of relaxation methods applied to linear systems, *Proc. Roy. Soc.* **169** A, 476 (1939).
4. L. F. RICHARDSON, The approximate arithmetical solution by finite differences of physical problems involving differential equations, with an application to the stresses in a masonry dam, *Phil. Trans.* **210** A, 307 (1911).
   How to solve differential equations approximately by arithmetic, *Math. Gaz.* **12**, 415 (1925).
5. H. LIEBMANN, Die angenäherte Ermittelung harmonischer Funktionen und konformer Abbildungen, *S. B. Bayer. Akad. Wiss., Math. Phys. Klasse*, 385–416 (1918).
6. S. P. FRANKEL, Convergence rates of iterative treatments of partial differential equations, *Math. Tab., Wash.* IV, 30, 65 (1950).
7. D. M. YOUNG, Iterative methods for solving partial differential equations of elliptic type, *Trans. Am. Math. Soc.* **76**, 92 (1954).
8. P. R. GARABEDIAN, Estimation of the relaxation factor for small mesh size, *Math. Tables Aids Comput.* **10**, 183 (1956).
9. H. E. KULSRUD, A practical technique for the determination of the optimum convergence factor of of the successive over-relaxation method, *Comm. Assoc. Comp. Mach.* **4**, 184 (1961).
10. B. A. CARRÉ, The determination of the optimum accelerating factor for successive over-relaxation, *Comp. J.* **4**, 73 (1961).
11. D. M. YOUNG, Ordvac solutions of the Dirichlet problem, *J. Assoc. Comp. Mach.*, July 1965.
12. R. J. ARMS, L. D. GATES, and B. ZONDEK, A method of block iteration, *J. Soc. Industr. Appl. Math.* **4**, 220 (1956).
13. D. W. PEACEMAN and H. H. RACHFORD, The numerical solution of parabolic and elliptic differential equations, *J. Soc. Industr. Appl. Math.* **3**, 28 (1955).
14. D. M. YOUNG and L. EHRLICH, Some numerical studies of iterative methods for solving elliptic difference equations, *Boundary Problems in Differential Equations*, Wisconsin, Madison, 1960.
15. H. B. KELLER, On some iterative methods for solving elliptic differential equations, *Qt. Appl. Math.* **16**, 3(1958).
16. J. HELLER, Simultaneous, successive and alternating direction iteration schemes, NY O-8675, AEC Computing and Applied Maths. Center, Institute of Math. Science, New York University (1958).

17. G. Shortley, Use of Chebyscheff-polynomial operators in the numerical solution of boundary-value problems, *J. Appl. Phys.* **24** (4), 392 (1953).
18. L. A. Hageman, Block iterative methods for two cyclic matrix equations, WAPD-TM-327, Westinghouse Bottis Atomic Power Labs., Pittsburgh, Pa., 1962.
19. H. S. Price and R. S. Varga, Recent numerical experiments comparing SOR and ADI methods, Rep. 91, Gulf Research and Development Co., 1962.
20. R. W. Hockney, A fast direct solution of Poisson's equation using Fourier analysis, *J. Assoc. Comp. Mach.* **12**, 95 (1965).
21. H. W. Milnes and R. B. Potts, Boundary contraction solution of Laplace's differential equation, *J. Assoc. Comp. Mach.* **6** (2), 226 (1959).
22. H. W. Milnes and R. B. Potts, Numerical solution of partial differential equations by boundary contraction, *Quart. Appl. Math.* **18** (1), 1 (1960).
23. S. V. Ahamed, Accelerated convergence of numerical solution of linear and non-linear vector field problems, *Comp. J.* **8**, 73 (1965).
24. P. Reece, Some experience with the application of block relaxation to the automatic iteration of finite difference equations, *Instn. Elect. Engrs. Colloquium: Applications of Computers to Field Analysis* **4** (1967).
25. R. L. Stoll, Numerical method of calculating eddy-currents in non-magnetic conductors, *Proc. Instn. Elect. Engrs.* **144**, 775 (1967).
26. L. Fox, Potential problems with mixed boundary conditions, *Q. Appl. Math.* **2**, 251 (1944).
27. H. McKibbin, Plotting magnetic fields produced by conductors housed in slots in iron, Ph.D. thesis, Queen's University, Belfast (1955).
28. L. F. Richardson and J. A. Gaunt, The deferred approach to the limit, *Phil. Trans.* **226** A (1926–7.)
29. L. Fox, Some improvements in the use of relaxation methods for the solution of ordinary and partial differential equations, *Proc. Roy. Soc.* **190** A, 31 (1947).
30. L. Fox, The numerical solution of elliptic differential equations when the boundary conditions involve a derivative, *Phil. Trans.* **242** A, 345 (1950).
31. W. G. Bickley, Formulae for numerical differentiation, *Math. Gaz.* **25**, 19 (1941).
32. F. C. Trutt, E. A. Erdélyi, and R. F. Jackson, The non-linear potential equation and its numerical solution for highly saturated electrical machines, *Trans. Inst. Elec. Electr. Engrs.* AS-1, 1 (1963).
33. F. C. Trutt and E. A. Erdélyi, No-load flux distribution in saturated homopolar generators, *Trans. Inst. Elec. Electr. Engrs.*, AS-1 (August, 1963).
34. E. A. Erdélyi, S. V. Ahamed and R. D. Burtness, Flux distribution in saturated DC machines at no-load, *Trans. Inst. Elec. Electr. Engrs.* PAS-**84**, 375 (1965).
35. S. V. Ahamed and E. A. Erdélyi, Nonlinear theory of salient-pole machines, *Trans. Inst. Elec. Electr. Engrs.* PAS-**85**, 61 (1966).
36. E. A. Erdélyi and E. F. Fuchs, Nonlinear magnetic field analysis of DC machines, *Trans. Inst. Elec. Electr. Engrs.* PAS-**89**, 1546 (1970).
37. P. Silvester and M. V. K. Chari, Finite element solution of saturable magnetic field problems, *Trans. Inst. Elec. Electr. Engrs.* PAS-**89**, 1642 (1970).
38. M. V. K. Chari and P. Silvester, Analysis of turbo alternator magnetic fields by finite elements, *Trans. Inst. Elec. Electr. Engrs.* PAS-**90**, 454 (1971).
39. P. Silvester and M. S. Hseih, Finite element solution of 2-dimensional exterior-field problems, *Proc. Instn. Elect. Engrs.* **118**, 1743 (1971).
40. M. J. L. Hussey, R. W. Thatcher, and M. J. M. Bernal, On the construction and use of finite elements, *J. Inst. Maths. Applics.* 263–82 (1970).

Additional References

Bellar, F. J., An iterative solution of large scale systems of simultaneous linear equations, *J. Soc. Indust. Appl. Math.* **9**, 189 (1961).
Brechna, H., and Gordon, H. S. (editors), *Proceedings of the International Symposium on Magnet Technology, Stanford Linear Accelerator Centre, Stanford University, California, September, 1965.*
Browne, B. T., Numerical solution of eddy-current fields, Ph.D. thesis, University of Leeds, 1971.
Davies, J. B., Review of methods for numerical solutions of the hollow waveguide problem, *Proc. Instn. Elect. Engrs.*, **119**, 35 (1972).

DAVIES, J. B. and MUILWYK, C. A., Numerical solution of uniform hollow waveguides with boundaries of arbitrary shape, *Proc. Instn. Elect. Engrs.* **113,** 277 (1966).

GOLUB, G. H., and VARGA, R. S., Chebyshev semi-iterative methods, successive over-relaxation iterative methods..., *Numerische Math.* **3,** Parts I and II, 1947 (1961).

HANNAFORD, C. D. The finite difference/variational method for inhomogeneous isotropic and anisotropic waveguides, Ph.D. thesis, University of Leeds, 1967.

KAHAN, W., The rate of convergence of the extrapolated Gauss–Seidel iteration, *Conference on Matrix Computations, Wayne State University,* 4, September, 1957.

MAMAK, R. S., and LAITHWAITE, E. R., Numerical evaluation of inductance and A.C. resistance, *Proc. Instn. Elect. Engrs.* **108** C, 252 (1961).

METCALF, W. S., Characteristic impedance of rectangular transmission lines, *Proc. Instn. Elect. Engrs.* **112,** 2033 (1965).

PARTER, S. V., A two-line iterative method for the Laplace and biharmonic difference equations, *Numerische Math.* **1,** 208 (1959).

PARTER, S. V., Multi-line iterative methods for elliptic difference equations and fundamental frequencies, *Numerische Math.* **3,** 309 (1961).

PARTER, S. V., On estimating the rates of convergence of iterative methods for elliptic difference equations, Tech. Rep. 28, Appl. Math. and Stat. Labs., Stanford University, Palo Alto, California (1963).

PARTER, S. V., Mildly non-linear elliptic partial differential equations and their numerical solutions, I. Tech. Rept. 470, Math. Res. Ctr., Madison, Wisconsin (1964).

STOLL, R. L., Solution of linear steady-state eddy-current problems by complex successive over-relaxation, *Proc. Instn. Elect. Engrs.* **117,** 1317 (1970).

WOODGER, M., and ALWAY, G. G., Solution of a two-dimensional heat conduction problem with mixed boundary conditions, *Deuce News,* No. 45, sheet No. 23 (1959).

CHAPTER 12

THE MONTE CARLO METHOD

12.1. Introduction

In the previous chapter it is shown how Poisson's equation can be replaced by a set of simultaneous finite-difference equations, and their solution by relaxation and iteration is described. In this chapter a statistical method of approximating to the solution of these equations is discussed. The method is extremely simple to apply (for both Laplacian and Poissonian fields) but, since the calculation is very long, its use requires the aid of a digital computer. The form of solution differs from that of an iteration solution in that for each application of the method the potential at one point only is obtained.

In general the Monte Carlo method is inferior to relaxation for the solution of problems lying within the scope of this book. It is laborious and relatively inefficient, and it should be realized that the main importance of the Monte Carlo method lies in the solution of other problems occurring, e.g. in operational research and nuclear physics. However, a very brief discussion of the application of this method of the solution of Laplacian and Poissonian fields is included and it was to the solution of these problems that much of the earliest work on the Monte Carlo method was applied.[1]

For general reading on the Monte Carlo method the reader should consult the recent book by Hammersley and Handscomb.

12.2. The method

The calculation. As with the methods of the previous chapter, the field region enclosed by the boundaries is replaced by a mesh, normally square, and Poisson's equation is replaced by the difference equations relating values of potential at adjacent points on the mesh. Then the coefficients in the difference eqn. (11.8), which can be written

$$A_0 = \tfrac{1}{4} \sum_{n=1}^{4} A_n + \tfrac{1}{4} \mu \mu_0 J h^2, \tag{12.1}$$

are interpreted as the probabilities of transition of a particle from one point to a neighbouring point. In eqn. (12.1) the potential at node 0 is dependent equally upon the potentials at the adjacent nodes 1, 2, 3, and 4, and this is interpreted by making equal the probabilities of transition from node 0 to each of the other nodes. This means that the probability of transition to any of the neighbouring points is made $\tfrac{1}{4}$, which corresponds with random movement from node 0.

Consider now a particle starting at any node on the mesh and performing a series of random movements, or a "walk", from node to node. Further, let the walk be terminated when the particle reaches a node m on the boundary, after making S steps. Then it can be shown that the most likely value of the potential T at the node at which the walk began is

$$T = A_m - \sum \tfrac{1}{4} h^2 \mu \mu_0 J, \tag{12.2}$$

where A_m is the value of potential at the point m, and $\mu\mu_0$ and J are the values of permeability and current density at each node passed through, including the starting point excluding the boundary point. Thus the potential at a point is given by the mean value of T for many walks starting at the point, that is

$$A = \sum_{n=1}^{N} \frac{T_n}{N}, \tag{12.3}$$

where N is the total number of walks. The proof of the above is to be found, for example, in the book edited by Beckenbach.

Boundary conditions. Problems having equipotential and flux line boundaries are simply treated. When an equipotential boundary is reached the walk is terminated. When a flux line boundary is reached, the walk is continued in the usual way until an instruction to cross the boundary occurs, and then the particle is moved in the opposite direction to that instructed. For example, if a particle at node b on a flux line boundary (Fig. 12.1), were instructed to move to node a, it would actually be moved to node c. Thereafter the walk proceeds in the usual manner.

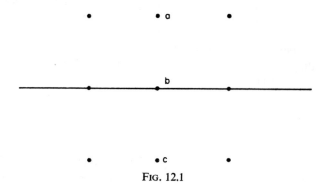

FIG. 12.1

Use of the Monte Carlo method for boundaries with other gradient conditions has been briefly examined by Kac.[2]

Use of random numbers. The sequence of instructions which is used to order the motion of the fictitious particle is derived from a sequence of random numbers. Methods of generation of these numbers are not described here, but it should be noted that their rapid generation by a digital computer is a routine matter; for a discussion of available methods the reader should consult Hammersley and Handscomb or the book edited by Meyer. The numbers can be used in a variety of ways to order a random "walk" generated into four equal parts and selecting the direction of motion according to the part in which the number lies.

12.3. Example

Consider the determination of the potential at the point P in the Laplacian field with the boundaries, a plane one opposite to a slotted one, as shown in Fig. 12.2. Let the plane boundary lie at a potential of unity and the slotted one at zero potential. Also, let the lines aa' and bb' be flux lines. The computation is carried out as described above and it is merely

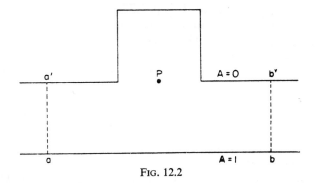

FIG. 12.2

necessary to note that since $J = 0$, eqns. (12.1) and (12.3) reduce to

$$A_p = \sum_1^N \frac{A_m}{N}. \tag{12.4}$$

Figure 12.3 shows the manner of convergence to the solution for a particular sequence of random numbers. The estimate of the solution taken after 2000 walks differs from the exact solution of the difference equation by 0·01. The total time taken for the computation using a fast computer is 11 minutes. This compares with a time of 3 minutes for the solution by iteration for all nodes.

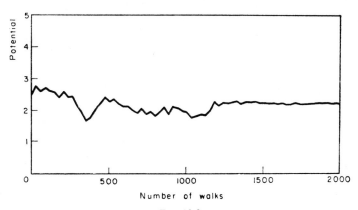

FIG. 12.3

12.4. Some general points

In this chapter is given a very brief outline of the use of these methods; a more detailed discussion is given by Curtiss.[3] Muller[4] has proposed a method having considerable advantage in that it does not resort to the difference approximation.

It is obviously useful to be able to estimate the length of time to achieve a given accuracy of solution, and this clearly involves both the probable duration of a random walk and the number of random walks required. It is possible to give estimates of these quantities together with an indication of their probable accuracies. The expected duration of a random walk is discussed by Ehrlich,[5] who gives a table of expected durations for various numbers of nodes and shapes of boundary. All walks are known to be of finite duration and a method of assessing the accuracy of solution is given by Kahn and Marshall.[6]

References

1. R. Courant, K. Friedrichs, and H. Lewy, Über die partiellen Differenzengleichungen der mathematischen Physik. *Math. Annalen.* **100,** 32 (1928).
2. M. Kac, Application of statistical methods to differential and integral equations, Lecture Notes Mass. Inst. Tech., 1949.
3. J. H. Curtiss, Sampling methods applied to differential and difference equations, *Proc. Sem. Sci. Comp.,* IBM Corporation, 1949.
4. M. E. Muller, Some continuous Monte Carlo methods for the Dirichlet problem, *Ann. Math. Stats.* **27,** 569 (1956).
5. L. W. Ehrlich, Monte Carlo solutions of boundary value problems, *J. Ass. Comp. Mach.* **6** (2), 204 (1959).
6. H. Kahn and A. W. Marshall, Methods of reducing sample size in Monte Carlo computations, *Res. Soc. Am.* **1,** 263 (1953).

Additional References

McCracken, D. D., The Monte Carlo method, *Scient. Am.* **5,** 90 (1955).
Wasow, W., Random walks and the eigenvalues of elliptic difference equations, *J. Res. Nat. Bur. Stds.* **46,** 65 (1951).
Yowell, E. C., A Monte Carlo method of solving Laplace's equation, *Proc. Sem. Sci. Comp.,* December 1949.

THE SUMS OF CERTAIN FOURIER SERIES

It is shown by Roth and Kouskoff, in reference 20 of Chapter 5, that certain of the Fourier series arising in Roth's method of solution for fields of distributed currents, section 5.6, can be expressed as finite functions. Use of these functions can greatly reduce the amount of computation and the more important of them are given here. Each falls into one of two groups, depending upon whether the angle, $m_h a$ is an even or an odd multiple of $\pi/2$. The relationships are given in terms of m_h, a, a_j and x, but it should be noted that there are also equivalent relationships in terms of n_k, b, b_j and y.

$m_h a$ is an odd multiple of $\pi/2$

$$\sum_{h=1}^{\infty} \frac{\sin m_h a_j \cdot \cos m_h x}{m_h(m_h^2 + n_k^2)} = \frac{a}{2}\left[\frac{\cosh n_k a - \cosh n_k(a - a_j) \cdot \cosh n_k x}{n_k^2 \cosh n_k a}\right], \quad x < a_j,$$

$$= \frac{a}{2}\left[\frac{\sinh n_k a_j \cdot \sinh n_k(a - x)}{n_k^2 \cosh n_k a}\right], \quad x > a_j;$$

$$\sum_{h=1}^{\infty} \frac{\sin m_h a_j \cdot \cos m_h x}{m_h^3} = \frac{a}{2}\left[aa_j - \frac{a_j^2 + x^2}{2}\right], \quad x < a_j,$$

$$= \frac{aa_j}{2}(a - x), \quad x > a_j.$$

$m_h a$ is an even multiple of $\pi/2$

$$\sum_{h=2}^{\infty} \frac{\sin m_h a_j \cdot \cos m_h x}{m_h(m_h^2 + n_k^2)} = \frac{a}{2n_k^2}\left[\frac{a - a_j}{a} - \frac{\cosh n_k x \cdot \sinh n_k(a - a_j)}{\sinh n_k a}\right], \quad x < a_j,$$

$$= \frac{a}{2n_k^2}\left[\frac{\sinh n_k a_j \cdot \cosh n_k(a - x)}{\sinh n_k a} - \frac{a_j}{a}\right], \quad x > a_j;$$

$$\sum_{h=2}^{\infty} \frac{\sin m_h a_j \cdot \cos m_h x}{m_h^3} = \frac{a - a_j}{12}[2aa_j - a_j^2 - 3x^2], \quad x < a_j,$$

$$= \frac{a_j}{12}[2a^2 - 6ax + 3x^2 + a_j^2], \quad x > a_j.$$

APPENDIX II

SERIES EXPANSIONS OF ELLIPTIC FUNCTIONS

THIS is a list of series which can be used to evaluate elliptic integrals and functions. The series proposed by King follow the other better-known series. However, it is to be noted that they converge more rapidly and are well suited to evaluation by digital computer.

Complete elliptic integral of the first kind

$$K(k) = \frac{\pi}{2}\left(1+\frac{1}{4}k^2+\frac{9}{64}k^4+\frac{25}{256}k^6+\frac{1225}{16\,384}k^8+\ \cdots\right)$$

$$= \frac{\pi}{2}\sum_{m=0}^{\infty}\left[\frac{(\frac{1}{2})_m k^m}{m!}\right]^2,$$

where $(\frac{1}{2})_m$ denotes m products of the form $(\frac{1}{2})(\frac{1}{2}+1)(\frac{1}{2}+2)\ldots$

Complete elliptic integral of the second kind

$$E(k) = \frac{\pi}{2}\left(1-\frac{1}{4}k^2-\frac{3}{64}k^4-\frac{5}{256}k^6-\frac{175}{16\,384}k^8-\ \cdots\right)$$

$$= \frac{\pi}{2}\sum_{m=0}^{\infty}\frac{1}{1-2m}\binom{-\frac{1}{2}}{m}^2 k^{2m},$$

where $\binom{-\frac{1}{2}}{m}$ denotes $\dfrac{(-\frac{1}{2})(-\frac{1}{2}-1)\ldots(-\frac{1}{2}-m+1)}{m!}$.

Jacobi's nome

$$q = \frac{k^2}{16}\left[1+2\left(\frac{k}{4}\right)^2+15\left(\frac{k}{4}\right)^4+150\left(\frac{k}{4}\right)^6+1707\left(\frac{k}{4}\right)^8+\ \cdots\right]^4 \qquad (k^2 < 1).$$

Incomplete elliptic integral of the first kind

$$F(\varphi, k) = \sum_{m=0}^{\infty}\binom{-\frac{1}{2}}{m}(-k^2)^m t_{2m}(\varphi) \qquad \left(0 < \varphi < \frac{\pi}{2},\ k^2 < 1\right),$$

where
$$t_0(\varphi) = \varphi,$$
$$t_2(\varphi) = \tfrac{1}{2}(\varphi-\sin\varphi\cos\varphi),$$
$$t_4(\varphi) = \tfrac{1}{8}[3\varphi-\sin\varphi\cos\varphi(3+2\sin^2\varphi)],$$
$$t_{2m}(\varphi) = \frac{2m-1}{2m}t_{2(m-1)}(\varphi)-\frac{1}{2m}\sin^{2m-1}\varphi\cos\varphi.$$

Incomplete elliptic integral of the second kind

$$E(\varphi, k) = \sum_{m=0}^{\infty} \binom{\frac{1}{2}}{m} (-k^2)^m t_{2m}(\varphi) \qquad \left(0 < \varphi < \frac{\pi}{2}, \; k^2 < 1\right),$$

where $t_{2m}(\varphi)$ is as given above.

Jacobian zeta function

$$Z(u, k) = \left(1 - \frac{E}{K}\right) u - 2k^2 \frac{u^3}{3!} + 8k^2(k^2+1) \frac{u^5}{5!}$$

$$- 16k^2(2k^4+13k^2+2) \frac{u^7}{7!} + 128k^2(k^6+30k^4+30k^2+1) \frac{u^9}{9!} - \dots \quad (|u| < K').$$

Jacobian elliptic functions

$$\text{sn}\, u = u - (1+k^2) \frac{u^3}{3!} + (1+14k^2+k^4) \frac{u^5}{5!} - (1+135k^2+135k^4+k^6) \frac{u^7}{7!} + \dots$$

$$\text{sn}\, u = \frac{2\pi}{kK} \sum_{m=0}^{\infty} \frac{q^{m+\frac{1}{2}}}{1-q^{2m+1}} \sin\left[(2m+1)\frac{\pi u}{2K}\right], \quad [|\,\text{Im}\,(u/K)| < \text{Im}\,(jK'/K)].$$

$$\text{cn}\, u = 1 - \frac{u^2}{2!} + (1+4k^2) \frac{u^4}{4!} - (1+44k^2+16k^4) \frac{u^6}{6!}$$

$$+ (1+408k^2+912k^4+64k^6) \frac{u^8}{8!} - \dots \quad (|u| < K').$$

$$\text{cn}\, u = \frac{2\pi}{kK} \sum_{m=0}^{\infty} \frac{q^{m+\frac{1}{2}}}{1+q^{2m+1}} \cos\left[(2m+1)\frac{\pi u}{2K}\right], \quad [|\,\text{Im}\,(u/K)| < \text{Im}\,(jK'/K)].$$

$$\text{dn}\, u = 1 - k^2 \frac{u^2}{2!} + (4+k^2)k^2 \frac{u^4}{4!} - (16+44k^2+k^4)k^2 \frac{u^6}{6!}$$

$$+ (64+912k^2+408k^4+k^6)k^2 \frac{u^8}{8!} - \dots \quad (|u| < K').$$

$$\text{dn}\, u = \frac{\pi}{2K} + \frac{2\pi}{K} \sum_{m=0}^{\infty} \frac{q^{m+1}}{1+q^{2(m+1)}} \cos\left[(m+1)\frac{\pi u}{K}\right], \quad [|\,\text{Im}\,(u/K)| < \text{Im}\,(jK'A/K)].$$

Evaluation of elliptic functions and integrals by the method of L. V. King

An arithmetico-geometric scale is formed as follows starting with $a_0 = 1$, $b_0 = k$, and $c_0 = k$.

$a_0 = 1$	$b_0 = k$	$c_0 = k$
$a_1 = \frac{1}{2}(a_0+b_0)$	$b_1 = \sqrt{a_0 b_0}$	$c_1 = \frac{1}{2}(a_0-b_0)$
$a_2 = \frac{1}{2}(a_1+b_1)$	$b_2 = \sqrt{a_1 b_1}$	$c_2 = \frac{1}{2}(a_1-b_1)$
$a_n = \frac{1}{2}(a_{n-1}+b_{n-1})$	$b_n = \sqrt{a_{n-1} b_{n-1}}$	$c_n = \frac{1}{2}(a_{n-1}-b_{n-1})$

The table of values is continued until c_n is negligibly small (say less than 10^{-8}).

For the calculation of the elliptic integrals, a series $\varphi_0, \varphi_1, \ldots, \varphi_n$ is formed, making $\varphi_0 = \varphi$ and using the recurrence relation

$$\varphi_{n+1} = \varphi_n + \tan^{-1}\left[\left(\frac{b_n}{a_n}\right)\tan\varphi_n\right].$$

Complete elliptic integral of the first kind

$$K(k) = \pi/2a_n.$$

Complete elliptic integral of the second kind

$$E(k) = \frac{\pi}{4a_n}(2 - c_0^2 - 2c_1^2 - \ldots - 2^n c_n^2).$$

Incomplete elliptic integral of the first kind

$$F(\varphi, k) = \frac{\varphi_n}{2^n a_n}.$$

When $k = 1$, use

$$F(\varphi, 1) = \frac{1}{2}\log_e\left(\frac{1+\sin\varphi}{1-\sin\varphi}\right).$$

Incomplete elliptic integral of the second kind

$$E(\varphi, k) = \frac{\varphi_n}{2^n a_n} + (c_1\sin\varphi_1 + \ldots + c_n\sin\varphi_n)$$

$$- \frac{\varphi_n}{2^{n+1}a_n}(c_0^2 + 2c_1^2 + 4c_2^2 + \ldots + 2^n c_n^2).$$

When $k = 1$, use $E(\varphi, 1) = \sin\varphi$.

For the evaluation of the elliptic functions, φ_n is first found from

$$\varphi_n = 2^n a_n u.$$

Then the recurrence formula

$$\sin(\varphi_{n-1} - \varphi_n) = \frac{c_n}{a_n}\sin\varphi_n$$

is used to give φ_{n-1} to φ_0.

The elliptic functions are then found as follows:

$$s_n u = \sin\varphi_0,$$

$$c_n u = \cos\varphi_0,$$

$$d_n u = \frac{\cos\varphi_0}{\cos(\varphi - \varphi_0)},$$

$$Z(u, k) = c\sin\varphi_1 + c\sin\varphi_2 + \ldots + c_n\sin\varphi_n.$$

For guidance on the use of these series the reader should consult the book by King or reference 8, Chapter 9.

APPENDIX III

TABLE OF TRANSFORMATIONS

This is a list of transformation equations and corresponding boundary shapes, which are of particular interest in the solution of electric and magnetic field problems. The transformations relate the planes of z and t, where

$$z = x+jy$$

and

$$t = u+jv.$$

In the case of transformations to the upper half-plane (which throughout is taken to be that of t) only the z-plane is shown, but the corresponding points of the t-plane are marked *inside* the mapped region.

Transformations to the unit circle can be obtained by combining those to the infinite straight line with the bilinear transformation.

1. Transformations to the upper half-plane

1.1. *Circles*

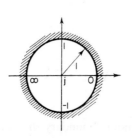

$$z = \frac{j-t}{j+t}.$$

(Exterior of circle on lower half-plane)

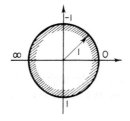

$$z = \frac{j+t}{j-t}.$$

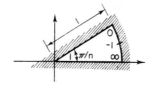

$$z = \frac{\sqrt{t}-1}{\sqrt{t}+1}.$$

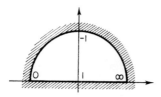

$$z = \left(\frac{\sqrt{t}-1}{\sqrt{t}+1}\right)^{1/n}$$

1.2. *Straight-line segments*

1.2.1. *One defining vertex*

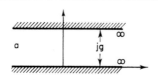

$$z = \frac{g}{\pi} \log(t-a).$$

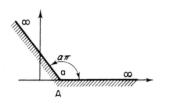

$$z + A = (t - a)^{\alpha}, \quad \alpha \neq 0,$$

1.2.2. *Two defining vertices*

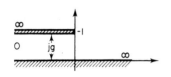

$$z = \frac{g}{\pi}(1 + t + \log t).$$

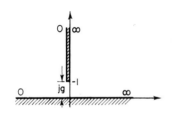

$$z = \frac{g}{2}(t^{1/2} - t^{-1/2}).$$

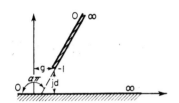

$$z = S\left(\frac{t^{1-\alpha}}{1-\alpha} - \frac{t^{-\alpha}}{\alpha}\right), \quad \begin{array}{c} \alpha \neq \pi, \\ \neq 0. \end{array}$$

S and position of origin evaluated from equivalence of $z = g + jd$ and $t = -1$.

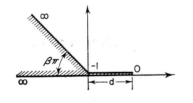

$$z = d[2 - \beta)(t + 1)^{1-\beta}$$
$$- (1 - \beta)(t + 1)^{2-\beta}$$

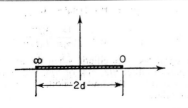

$$z = d\frac{1 - t^2}{1 + t^2}.$$

($t = j$ corresponds with $z = \infty$).

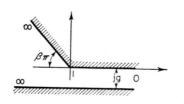

$$z = -\frac{g}{\pi} \int \frac{(1-t)^\beta}{t}\, dt.$$

Representable in elementary functions if $\beta = p/q$, where $0 < p < q$, and p and q are integers.

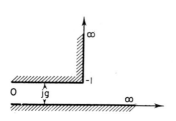

$$z = \frac{g}{\pi} \{2(t+1)^{1/2} - 2\log [(t+1)^{1/2} + 1] + \log t\}.$$

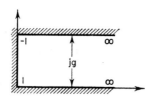

$$z = \frac{g}{\pi} \cosh^{-1} t.$$

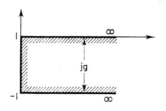

$$z = \frac{g}{\pi} \left[t\sqrt{t^2-1} - \log\left(t + \sqrt{t^2-1}\right)\right].$$

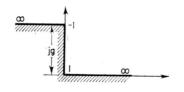

$$z = \frac{g}{\pi} \left(\cosh^{-1} t - \sqrt{t^2-1}\right).$$

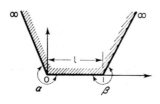

$$z = S \int t^{[(\alpha/\pi)-1]}(t-1)^{[(\beta/\pi)-1]}\, dt;$$

$$S = \frac{l}{(-1)^{[(\beta/\pi)-1]}} \cdot \Gamma \frac{(\alpha+\beta)/\pi}{\Gamma(\alpha/\pi)\,\Gamma(\beta/\pi)}.$$

The integral becomes elliptic when the angles are multiples of either $\pi/3$, $\pi/4$, or $\pi/6$. When all the vertices occur in the finite region, the boundary is triangular.

1.2.3. *Three defining vertices*

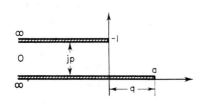

$$z = S\left[\tfrac{1}{2}t^2+(1-a)t-a\log t\right]+k.$$
S and k are determined graphically.

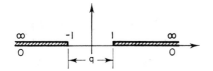

$$z = \tfrac{1}{4}q(t+t^{-1}).$$

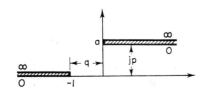

$$z = S\left[t+(1-a)\log t+\frac{a}{t}\right]+k.$$
S and k are determined graphically.
See p. 171.

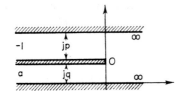

$$z = \frac{t^{2-\beta}}{2-\beta}+\frac{(a-1)}{(1-\beta)}\,t^{1-\beta}+\frac{a}{\beta}\,t^{-\beta}.$$

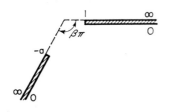

$$z = \frac{p}{\pi}\left[\log\,(t+1)+\frac{q}{p}\log\left(t-\frac{q}{p}\right)\right]-\frac{q}{\pi}\log\frac{q}{p}\,.$$

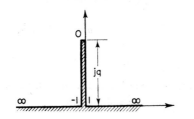

$$z = q\sqrt{t^2-1}.$$

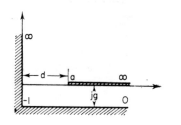

$$z = \frac{g}{\pi}\left[\frac{2}{a}\sqrt{t+1}+\log\frac{\sqrt{t+1}+1}{\sqrt{t+1}-1}\right];$$

a is determined graphically from

$$\frac{d}{g}\pi = \frac{2}{a}\sqrt{a+1}+\log\frac{\sqrt{a+1}+1}{\sqrt{a+1}-1}.$$

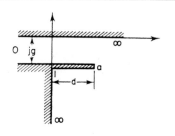

$$z = \frac{g}{\pi}\left[\frac{2j}{a}\sqrt{t-1}-\log\frac{\sqrt{t-1}-j}{\sqrt{t-1}+j}\right].$$

a is determined from the equivalence of
$t = a$ and $z = d-jg.$

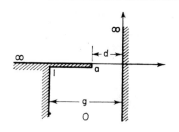

$$z = \frac{g}{\pi}\left[\frac{2}{a}\sqrt{t-1}+j\log\frac{\sqrt{t-1}-j}{\sqrt{t+1}+j}\right].$$

a is determined from the equivalence of
$t = a$ and $z = -d.$

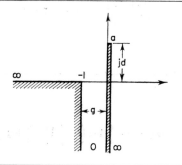

$$z = \frac{jg}{\pi}\left[\log\frac{\sqrt{t+1}-1}{\sqrt{t+1}+1}-\frac{2}{a}\sqrt{t+1}\right].$$

a is determined from the equivalence of
$t = a$ and $z = jd.$

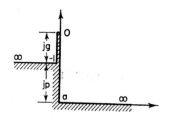

$$z = \frac{p}{\pi}\frac{2}{(a-1)}\left\{\sqrt{(t+1)\,(t-a)}+\frac{(a-1)}{2}\right.$$
$$\left.\times[\log(2\sqrt{(t+1)\,(t-a)}+2t+1-a)]\right\}$$
$$-\frac{p}{\pi}\log(1+a).$$

a is determined from the equivalence of
$t = 0$ and $z = j(p+g).$

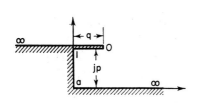

$$z = \frac{p}{\pi} \frac{2}{(a-1)} \left\{ \sqrt{(t-1)(t-a)} + \frac{(a-1)}{2} \right.$$
$$\left. \times [\log (2\sqrt{(t-1)(t-a)} + 2t - 1 - a))] \right\}$$
$$- \frac{p}{\pi} \log (a-1).$$

a is determined from the equivalence of
$$t = 0 \text{ and } z = q + jp.$$

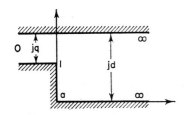

$$z = \frac{d}{\pi} \left[\cosh^{-1} \frac{2t - (a+1)}{a-1} - \cosh^{-1} \frac{(a+1)t - 2a}{(a-1)t} \right]$$
$$a = \left(\frac{d}{g} \right)^2.$$

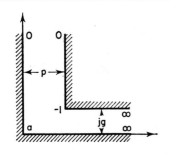

$$z = \frac{2g}{\pi} \left[\frac{p}{g} \tan^{-1} \frac{pu}{g} + \frac{1}{2} \log \left(\frac{1+u}{1-u} \right) \right];$$
$$u^2 = \frac{1 - (g/p)^2}{t+1}.$$

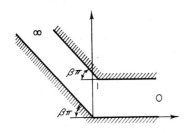

$$z = S \int \frac{1}{t} \left(\frac{1-t}{a+t} \right)^\beta dt.$$

Representable in terms of elementary functions if $\beta = p/q$, where $0 < p < q$ and p and q are integers. See Köber, p. 156.

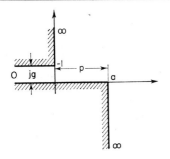

$$z = -\frac{jg}{\pi \sqrt{a}} \left[\frac{(a+1)R}{R^2 - 1} + (1-a) \tanh^{-1} R \right.$$
$$\left. + j\sqrt{a} \log \frac{(R\sqrt{a} - j)}{(R\sqrt{a} + j)} \right]; \quad R = \sqrt{\frac{t+1}{t-a}},$$
$$a = 1 + 2p/g \pm \sqrt{(1 + 2p/g)^2 - 1}.$$

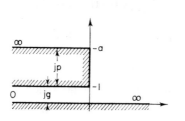

$$z = \frac{g}{\pi}\left[\frac{(a+1)}{\sqrt{a}}\tanh^{-1}R + \frac{(a-1)}{\sqrt{a}}\right.$$

$$\left. \times\frac{R}{(1-R^2)} + \log\frac{(R\sqrt{a}-1)}{(R\sqrt{a}+1)}\right];$$

$$R = \sqrt{\frac{t+1}{t+a}},$$

$$a = -1+2k^2 \pm 2k\sqrt{k^2-1}, \quad k = 1+p/g.$$

1.2.4. *Four defining vertices*

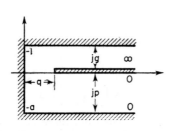

$$z = \frac{g}{\pi}\left[ab\log t + (1-ab)\log(t-1)\right.$$

$$\left. +(1+a+b+ab)\frac{1}{(t-1)}\right] + k;$$

$$ab = p/g,$$

If $p = g$, then $ab = 1$, $k = -jg$ and a
is determined graphically from

$$\log a + \frac{(a+1)^2}{a(a-1)} = -\frac{q}{g}\pi.$$

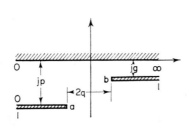

$$z = \frac{1}{\pi}\left\{g\log\left[2\sqrt{(t+a)(t+1)}+2t+a+1\right]\right.$$

$$\left. +p\log\left[\frac{2\sqrt{a}\sqrt{(t+a)(t+1)}}{t}+\frac{2a}{t}+a+1\right]\right\}$$

$$-\frac{1}{\pi}(p+g)\log(1-a);$$

$gb = \sqrt{a}\,p$, and a is determined from the
equivalence of $t = b$, $z = q$.

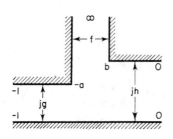

$$z = -\frac{jf}{\pi}\left\{\cosh^{-1}\left[\frac{2t+a-b}{a+b}\right]\right.$$

$$+\frac{h}{f}\cos^{-1}\left[\frac{(a-b)t-2ab}{(a+b)t}\right]$$

$$\left. +\frac{g}{f}\cos^{-1}\left[\frac{2(a-1)(b+1)+(b-a+2)(t+1)}{(a+b)(t+1)}\right]\right\};$$

$$\frac{h}{f} = \sqrt{ab}, \quad \text{and} \quad \frac{g}{f} = \sqrt{(b+1)(a-1)}.$$

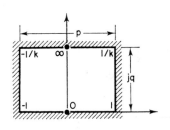

$$t = \text{sn}\left(\frac{2K}{p} z, k\right);$$

$$\frac{K'}{K} = \frac{2q}{p}.$$

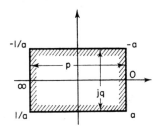

$$z = \frac{p}{E - k'^2 K}$$

$$\times \left\{ E' - k^2 K - j \left[Z(\beta) + \frac{\beta}{K} (E - k'^2 K) \right] \right\};$$

$$k \, \text{sn} \, \beta = -\sin \left[2 \tan^{-1} (-t) \right]$$

$$\frac{p}{g} = \frac{E - k'^2 K}{E' - k^2 K'}.$$

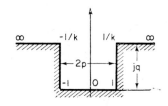

$$z = \frac{p}{E} E(t, k);$$

$$\frac{q}{p} = \frac{K' - E'}{E}.$$

1.2.5. Five defining vertices

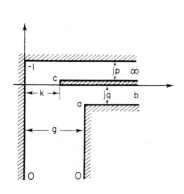

$$z = \frac{1}{\pi} \left\{ p \cosh^{-1} \left[\frac{2t + 1 - a}{1 + a} \right] \right.$$

$$+ q \cos^{-1} \left[\frac{(1 - a)t - 2a}{(1 + a)t} \right]$$

$$\left. + q \cosh^{-1} \left[\frac{2(1 + b)(b - a) + (1 + 2b - a)(t - b)}{(1 + a)(t - b)} \right] \right\};$$

$$\frac{c \sqrt{a}}{b} = \frac{g}{p},$$

$$\frac{c - b}{b} \sqrt{\frac{b - a}{1 + b}} = \frac{q}{p},$$

and the third relationship between the constants is determined by substituting for $t = c$, $z = k$.

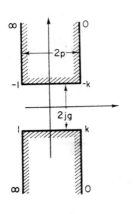

$$z = \frac{jp}{2E - k'^2 K}\left[U\left(k'^2 - \frac{2E}{K}\right) - 2Z(U)\right.$$
$$\left. - \frac{\operatorname{cn} U \operatorname{dn} U}{\operatorname{sn} U}\right] - g;$$

$$t = \frac{1}{\operatorname{sn} U},$$

$$\frac{p}{g} = \frac{K'k'^2 - 2K' + 2E'}{2(Kk'^2 - 2E)}.$$

1.2.6. *Six defining vertices*

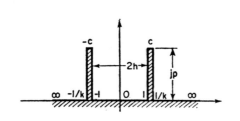

$$z = \frac{2hK'}{\pi}\left[Z(w) + \frac{\pi w}{2KK'}\right];$$

$t = \operatorname{sn} w$ to modulus k,

$$\frac{p}{h} = \frac{2K'}{\pi}\left[jZ(K + jv) + \frac{\pi v}{2KK'}\right],$$

$$c = \frac{1}{K}\sqrt{\frac{E'}{K'}} = \operatorname{sn}(K + jv).$$

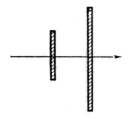

$$z = S\left[Z(w) + \frac{\pi w}{2KK'} + \frac{\operatorname{cn} w \operatorname{dn} w}{c + \operatorname{sn} w}\right] + \text{const.};$$

$$t = \operatorname{sn} w.$$

For the determination of the constants see Love, Additional references of Chapter 10.

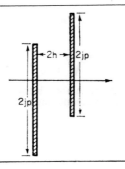

$$z = \frac{2hK'}{\pi}\left[Z(w) + \frac{\pi w}{2KK'} + jBk \operatorname{sn} w\right];$$

$$t = \operatorname{sn} w \text{ to modulus } k.$$

For the determination of the constants see Love, Additional references of Chapter 10.

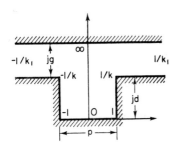

$$z = -\frac{2g}{\pi}\left[\frac{\operatorname{sn}\alpha\,\operatorname{dn}\alpha}{\operatorname{cn}\alpha}\,U + \Pi(U,\alpha,k)\right];$$

$$t = \operatorname{sn}\alpha \text{ to modulus } k,$$

$$k_1 = k\operatorname{sn}\alpha,$$

$$\frac{p}{g} = \frac{4K}{\pi}\left[Z(\alpha) - \frac{\operatorname{sn}\alpha\,\operatorname{dn}\alpha}{\operatorname{cn}\alpha}\right],$$

$$\frac{d}{g} = \frac{2K'}{\pi}\left[Z(\alpha) - \frac{\operatorname{sn}\alpha\,\operatorname{dn}\alpha}{\operatorname{cn}\alpha}\right] + \frac{\alpha}{K}.$$

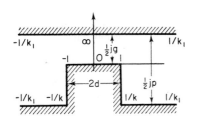

$$z = \frac{p}{\pi}\frac{k^2\operatorname{sn}\alpha\,\operatorname{cn}\alpha}{\operatorname{dn}\alpha}\left[U - \frac{\operatorname{dn}\alpha}{k^2\operatorname{sn}\alpha\,\operatorname{cn}\alpha}\,\Pi(U,\alpha)\right];$$

$$t = \operatorname{sn}U \text{ to modulus } k,$$

$$k_1 = k\operatorname{sn}\alpha,$$

$$\frac{g}{p} = \frac{2K'}{\pi}\left[\frac{k^2\operatorname{sn}\alpha\,\operatorname{cn}\alpha}{\operatorname{dn}\alpha} - Z(\alpha)\right] + 1 - \frac{\alpha}{K}$$

$$\frac{d}{p} = \frac{K}{\pi}\left[\frac{k^2\operatorname{sn}\alpha\cdot\operatorname{cn}\alpha}{\operatorname{dn}\alpha} - Z(\alpha)\right],$$

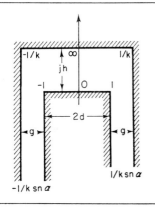

$$z = U - \frac{\operatorname{dn}\alpha}{k^2\operatorname{sn}\alpha\,\operatorname{cn}\alpha}\,\Pi(U,\alpha,k);$$

$$t = \operatorname{sn}U,$$

$$\alpha = K - \tfrac{1}{2}jK',$$

$$d = K\left(\frac{1+k}{2k} - \frac{\pi}{4kK}\right),$$

$$g = \frac{\pi}{2k},$$

$$h = \frac{K'(1+k)}{2k}.$$

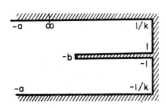

$$z = kC(1 - bk\sin\delta)\frac{\operatorname{sn}\delta}{\operatorname{cn}\delta\,\operatorname{dn}\delta}\,\Pi(\gamma,\delta,k)$$

$$+ \frac{kbC\lambda}{a} - \frac{jC(a-b)k^2\operatorname{sn}^2\delta}{\operatorname{cn}\delta\,\operatorname{dn}\delta}$$

$$\times\sin^{-1}\sqrt{\left(\frac{a^2-1}{1-k^2}\frac{1-k^2t^2}{a^2-t^2}\right)}.$$

For the evaluation of the constants see E.P. Adams, Electric distributions on cylinders, *Proc. Am. Phil. Soc.* **125**, 11 (1936).

1.2.7. *n defining vertices*

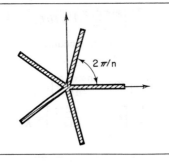

n equally spaced, equal, finite intersecting plates terminating on the unit circle.

$$z = [\cos(n \tan^{-1} t)]^{2/n}.$$

The points corresponding to the ends of the plates are

$$t = \tan \frac{(2k-1)\pi}{2n}, \qquad k = 1, 2, 3, \ldots n.$$

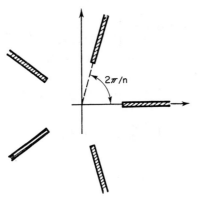

n equally spaced, semi-infinite plates, terminating on the unit circle.

$$z = [\cos(n \tan^{-1} t)]^{-2/n}.$$

The points corresponding to the ends of the plates are

$$t = \tan \frac{(2k-1)\pi}{2n}, \qquad k = 1, 2, 3, \ldots n.$$

1.3. *Straight-line segments and curves*

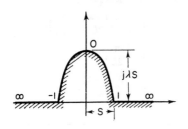

Semi-ellipse and straight line.

$$z = S\left(t + \lambda \sqrt{t^2 - 1}\right).$$

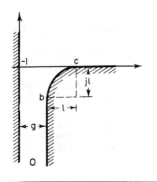

$$z = \frac{-2g}{\pi\left(\sqrt{b} + \lambda\sqrt{c}\right)}\left\{\sqrt{t-b}\right.$$

$$-\sqrt{b}\,\tan^{-1}\sqrt{\left(\frac{t-b}{b}\right)} + \lambda\left[\sqrt{t-c}\right.$$

$$\left.\left.-\sqrt{c}\,\tan^{-1}\sqrt{\left(\frac{t-c}{c}\right)}\right]\right\}.$$

b, c, and λ are determined from the equivalence of $t = -1$ and $z = 0$ and from the desired shape of the rounded corner.

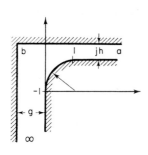

$$2 = -j\sqrt{\frac{a+1}{b-a}}\,\log\frac{1+\sqrt{\left(\frac{b-a}{a+1}\right)\left(\frac{2}{b-1}\right)}}{1-\sqrt{\left(\frac{b-a}{a+1}\right)\left(\frac{2}{b-1}\right)}}$$

$$+2j\tan^{-1}\sqrt{\frac{2}{b-1}}-\left(\frac{b+1}{b-1}\right)$$

$$\times\log\frac{1+\sqrt{\frac{2}{b+1}}}{1-\sqrt{\frac{2}{b+1}}}+2\sqrt{\left(\frac{b+1}{b-1}\right)\left(\frac{a-1}{b-a}\right)}$$

$$\times\tan^{-1}\sqrt{\left(\frac{b-a}{a-1}\right)\left(\frac{2}{b+1}\right)};$$

$$h = \pi\left\{\sqrt{\frac{a+1}{b-a}}+\sqrt{\left(\frac{b+1}{b-1}\right)\left(\frac{a-1}{b-a}\right)}\right\},$$

$$g = \pi\left\{1+\sqrt{\frac{b+1}{b-1}}\right\}.$$

Choice of constants gives a rounded corner with almost uniform curvature.

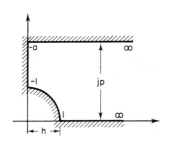

$$z = \frac{2p}{\pi(1+\lambda)}\left[\tan^{-1}\sqrt{\left(\frac{t-1}{t+a}\right)}\right.$$

$$\left.+\lambda\tanh^{-1}\sqrt{\left(\frac{t+1}{t+a}\right)}\right].$$

λ and a are determined from the equivalence of $t = -1$ and $z = jb$, and $t = 1$, $z = b$. The curve approximates closely to a circle.

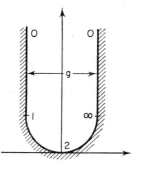

$$z = jg\frac{(K'-E')}{E}+\frac{1}{2}g(1-j);$$

$$k = \sqrt{t}.$$

The curve is a semicircle.

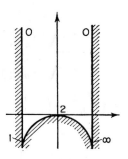

$$z = jg\frac{K'}{K} - \frac{1}{2}g(1+j);$$

$$k = \sqrt{t}.$$

The curve is a semicircle.

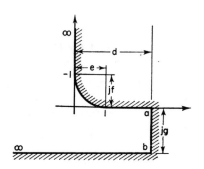

$$z = S\int \frac{\lambda(t+1)+\sqrt{t-1}}{\sqrt{[(t-a)(t-b)]}}\,dt;$$

$$g = 2s\left[\lambda\sqrt{b+1}\,E\left(\frac{\pi}{2}, k\right)+\sqrt{(b-1)}\right.$$
$$\left.\times E\left(\frac{\pi}{2}, k'\right)\right],$$

$$d = 2s\left\{\lambda(b+1)\left[F\left(\frac{\pi}{2}, k_1\right)-E\left(\frac{\pi}{2}, k_1\right)\right]\right.$$
$$\left.+\sqrt{(b-1)}\left[F\left(\frac{\pi}{2}, k_1'\right)-E\left(\frac{\pi}{2}, k_1'\right)\right]\right\},$$

$$e = 2s\lambda\sqrt{b+1}\,[F(\varphi, k_1)-E(\varphi, k_1,)],$$

$$f = 2s\sqrt{b-1}\left\{\sqrt{\left[\frac{2(b+1)}{(b-1)(a+1)}\right]}\right.$$
$$\left.-E(\varphi, k')\right\},$$

$$k = \sqrt{\frac{b-a}{b+1}}, \qquad k' = \sqrt{\frac{b-a}{b-1}},$$

$$k_1 = \sqrt{1-k^2}, \qquad k_1' = \sqrt{1-k'^2},$$

$$\sin\varphi = \sqrt{\frac{2}{a+1}}.$$

The curve approximates to a circular arc.

2. Other transformations

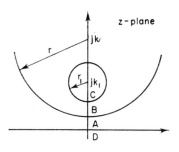

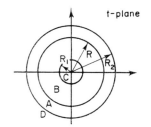

Non-concentric circles (interior) and straight line on concentric circles.

$$2 = \frac{jp_1(R_2+t)}{R_2-t};$$

$$p_1 = \sqrt{k_1^2-r_1^2}, \quad k_1 > r_1 > 0.$$

The following correspond:

The line $y = 0$ and the circle $|t| = R_2$.
The circle $|z-jk_1| = r_1$ and the circle

$$|t| = R_1 = \frac{R_2 r_1}{|k_1|+\sqrt{k_1^2-r_1^2}}.$$

The circle $|z-jk| = r$ and the circle $|t| = R$

where $k = p_1 \dfrac{R_2^2+R^2}{R_2^2-R^2}$ and $r = \dfrac{2p_1 R_2 R}{|R_2^2-R^2|}$.

The regions A, B, C, and D in the two planes.

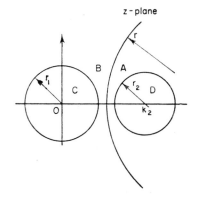

Non-concentric circles (exterior) on concentric circles.

t-plane as above. $z = \dfrac{a(r_1 t-bR_1)}{r_1 t-aR_1};$

$$ab = r_1^2, \quad (k_2-a)(k_2-b) = r_2^2 \quad \text{and}$$

$$\frac{R_2}{R_1} = \frac{r_2}{r_1}\frac{b}{(k_2-b)}, \quad k_2 > 0.$$

The following correspond:

The circle $|z| = r_1$ and the circle $|t| = R_1$.
The circle $|z-k_2| = r_2$ and the circle $|t| = R_2$.
The circle $|z-k| = r$ and the circle $|t| = R$,

$$\text{where } k = \frac{r_1^2(R_1^2-R^2)}{bR_1^2-aR^2},$$

$$\text{and } r = \frac{r_1 R_1 R|b-a|}{|aR^2-bR^2|}.$$

The regions A, B, C, and D in the two planes.

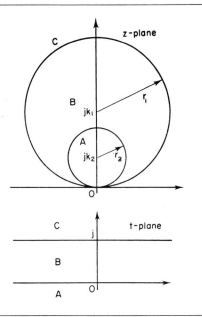

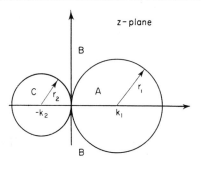

Circles with interior contact on straight lines.

$$z = \frac{2r_1r_2}{(r_2-r_1)t - r_2 j}.$$

The following correspond:

The circle $|z - jk_1| = r_1$ and the line $v = 0$.
The circle $|z - jk_2| = r_2$ and the line $v = j$.
The line $x = 0$ and the line $u = 0$.
The line $y = 0$ and the line $v = jr_2/(r_2 - r_1)$

The regions A, B, and C in the two planes.

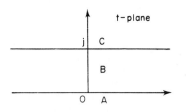

Circles with exterior contact on straight lines.

$$z = \frac{2r_1r_2}{j(r_1+r_2)t + r_2}.$$

The following correspond:

The circle $|z - k_1| = r_1$ and the line $v = 0$.
The circle $|z + k_2| = r_2$ and the line $v = j$.
The line $x = 0$ and the line $v = r_2/(r_1+r_2)$.

The line $y = 0$ and the line $u = 0$.

The regions A, B and C in the two planes.

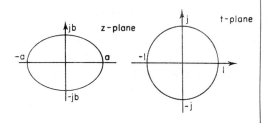

Interior of ellipse on interior of unit circle:

$$t = \sqrt{k}\, \mathrm{sn}\left(\frac{2K}{\pi}\sin^{-1}\frac{z}{\sqrt{a^2 - b^2}}\right)$$

to modulus k;

$$\left(\frac{a-b}{a+b}\right)^2 = \exp\left(-\frac{\pi K'}{K}\right).$$

APPENDIX IV

BIBLIOGRAPHIES

1. Bibliography

ABRAHAM, M., and BECKER, R., *The Classical Theory of Electricity and Magnetism*, Blackie, London and Glasgow, 1937.

ALLEN, D. N. DE G., *Relaxation Methods*, McGraw-Hill, London, New York, and Toronto, 1954.

BATEMANN, H., *Partial Differential Equations of Mathematical Physics*, Dover, New York, 1944.

BECKENBACH, E. F., *Modern Mathematics for the Engineer*, McGraw-Hill, New York, 1956.

BEWLEY, L. V., *Two-dimensional Fields in Electrical Engineering*, Dover, New York, 1963.

BOWMAN, F., *Introduction to Elliptic Functions with Applications*, English Universities Press, Suffolk, 1953.

BUCHHOLZ, H., *Elektrische und magnetische Potentialfelder*, Springer, Berlin, 1957.

BUCKINGHAM, R. A., *Numerical Methods*, Pitman, London, 1957.

BYERLY, W. E., *An Elementary Treatise on Fourier Series and Spherical, Cylindrical and Ellipsoidal Harmonics*, Ginn, Boston, 1893.

BYRD, P., and FRIEDMAN, M. D., *Handbook of Elliptic Integrals for Engineers and Physicists*, Springer, Berlin, 1954.

CARTER, G. W., *The Electromagnetic Field in its Engineering Aspect*, Longmans, London, New York, and Toronto, 1954.

CHURCHILL, R. V., *Fourier Series and Boundary Value Problems*, McGraw-Hill, New York, 1941.

COPSON, E. T., *An Introduction to the Theory of Functions of a Complex Variable*, Clarendon Press, Oxford, 1935.

EDWARDS, J., *A Treatise on the Integral Calculus*, Macmillan, London, 1921.

FERRARO, V. C. A., *Electromagnetic Theory*, Athlone Press, London, 1956.

FORSYTHE, G. E., and WASOW, W. R., *Finite Difference Methods for Partial Differential Equations*, Wiley, New York and London, 1960.

FRANK, P., and MISES, R. V., *Die Differential- und Integralgleichungen der Mechanik und Physik*, 2 vols., Vieweg, Braunschweig, 1935.

GIBBS, W. J., *Conformal Transformations in Electrical Engineering*, Chapman and Hall, London, 1958.

GRINTER, L. E., *Numerical Methods of Analysis in Engineering*, Macmillan, New York, 1949.

GROVER, F. W., *Inductance Calculations*, van Nostrand, New York, 1946.

HAGUE, B., *The Principles of Electromagnetism Applied to Electrical Machines*, Dover, London, 1962.

JEANS, J., *The Mathematical Theory of Electricity and Magnetism*, University Press, Cambridge, 1951.

JENNINGS, W., *First Course in Numerical Methods*, Macmillan, London, 1964.

KANTOROVICH, L. V. and KRYLOV, V. I. *Approximate Methods of Higher Analysis*, Groningen, 1958.

KARMAN, T. V., and BURGERS, J. M., *General Aerodynamic Theory*, vol. II, Springer, Berlin, 1935.

KELVIN, LORD, *Reprint of Papers on Electrostatics and Magnetism*, Macmillan, London, 1872.

KING, L. V., *Numerical Evaluation of Elliptic Functions and Elliptic Integrals*, Cambridge, 1924.

KÖBER, H., *Dictionary of Conformal Representation*, Dover, New York, 1952.

KOPPENFELS, W. VON, and STALLMANN, F., *Praxis der konformen Abbildung*, Springer, Berlin, 1959.

LANCZOS, C., *Applied Analysis*, Pitman, London, 1957.

LEGENDRE, A. M., *Mémoires sur les Transcendantes élliptiques*, 1793.

LEGENDRE, A. M., *Exercices de calcul intégral*, 1811.

MAXWELL, J. C., *A Treatise on Electricity and Magnetism*, 2 vols., Clarendon Press, Oxford, 1892.

MEYER, H. A., *Symposium on Monte Carlo Methods*, Chapman and Hall, London, 1956.

MILNE, W. E., *Numerical Solution of Differential Equations*, Wiley, New York; Chapman and Hall, London, 1953.

MILNE-THOMSON, L. M., *Theoretical Hydrodynamics*, Macmillan, London, 1938.

MOON, P., and SPENCER, D. E., *Field Theory for Engineers*, van Nostrand, New York, 1961.

MORSE, P. M., and FESHBACH, H., *Methods of Theoretical Physics*, 2 vols., McGraw-Hill, New York, 1953.

MOULLIN, E. B., *Principles of Electromagnetism*, University Press, Oxford, 1955.

OLLENDORF, F., *Technische Elektrodynamik*, Band 1, *Berechnung magnetischer Felder*, Springer, Vienna, 1952.

RALSTON, A., and WILFE, H. S., *Mathematical Methods for Digital Computers*, Wiley, New York, London, 1960.

RICHTMEYER, R. D., *Difference Methods for Initial Value Problems*, Wiley–Interscience, New York, 1957.

SEELY, S., *Introduction to Electromagnetic Fields*, McGraw-Hill, New York, 1958.

SHAW, F. S., *An Introduction to Relaxation Methods*, Dover, New York, 1953.

SMITH, G. D., *Numerical Solution of Partial Differential Equations*, Oxford, 1965.

SMYTHE, W. R., *Static and Dynamic Electricity*, McGraw-Hill, New York, 1939.

SOKOLNIKOFF, I. S. and E. S., *Higher Mathematics for Engineers and Physicists*, McGraw-Hill, New York and London, 1941.

SOUTHWELL, R. V., *Relaxation Methods in Theoretical Physics*, University Press, Oxford, 1946.

STRATTON, J. A., *Electromagnetic Theory*, McGraw-Hill, New York, 1941.

TANNERY, J., and MOLK, J., *Elements de la théorie des fonctions élliptiques*, Gauther-Villars, Paris, 1893.

TODD, J., *Survey of Numerical Analysis*, McGraw-Hill, New York, 1968.

VARGA, R. S., *Matrix Iterative Analysis*, Prentice-Hall, London, 1962.

VITKOVITCH, D. (editor), *Field Analysis: Experimental and Computational Methods*, Van Nostrand, London, 1966.

WACHSPRESS, E. L., *Iterative Solution of Elliptic Systems*, Prentice-Hall, 1966.

WALSH, J., *Numerical Analysis: An Introduction*, Academic Press, London, New York, 1966.

WEBER, E., *Electromagnetic Fields*, vol. 1, *Mapping of Fields*, Wiley, New York, London, 1960.

WHITTAKER, E. T., and WATSON, G. N., *A Course of Modern Analysis*, University Press, Cambridge, 1920.

ZWORYKIN, V. K., MORTON, G. A., RAMBERG, E. G., HILLIER, J., and VANCE, A. W., *Electron Optics and the Electron Microscope*, Wiley, New York, 1945.

2. Tables of elliptic functions

BYRD, P. F., and FRIEDMAN, M. D., *Handbook of Elliptic Integrals for Engineers and Physicists*, Springer, Berlin, Göttingen, and Heidelberg, 1954.

DWIGHT, H. B., *Tables of Integrals and other Mathematical Data*, Macmillan, New York, 1934.

JAHNKE, E., and EMDE, F., *Tables of Functions with Formulas and Curves*, Dover, New York, 1945.

MILNE-THOMSON, L. M., *Jacobian Elliptic Function Tables*, Dover, New York, 1950.

PEARSON, K., *Tables of Complete and Incomplete Elliptic Integrals*, London, 1934.

PEIRCE, B. O., *A Short Table of Integrals*, Ginn, Boston, 1929.

SPENCELEY, G. W., and SPENCELEY, R. M., *Smithsonian Elliptic Function Tables*, Smithson, Washington, 1947.

3. Additional bibliography

ABRAHAM, M., and BECKER, R., *The Classical Theory of Electricity*, Blackie, London, 1932.

BIEBERBACH, L., *Einführung in die konforme Abbildung* (Sammlung Göschen 768), 5 Aufl., Berlin, 1956.

BIEBERBACH, L., *Lehrbuch der Funtionentheorie*, 2 vols.; reprint by Chelsea Publishing Co., New York, 1945; originally published by Teubner, Leipzig.

BYERLY, W. E., *Fourier's Series and Spherical, Cylindrical and Ellipsoidal Harmonics*, Ginn, Boston, 1902.

CARATHEODORY, C., *Conformal Representation*, Cambridge, 1932.

CARSLAW, H. S., and JAEGER, J. C., *Conduction of Heat in Solids*, Oxford University Press, New York, 1947.

CHAFFEE, E. L., *Theory of Thermionic Vacuum Tubes*, McGraw-Hill, New York, 1933.

COURANT, R., and HILBERT, D., *Methoden der mathematischen Physik*, vol. I, 1931, vol. II, 1937, Springer, Berlin.

CRANDALL, S. H., *Engineering Analysis*, McGraw-Hill, New York, 1956.

CULLWICK, E. G., *The Fundamentals of Electromagnetism*, Macmillan, New York, 1939.

Eck, B., *Einführung in die technische Strömungslehre*, vol. I, *Theory*, 1935; vol. II, *Laboratory Methods*, 1936, Springer, Berlin.

Evans, G. C., *The Logarithmic Potential, Discontinuous Dirichlet, and Newmann Problems*, American Mathematical Society, vol. VI, New York, 1927.

Fourier, J. B. J., *Théorie analytique de la chaleur* (translated by A. Freeman), Dover, New York, 1955.

Goursat, E., *Cours d'analyse mathématique*, A. Hermann, Paris, 1910, 1911.

Hammond, P., *Applied Electromagnetism*, Pergamon, Oxford, 1971.

Handbuch der Physik, vol. 12, *Theorien der Elektrizität, Elektrostatik*, 1927; vol. 15, *Magnetismus, elektromagnetisches Feld*, 1927, Springer, Berlin.

Heine, E., *Anwendungen der Kugelfunktionen*, Berlin, 1881.

Hobson, E. W., *Spherical and Ellipsoidal Harmonics*, Cambridge, 1931.

Hurwitz, A., *Vorlesungen über allgemeine Funktionentheorie*, Springer, Berlin, 1929.

Karapetoff, V., *The Electric Circuit*, McGraw-Hill, New York, 1910.

Karapetoff, V., *The Magnetic Circuit*, McGraw-Hill, New York, 1910.

Karman, Th. V., and Burgers, J. M., *General Aerodynamic Theory, Perfect Fluids*, vol. II of *Aerodynamic Theory* (edited by W. F. Durand), Springer, Berlin, 1935.

Kellogg, O. D., *Foundations of Potential Theory*, Springer, Berlin, 1929.

Korn, A., *Lehrbuch der Potentialtheorie*, Dummler, Berlin, 1899.

Lamb, H., *Hydrodynamics*, 6th edn., Cambridge University Press, England, 1932.

Lance, G. N., *Numerical Methods for High Speed Computers*, Iliffe, London, 1960.

Livens, G. H., *The Theory of Electricity*, Cambridge University Press, London, 1926.

Macrobert, T. M., *Spherical Harmonics*, E. P. Dutton, New York, 1927.

Macrobert, T. M., *Functions of a Complex Variable*, Macmillan, London, 1954.

Mason, M., and Weaver, W., *The Electromagnetic Field*, Chicago University Press, 1929.

Miller, K. S., *Partial Differential Equations in Engineering Problems*, Prentice-Hall, New York, 1953.

Milne-Thomson, L. M., *Theoretical Aerodynamics*, Macmillan, London, 1952.

Murnaghan, F. D., *Introduction to Applied Mathematics*, Wiley, New York, 1948.

Muskhelishvili, N. G., *Some Basic Problems of the Mathematical Theory of Elasticity* (translated by J. R. M. Radok), P. Neordhoff Ltd., Groningen-Holland, 1953.

Myers, L. M., *Electron Optics*, Van Nostrand, New York, 1939.

Nehari, Z., *Conformal Mapping*, McGraw-Hill, New York, Toronto, and London, 1952.

Neville, H. H., *Jacobian Elliptic Functions*, Clarendon Press, Oxford, 1951.

Ollendorff, F., *Potentialfelder der Elektrotechnik*, Springer, Berlin, 1932.

Peirce, B. O., *Newtonian Potential Function*, Ginn, Boston, 1902.

Pierpoint, J., *Functions of a Complex Variable*, Ginn, Boston, 1914.

Ramsay, A. S., *A Treatise of Hydromechanics*, Bell, 1935.

Richter, R., *Elektrische Maschinen*, vol. 1, *Fundamentals and D.C. Machines*, 1924; vol. 2, *Synchronous Machines and Converters*, 1930; vol. 3, *Transformers*, 1932; vol. 4, *Asynchronous Machines*, 1936, Springer, Berlin.

Rothe, R., Ollendorff, F., and Pohlhausen, K., *Theory of Functions as Applied to Engineering Problems*, Technology Press, Cambridge, Mass., 1933.

Spangenberg, K. R., *Vacuum Tubes*, McGraw-Hill, New York, 1948.

Sternberg, W., *Potentialtheorie*, W. de Gruyter, Leipzig, 1925.

Strutt, M. J. O., *Moderne Mehrgitter-Elektronenröhren*, vol. 2, Springer, Berlin, 1928.

Thom, A., and Apelt, C. J., *Field Computations in Engineering and Physics*, Van Nostrand, New York, 1961.

Titchmarsh, E. C., *Theory of Functions*, Oxford, 1932.

Walker, M., *Conjugate Functions for Engineers*, Oxford, 1933.

Webster, A. G., *Partial Differential Equations*, Teubner, Leipzig, 1927.

INDEX